Plan B 2.0

Other Norton Books by Lester R. Brown

Outgrowing the Earth: The Food Security Challenge in an Age of Falling Water Tables and Rising Temperatures

Plan B: Rescuing a Planet under Stress and a Civilization in Trouble

The Earth Policy Reader
with Janet Larsen and Bernie Fischlowitz-Roberts

Eco-Economy: Building an Economy for the Earth

State of the World 1984 through *2001*
annual, with others

Vital Signs 1992 through *2001*
annual, with others

Beyond Malthus
with Gary Gardner and Brian Halweil

The World Watch Reader 1998
editor with Ed Ayres

Tough Choices

Who Will Feed China?

Full House
with Hal Kane

Saving the Planet
with Christopher Flavin and Sandra Postel

Building a Sustainable Society

Running on Empty
with Colin Norman and Christopher Flavin

The Twenty-Ninth Day

In the Human Interest

Earth Policy Institute® is a nonprofit environmental research organization providing a vision of an eco-economy and a roadmap on how to get from here to there. It seeks to reach a global constituency through the media and the Internet. Its primary publications are *Outgrowing the Earth*, *Plan B*, *Eco-Economy: Building an Economy for the Earth*, *The Earth Policy Reader*, and a series of four-page Eco-Economy Updates that assess progress in building an eco-economy. All of these can be downloaded at no charge from the EPI Web site.

Web site: www.earthpolicy.org

Plan B 2.0

Rescuing a Planet Under Stress and a Civilization in Trouble

Lester R. Brown

EARTH POLICY INSTITUTE

W · W · NORTON & COMPANY

NEW YORK LONDON

Printed in the United States of America
First Edition

The text of this book is composed in Sabon. Composition by Elizabeth Doherty; manufacturing by the Maple-Vail Book Manufacturing Group.

Library of Congress Cataloging-in-Publication Data

Brown, Lester Russell, 1934–
Plan B 2.0 : rescuing a planet under stress and a civilization in trouble / Lester R. Brown.
p. cm.
Rev. ed. of: Plan B : rescuing a planet under stress and a civilization in trouble. 1st ed. 2003.
Includes bibliographical references and index.
ISBN 0-393-06162-0 (hardcover)—ISBN 0-393-32831-7 (pbk.)
1. Environmental policy--Economic aspects. 2. Natural resources—Management.
3. Economic development—Environmental aspects.
I. Title.
HC79.E5B7595 2006
333.7—dc22

2005031835

W. W. Norton & Company, Inc., 500 Fifth Avenue,
New York, N.Y. 10110
www.wwnorton.com

W. W. Norton & Company, Ltd., Castle House, 75/76 Wells Street,
London W1T 3QT

1 2 3 4 5 6 7 8 9 0

This book is printed on recycled paper.

Contents

Preface

Plan A, business-as-usual, has the world on an environmental path that is leading toward economic decline and eventual collapse. If our goal is to sustain economic progress, we have no choice other than move onto a new path—Plan B. This is why I wrote the original *Plan B* in 2003.

There are many reasons why we have updated and expanded this 2003 edition into *Plan B 2.0*. Most fundamentally, there still is no widely shared sense that we need to build a new economy—and even less, a vision of what it might look like. The purpose of this book is to make a convincing case for building the new economy, to offer a more detailed vision of what it would look like, and to provide a roadmap of how to get from here to there.

There are several other reasons for this new edition. One, there is strong new evidence that the western economic model will not work for China. Two, the tightening oil supply raises challenging new issues that deserve attention. Three, since poverty cannot be eradicated if the economy's natural support systems continue to deteriorate, we have also included here an earth restoration budget to complement the poverty eradication budget in the first edition. Four, technological advances in the last few years offer exciting new possibilities for reversing the environmental trends that are undermining our future. And, five, we wanted to do a new edition simply because of the unexpectedly enthusiastic response to the first edition.

To elaborate on the first of these points, China has now overtaken the United States in the consumption of most basic resources. Among the leading commodities in the food sector (grain and meat), in the energy sector (oil and coal), and in the industrial economy (steel), China now leads the United States in consumption of all except oil.

What if China catches up to the United States in consumption per person? If China's economy continues to expand at 8 percent per year, its income per person will reach the current U.S. level in 2031. If we assume that Chinese consumption levels per person in 2031 are the same as those in the United States today, then the country's projected population of 1.45 billion would consume an amount of grain equal to two thirds of the current world grain harvest, its paper consumption would be double current world production, and it would use 99 million barrels of oil per day—well above current world production of 84 million barrels.

The western economic model is not going to work for China. Nor will it work for India, which by 2031 is projected to have a population even larger than China's, or for the other 3 billion people in developing countries who are also dreaming the "American dream." And in an increasingly integrated world economy, where all countries are competing for the same oil, grain, and mineral resources, the existing economic model will not work for industrial countries either. The days of the fossil-fuel-based, automobile-centered, throwaway economy are numbered.

Closely related to China's expanding resource consumption is the world's fast-changing oil outlook and the new issues it generates. For example, we have long been concerned about the effect of rising oil prices on food production costs, but of even more concern is the effect on the demand for food commodities. Since virtually everything we eat can be converted into automotive fuel either in ethanol distilleries or biodiesel refineries, high oil prices are opening a vast new market for farm products. Those buying commodities for fuel producers are competing directly with food processors for supplies of wheat, corn, soybeans, sugarcane, and other foodstuffs. In effect, supermarkets and service stations are now competing for the same commodities.

The price of oil is setting the price for food simply because if the fuel value of a commodity exceeds its value as food, it will be converted into fuel. As more and more ethanol distilleries and biodiesel refineries are built, the world's affluent automobile owners will be competing with the world's poor for the same commodities.

In the original *Plan B*, we had a budget for eradicating poverty, but if the economy's environmental support systems are collapsing, poverty eradication will not be possible. If croplands are eroding and harvests are shrinking, if water tables are falling and wells are going dry, if rangelands are turning to desert and livestock are dying, if fisheries are collapsing, if forests are shrinking, and if rising temperatures are scorching crops, a poverty eradication program—no matter how carefully crafted and well implemented—will not succeed.

For this reason, we have added an earth restoration budget to restore the earth's productive health that parallels the budget for poverty eradication. It includes the costs of protecting and restoring soils, forests, rangelands, and oceanic fisheries, plus conserving the earth's biological diversity. It also means halting advancing deserts that threaten to displace millions of people.

And finally, the good news—and another reason for updating *Plan B*—is that new technologies offer hope in dealing with the mounting challenges we face on the environmental front. For example, advances in gas-electric hybrid cars and in wind turbine design have set the stage for the evolution of a new automotive fuel economy. Using gas-electric hybrids with an extra storage battery plus a plug-in capacity enables us to do our short-distance driving largely with electricity. If we combine this with investment in wind farms to feed cheap electricity into the grid, we can largely power automobiles with wind energy. Using cheap wind-generated electricity to recharge batteries during off-peak hours costs the equivalent of 50¢-a-gallon gasoline! This is but one example of the possibilities for building a new economy, one that can sustain economic progress while saving money, reducing oil dependence, and cutting carbon emissions.

We were also inspired to do *Plan B 2.0* because of the extraordinary response to the first edition. In looking at our sales database several months after publication, we noticed that many individuals who had ordered a copy initially had returned to order 5, 10, 20, even 50 or more copies for distribution to colleagues, opinion leaders, political leaders, and others.

In response to this, we formed a Plan B Team of people who

ordered five or more copies. That team is now some 650 strong. Ted Turner, who purchased 3,569 copies to distribute to heads of state, cabinet members, Fortune 500 CEOs, the U.S. Congress, and others, was designated team captain. With the Plan B Team now in place as this revised, expanded revision comes out, we hope we can expand its membership so that before long there will be thousands of people actively promoting this plan to save our civilization.

There is a mounting tide of public concern about where the world is heading and a growing sense that we need to change course. The rising price of oil and growing competition for this resource are feeding this concern. So, too, are the various manifestations of climate change, such as ice melting and rising sea level. When Hurricane Katrina left in its wake a $200-billion bill—nearly seven times the cost of any previous storm—it sent a message to the entire world.

It is this rise in public concern that may soon start to drive the policymaking process in the right direction, a direction that will move the world onto an environmental path that will sustain economic progress.

This book can be downloaded without charge from our Web site. Permission for reprinting or excerpting portions of the manuscript can be obtained from Reah Janise Kauffman at Earth Policy Institute.

Lester R. Brown
October 2005

Earth Policy Institute
1350 Connecticut Ave. NW
Suite 403
Washington, DC 20036

Phone: (202) 496-9290
Fax: (202) 496-9325
E-mail: epi@earthpolicy.org
Web: www.earthpolicy.org

For additional information on the topics discussed in this book, see www.earthpolicy.org/Books/PB2/index.htm.

PLAN B 2.0

1

Entering a New World

Our global economy is outgrowing the capacity of the earth to support it, moving our early twenty-first century civilization ever closer to decline and possible collapse. In our preoccupation with quarterly earnings reports and year-to-year economic growth, we have lost sight of how large the human enterprise has become relative to the earth's resources. A century ago, annual growth in the world economy was measured in billions of dollars. Today it is measured in trillions.

As a result, we are consuming renewable resources faster than they can regenerate. Forests are shrinking, grasslands are deteriorating, water tables are falling, fisheries are collapsing, and soils are eroding. We are using up oil at a pace that leaves little time to plan beyond peak oil. And we are discharging greenhouse gases into the atmosphere faster than nature can absorb them, setting the stage for a rise in the earth's temperature well above any since agriculture began.

Our twenty-first century civilization is not the first to move onto an economic path that was environmentally unsustainable. Many earlier civilizations also found themselves in environmental trouble. As Jared Diamond notes in *Collapse: How Societies*

Choose to Fail or Succeed, some were able to change course and avoid economic decline. Others were not. We study the archeological sites of Sumerians, the Mayans, Easter Islanders, and other early civilizations that were not able to make the needed adjustments in time.[1]

Fortunately, there is a consensus emerging among scientists on the broad outlines of the changes needed. If economic progress is to be sustained, we need to replace the fossil-fuel-based, automobile-centered, throwaway economy with a new economic model. Instead of being based on fossil fuels, the new economy will be powered by abundant sources of renewable energy: wind, solar, geothermal, hydropower, and biofuels.

Instead of being centered around automobiles, future transportation systems will be far more diverse, widely employing light rail, buses, and bicycles as well as cars. The goal will be to maximize mobility, not automobile ownership.

The throwaway economy will be replaced by a comprehensive reuse/recycle economy. Consumer products from cars to computers will be designed so that they can be disassembled into their component parts and completely recycled. Throwaway products such as single-use beverage containers will be phased out.

The good news is that we can already see glimpses here and there of what this new economy looks like. We have the technologies to build it—including, for example, gas-electric hybrid cars, advanced-design wind turbines, highly efficient refrigerators, and water-efficient irrigation systems.

We can see how to build the new economy brick by brick. With each wind farm, rooftop solar panel, paper recycling facility, bicycle path, and reforestation program, we move closer to an economy that can sustain economic progress.

If, instead, we continue on the current economic path, the question is not whether environmental deterioration will lead to economic decline, but when. No economy, however technologically advanced, can survive the collapse of its environmental support systems.

The Nature of the New World

We recently entered a new century, but we are also entering a new world, one where the collisions between our demands and the earth's capacity to satisfy them are becoming daily events. It may be another crop-withering heat wave, another village abandoned because of invading sand dunes, or another aquifer pumped dry. If we do not act quickly to reverse the trends, these seemingly isolated events will come more and more frequently, accumulating and combining to determine our future.

Resources that accumulated over eons of geological time are being consumed in a single human lifespan. We are crossing natural thresholds that we cannot see and violating deadlines that we do not recognize. These deadlines, determined by nature, are not politically negotiable.

Nature has many thresholds that we discover only when it is too late. In our fast-forward world, we learn that we have crossed them only after the fact, leaving little time to adjust. For example, when we exceed the sustainable catch of a fishery, the stocks begin to shrink. Once this threshold is crossed, we have a limited time in which to back off and lighten the catch. If we fail to meet this deadline, breeding populations shrink to where the fishery is no longer viable, and it collapses.

We know from earlier civilizations that the lead indicators of economic decline were environmental, not economic. The trees went first, then the soil, and finally the civilization itself. To archeologists, the sequence is all too familiar.

Our situation today is far more challenging because in addition to shrinking forests and eroding soils, we must deal with falling water tables, more frequent crop-withering heat waves, collapsing fisheries, expanding deserts, deteriorating rangelands, dying coral reefs, melting glaciers, rising seas, more-powerful storms, disappearing species, and, soon, shrinking oil supplies. Although these ecologically destructive trends have been evident for some time, and some have been reversed at the national level, not one has been reversed at the global level.

The bottom line is that the world is in what ecologists call an "overshoot-and-collapse" mode. Demand has exceeded the sustainable yield of natural systems at the local level countless times in the past. Now, for the first time, it is doing so at the global level. Forests are shrinking for the world as a whole. Fishery collapses are widespread. Grasslands are deteriorating on every continent. Water tables are falling in many countries. Carbon dioxide (CO_2) emissions exceed CO_2 fixation everywhere.

In 2002, a team of scientists led by Mathis Wackernagel, who now heads the Global Footprint Network, concluded that humanity's collective demands first surpassed the earth's regenerative capacity around 1980. Their study, published by the U.S. National Academy of Sciences, estimated that global demands in 1999 exceeded that capacity by 20 percent. The gap, growing by 1 percent or so a year, is now much wider. We are meeting current demands by consuming the earth's natural assets, setting the stage for decline and collapse.[2]

In a rather ingenious approach to calculating the human physical presence on the planet, Paul MacCready, the founder and Chairman of AeroVironment and designer of the first solar-powered aircraft, has calculated the weight of all vertebrates on the land and in the air. He notes that when agriculture began, humans, their livestock, and pets together accounted for less than 0.1 percent of the total. Today, he estimates, this group accounts for 98 percent of the earth's total vertebrate biomass, leaving only 2 percent for the wild portion, the latter including all the deer, wildebeests, elephants, great cats, birds, small mammals, and so forth.[3]

Ecologists are intimately familiar with the overshoot-and-collapse phenomenon. One of their favorite examples began in 1944, when the Coast Guard introduced 29 reindeer on remote St. Matthew Island in the Bering Sea to serve as the backup food source for the 19 men operating a station there. After World War II ended a year later, the base was closed and the men left the island. When U.S. Fish and Wildlife Service biologist David Kline visited St. Matthew in 1957, he discovered a thriving population of 1,350 reindeer feeding on the four-inch-thick mat of lichen that covered the 332-square-kilometer (128-square-mile) island. In the absence of any predators, the population was exploding. By 1963, it had reached 6,000. He returned to St. Matthew in 1966 and discovered an island strewn with reindeer skeletons and not much lichen. Only 42 of the reindeer survived: 41 females and 1 not entirely healthy male. There were no fawns. By 1980 or so, the remaining reindeer had died off.[4]

Like the deer on St. Matthew Island, we too are overconsuming our natural resources. Overshoot leads sometimes to decline and sometimes to a complete collapse. It is not always clear which it will be. In the former, a remnant of the population or economic activity survives in a resource-depleted environment. For example, as the environmental resource base of Easter Island in the South Pacific deteriorated, its population declined from a peak of 20,000 several centuries ago to today's population of fewer than 4,000. In contrast, the 500-year-old Norse settlement in Greenland collapsed during the 1400s, disappearing entirely in the face of environmental adversity.[5]

As of 2005, some 42 countries have populations that are stable or declining slightly in size as a result of falling birth rates. But now for the first time ever, demographers are projecting population declines in some countries because of rising death rates, among them Botswana, Lesotho, Namibia, and Swaziland. In the absence of an accelerated shift to smaller families, this list of countries is likely to grow much longer in the years immediately ahead.[6]

The most recent mid-level U.N. demographic projections show world population increasing from 6.1 billion in 2000 to 9.1 billion in 2050. But such an increase seems highly unlikely, considering the deterioration in life-support systems now under way in much of the world. Will we not reach 9.1 billion because we quickly eradicate global poverty and lower birth rates? Or because we fail to do so and death rates begin to rise, as they are already doing in many African countries? We thus face two urgent major challenges: restructuring the global economy and stabilizing world population.[7]

Even as the economy's environmental support systems are deteriorating, the world is pumping oil with reckless abandon. Leading geologists now think oil production may soon peak and turn downward. This collision between the ever-growing demand for oil and the earth's finite resources is but the latest in a long series of collisions. Although no one knows exactly when oil production will peak, supply is already lagging behind demand, driving prices upward.[8]

In this new world, the price of oil begins to set the price of food, not so much because of rising fuel costs for farmers and food processors but more because almost everything we eat can be converted into fuel for cars. In this new world of high oil prices, supermarkets and service stations will compete in commodity markets for basic food commodities such as wheat, corn, soybeans, and sugarcane. Wheat going into the market can be converted into bread for supermarkets or ethanol for service stations. Soybean oil can go onto supermarket shelves or it can go to service stations to be used as diesel fuel. In effect, owners of the world's 800 million cars will be competing for food resources with the 1.2 billion people living on less than $1 a day.[9]

Faced with a seemingly insatiable demand for automotive fuel, farmers will want to clear more and more of the remaining tropical forests to produce sugarcane, oil palms, and other high-yielding fuel crops. Already, billions of dollars of private capital are moving into this effort. In effect, the rising price of oil is generating a massive new threat to the earth's biological diversity.

As the demand for farm commodities climbs, it is shifting the focus of international trade concerns from the traditional goal of assured access to markets to one of assured access to supplies. Countries heavily dependent on imported grain for food are beginning to worry that buyers for fuel distilleries may outbid them for supplies. As oil security deteriorates, so, too, will food security.

As the role of oil recedes, the process of globalization will be reversed in fundamental ways. As the world turned to oil during the last century, the energy economy became increasingly globalized, with the world depending heavily on a handful of countries in the Middle East for energy supplies. Now as the world turns to wind, solar cells, and geothermal energy in this century, we are witnessing the localization of the world energy economy.

The globalization of the world food economy will also be reversed, as the higher price of oil raises the cost of transporting food internationally. In response, food production and consumption will become much more localized, leading to diets based more on locally produced food and seasonal availability.

The world is facing the emergence of a geopolitics of scarcity, which is already highly visible in the efforts by China, India, and other developing countries to ensure their access to oil supplies. In the future, the issue will be who gets access to not only Middle Eastern oil but also Brazilian ethanol and North American grain. Pressures on land and water resources, already excessive in most of the world, will intensify further as the demand for biofuels climbs. This geopolitics of scarcity is an early manifestation of civilization in an overshoot-and-collapse mode, much like the one that emerged among the Mayan cities competing for food in that civilization's waning years.[10]

You do not need to be an ecologist to see that if recent environmental trends continue, the global economy eventually will come crashing down. It is not knowledge that we lack. At issue is whether national governments can stabilize population and restructure the economy before time runs out. Looking at what is happening in China helps us to see the urgency of acting quickly.

Learning from China

For many years environmentalists have pointed to the United States as the world's leading consumer, noting that 5 percent of the world's people were consuming nearly a third of the earth's resources. Although that was true for some time, it no longer is. China has replaced the United States as the leading consumer of basic commodities.[11]

Among the five basic food, energy, and industrial commodities—grain and meat, oil and coal, and steel—consumption in China has eclipsed that of the United States in all but oil. China has opened a wide lead with grain, consuming 380 million tons in 2005 versus 260 million tons in the United States. Among the big three grains, China leads in the consumption of both wheat and rice and trails the United States only in corn.[12]

Although eating hamburgers is a defining element of the U.S. lifestyle, China's 2005 meat consumption of 67 million tons is far above the 38 million tons eaten in the United States. While U.S. meat intake is rather evenly distributed between beef, pork, and poultry, in China pork totally dominates. Indeed, half the world's pigs are now found in China.[13]

With oil, the United States was still solidly in the lead in 2004, using more than three times as much as China—20.4 million barrels per day versus 6.5 million barrels. But U.S. oil use expanded by only 15 percent between 1994 and 2004, while use in China more than doubled. Having recently eclipsed Japan as an oil consumer, China now trails only the United States.[14]

Energy use in China also obviously includes coal, which supplies nearly two thirds of the country's energy. China's annual burning of 960 million tons easily exceeds the 560 million tons used in the United States. With this level of coal use and with oil and natural gas use also climbing fast, it is only a matter of time before China's carbon emissions match those of the United States. Then the world will have two major countries driving climate change.[15]

China's consumption of steel, a basic indicator of industrial development, is now nearly two and a half times that of the United States: 258 million tons to 104 million tons in 2003. As China has moved into the construction phase of development, building hundreds of thousands of factories and high-rise apartment and office buildings, steel consumption has climbed to levels never seen in any country.[16]

With consumer goods, China leads in the number of cell phones, television sets, and refrigerators. The United States still leads in the number of personal computers, though likely not for much longer, and in automobiles.[17]

That China has overtaken the United States in consumption of basic resources gives us license to ask the next question. What if China catches up with the United States in consumption per person? If the Chinese economy continues to grow at 8 percent a year, by 2031 income per person will equal that in the United States in 2004. If we further assume that consumption patterns of China's affluent population in 2031, by then 1.45 billion, will be roughly similar to those of Americans in 2004, we have a startling answer to our question.[18]

At the current annual U.S. grain consumption of 900 kilograms per person, including industrial use, China's grain consumption in 2031 would equal roughly two thirds of the current world grain harvest. If paper use per person in China in 2031 reaches the current U.S. level, this translates into 305 million tons of paper—double existing world production of 161 million tons. There go the world's forests. And if oil consumption per person reaches the U.S. level by 2031, China will use 99 million barrels of oil a day. The world is currently producing 84 million barrels a day and may never produce much more. This helps explain why China's fast-expanding use of oil is already helping to create a politics of scarcity.[19]

Or consider cars. If China one day should have three cars for every four people, as the United States now does, its fleet would total 1.1 billion vehicles, well beyond the current world fleet of 800 million. Providing the roads, highways, and parking lots for such a fleet would require paving an area roughly equal to China's land in rice, its principal food staple.[20]

The inevitable conclusion to be drawn from these projections is that there are not enough resources for China to reach U.S.

consumption levels. The western economic model—the fossil-fuel-based, automobile-centered, throwaway economy—will not work for China's 1.45 billion in 2031. If it does not work for China, it will not work for India either, which by 2031 is projected to have even more people than China. Nor will it work for the other 3 billion people in developing countries who are also dreaming the "American dream." And in an increasingly integrated world economy, where countries everywhere are competing for the same resources—the same oil, grain, and iron ore—the existing economic model will not work for industrial countries either.[21]

Learning from the Past

Our twenty-first century global civilization is not the first to face the prospect of environmentally induced economic decline. The question is how we will respond. We do have one unique asset at our command—an archeological record that shows us what happened to earlier civilizations that got into environmental trouble and failed to respond.

As Jared Diamond points out in *Collapse*, some of the early societies that were in environmental trouble were able to change their ways in time to avoid decline and collapse. Six centuries ago, for example, Icelanders realized that overgrazing on their grass-covered highlands was leading to extensive soil loss from the inherently thin soils of the region. Rather than lose the grasslands and face economic decline, farmers joined together to determine how many sheep the highlands could sustain and then allocated quotas among themselves, thus preserving their grasslands and avoiding what Garrett Hardin later termed the "tragedy of the commons."[22]

The Icelanders understood the consequences of overgrazing and reduced their sheep numbers to a level that could be sustained. We understand the consequences of burning fossil fuels and the resulting CO_2 buildup in the atmosphere. Unlike the Icelanders who were able to restrict their livestock numbers, we have not been able to restrict our CO_2 emissions.

Not all societies have fared as well as the Icelanders, whose economy continues to produce wool and to thrive. The early Sumerian civilization of the fourth millennium BC was an extraordinary one, advancing far beyond any that had existed before. Its carefully engineered irrigation system gave rise to a highly productive agriculture, one that enabled farmers to produce a food surplus, supporting formation of the first cities. Managing the irrigation system required a sophisticated social organization. The Sumerians had the first cities and the first written language, the cuneiform script.[23]

By any measure it was an extraordinary civilization, but there was an environmental flaw in the design of its irrigation system, one that would eventually undermine its food supply. The water that backed up behind dams built across the Euphrates was diverted onto the land through a network of gravity-fed canals. Some water was used by the crops, some evaporated, and some percolated downward. In this region, where underground drainage was weak, percolation slowly raised the water table. As the water climbed to within inches of the surface, it began to evaporate into the atmosphere, leaving behind salt. Over time, the accumulation of salt on the soil surface lowered its productivity.[24]

As salt accumulated and wheat yields declined, the Sumerians shifted to barley, a more salt-tolerant plant. This postponed Sumer's decline, but it was treating the symptoms, not the cause, of falling crop yields. As salt concentrations continued to build, the yields of barley eventually declined also. The resultant shrinkage of the food supply undermined the economic foundation of this once-great civilization. As land productivity declined, so did the civilization.[25]

Archeologist Robert McC. Adams has studied the site of ancient Sumer on the central floodplain of the Euphrates River, an empty, desolate area now outside the frontiers of cultivation. He describes how the "tangled dunes, long disused canal levees, and the rubble-strewn mounds of former settlement contribute only low, featureless relief. Vegetation is sparse, and in many areas it is almost wholly absent....Yet at one time, here lay the core, the heartland, the oldest urban, literate civilization in the world."[26]

The New World counterpart to Sumer is the Mayan civilization that developed in the lowlands of what is now Guatemala. It flourished from AD 250 until its collapse around AD 900. Like the Sumerians, the Mayans had developed a sophisticated, highly productive agriculture, this one based on raised plots of earth surrounded by canals that supplied water.[27]

As with Sumer, the Mayan demise was apparently linked to a failing food supply. For this New World civilization, it was deforestation and soil erosion that undermined agriculture. Changes in climate may also have played a role. Food shortages apparently triggered civil conflict among the various Mayan cities as they competed for food. Today this region is covered by jungle, reclaimed by nature.[28]

During the later centuries of the Mayan civilization, a new society was evolving on faraway Easter Island, some 166 square kilometers of land in the South Pacific roughly 3,200 kilometers west of South America and 2,200 kilometers from Pitcairn Island, the nearest habitation. Settled around AD 400, this civilization flourished on a volcanic island with rich soils and lush vegetation, including trees that grew 25 meters tall with trunks 2 meters in diameter. Archeological records indicate that the islanders ate mainly seafood, principally dolphins—a mammal that could only be caught by harpoon from large sea-going canoes.[29]

The Easter Island society flourished for several centuries, reaching an estimated population of 20,000. As its human numbers gradually increased, tree cutting exceeded the sustainable yield of forests. Eventually the large trees that were needed to build the sturdy canoes disappeared, depriving islanders of access to the dolphins and dramatically shrinking their food supply. The archeological record shows that at some point human bones became intermingled with the dolphin bones, suggesting a desperate society that had resorted to cannibalism. Today the island has some 2,000 residents.[30]

One unanswerable question about these earlier civilizations was whether they knew what was causing their decline. Did the Sumerians understand that the rising salt content in the soil from water evaporation was reducing their wheat yields? If they knew, were they simply unable to muster the political support needed to lower water tables, just as the world today is struggling unsuccessfully to lower carbon emissions?

These are just three of the many early civilizations that moved onto an economic path that nature could not sustain. We, too, are on such a path. Any one of several trends of environmental degradation could undermine civilization as we know it. Just as the irrigation system that defined the early Sumerian economy had a flaw, so too does the fossil fuel energy system that defines our modern economy. For them it was a rising water table that undermined the economy; for us it is rising CO_2 levels that threaten to disrupt economic progress. In both cases, the trend is invisible.

Whether it resulted from the salting of Sumer's cropland, the deforestation and soil erosion of the Mayans, or the depleted forests and loss of the distant-water fishing capacity of the Easter Islanders, collapse of these early civilizations appears to have been associated with a decline in food supply. Today the annual addition of more than 70 million people to a world population of over 6 billion at a time when water tables are falling, temperatures are rising, and oil supplies will soon be shrinking suggests that the food supply again may be the vulnerable link between the environment and the economy.[31]

The Emerging Politics of Scarcity

The first big test of the international community's capacity to manage scarcity may come with oil or it could come with grain. If the latter is the case, this could occur when China—whose grain harvest fell by 34 million tons, or 9 percent, between 1998 and 2005—turns to the world market for massive imports of 30 million, 50 million, or possibly even 100 million tons of grain per year. Demand on this scale could quickly overwhelm world grain markets. When this happens, China will have to look to the United States, which controls the world's grain exports of over 40 percent of some 200 million tons.[32]

This will pose a fascinating geopolitical situation. More than 1.3 billion Chinese consumers, who had an estimated $160-billion trade surplus with the United States in 2004—enough to buy the entire U.S. grain harvest twice—will be competing with Americans for U.S. grain, driving up U.S. food prices. In such a situation 30 years ago, the United States simply restricted exports. But China is now banker to the United States, underwriting much of the massive U.S. fiscal deficit with monthly purchases of U.S. Treasury bonds.[33]

Within the next few years, the United States may be loading one or two ships a day with grain for China. This long line of ships stretching across the Pacific, like an umbilical cord providing nourishment, will intimately link the two economies. Managing this flow of grain so as to simultaneously satisfy the food needs of consumers in both countries, at a time when ethanol fuel distilleries are taking a growing share of the U.S. grain harvest, may become one of the leading foreign policy challenges of this new century.

The way the world accommodates the vast projected needs of China, India, and other developing countries for grain, oil, and other resources will help determine how the world addresses the stresses associated with outgrowing the earth. How low-income, importing countries fare in this competition for grain will also tell us something about future political stability. And, finally, the U.S. response to China's growing demands for grain even as they drive up food prices for U.S. consumers will tell us much about the capacity of countries to manage the emerging politics of scarcity.

The most imminent risk is that China's entry into the world market, combined with the growing diversion of farm commodities to biofuels, will drive grain prices so high that many low-income developing countries will not be able to import enough grain. This in turn could lead to escalating food prices and political instability on a scale that will disrupt global economic progress.

Earlier civilizations that moved onto an economic path that was environmentally unsustainable did so largely in isolation. But in today's increasingly integrated, interdependent world economy, if we are facing civilizational decline, we are facing it together. The fates of all peoples are intertwined. This interdependence can be managed to our mutual benefit only if we recognize that the term "in the national interest" is in many ways obsolete.

Getting the Price Right

The question facing governments is whether they can respond quickly enough to prevent threats from becoming catastrophes. The world has precious little experience in responding to aquifer depletion, rising temperatures, expanding deserts, melting polar ice caps, and a shrinking oil supply. These new trends will fully challenge the capacity of our political institutions and leadership. In times of crisis, societies sometimes have a Nero as a leader and sometimes a Churchill.

The central challenge, the key to building the new economy, is getting the market to tell the ecological truth. The dysfunctional global economy of today has been shaped by distorted market prices that do not incorporate environmental costs. Many of our environmental travails are the result of severe market distortions.

One of these distortions became abundantly clear in the summer of 1998 when China's Yangtze River valley, home to 400 million people, was wracked by some of the worst flooding in history. The resulting damages of $30 billion exceeded the value of the country's annual rice harvest.[34]

After several weeks of flooding, the government in Beijing announced in mid-August a ban on tree cutting in the Yangtze River basin. It justified the ban by noting that trees standing are worth three times as much as trees cut. The flood control services provided by forests were three times as valuable as the lumber in the trees. In effect, the market price was off by a factor of three! With this analysis, no one could economically justify cutting trees in the basin.[35]

A similar situation exists with gasoline. In the United States, the gasoline pump price was over $2 per gallon in mid-2005. But this reflects only the cost of pumping the oil, refining it into gasoline, and delivering the gas to service stations. It does not include the costs of tax subsidies to the oil industry, such as the oil depletion allowance; the subsidies for the extraction, production, and use of petroleum; the burgeoning military costs of protecting access to oil supplies; the health care costs for treating respiratory illnesses ranging from asthma to emphysema; and, most important, the costs of climate change.[36]

If these costs, which in 1998 the International Center for Technology Assessment calculated at roughly $9 per gallon of gasoline burned in the United States, were added to the $2 cost of the gasoline itself, motorists would pay about $11 a gallon for gas at the pump. Filling a 20-gallon tank would cost $220. In reality, burning gasoline is very costly, but the market tells us it is cheap, leading to gross distortions in the structure of the economy. The challenge facing governments is to incorporate such costs into market prices by systematically calculating them and incorporating them as a tax on the product to make sure its price reflects the full costs to society.[37]

If we have learned anything over the last few years, it is that accounting systems that do not tell the truth can be costly. Faulty corporate accounting systems that leave costs off the books have driven some of the world's largest corporations into bankruptcy, costing millions of people their lifetime savings, retirement incomes, and jobs. Distorted world market prices that do not incorporate major costs in the production of various products and the provision of services could be even costlier. They could lead to global bankruptcy and economic decline.

Plan B—A Plan of Hope

Even given the extraordinarily challenging situation we face, there is much to be upbeat about. First, virtually all the destructive environmental trends are of our own making. All the problems we face can be dealt with using existing technologies. And almost everything we need to do to move the world economy onto an environmentally sustainable path has been done in one or more countries.

We see the components of Plan B—the alternative to business as usual—in new technologies already on the market. On the energy front, for example, an advanced-design wind turbine can produce as much energy as an oil well. Japanese engineers have designed a vacuum-sealed refrigerator that uses only one eighth as much electricity as those marketed a decade ago. Gas-electric hybrid automobiles, getting 55 miles per gallon, are easily twice as efficient as the average vehicle on the road.[38]

Numerous countries are providing models of the different components of Plan B. Denmark, for example, today gets 20 percent of its electricity from wind and has plans to push this to 50 percent by 2030. Similarly, Brazil is on its way to automotive fuel self-sufficiency. With highly efficient sugarcane-based ethanol supplying 40 percent of its automotive fuel in 2005, it could phase out gasoline within a matter of years.[39]

With food, India—using a small-scale dairy production model that relies almost entirely on crop residues as a feed source—has more than quadrupled its milk production since 1970, overtaking the United States to become the world's leading milk producer. The value of India's dairy production in 2002 exceeded that of the rice crop.[40]

On another front, fish farming advances in China, centered on the use of an ecologically sophisticated carp polyculture, have made China the first country where fish farm output exceeds oceanic catch. Indeed, the 29 million tons of farmed fish produced in China in 2003 was equal to roughly 30 percent of the world's oceanic fish catch.[41]

We see what a Plan B world could look like in the reforested mountains of South Korea. Once a barren, almost treeless country, the 65 percent of South Korea now covered by forests has checked flooding and soil erosion, returning a high degree of environmental stability to the Korean countryside.[42]

The United States—which retired one tenth of its cropland, most of it highly erodible, and shifted to conservation tillage practices—has reduced soil erosion by 40 percent over the last 20 years. At the same time, the nation's farmers expanded the grain harvest by more than one fifth.[43]

Some of the most innovative leadership has come at the urban level. Amsterdam has developed a diverse urban transport system; today 35 percent of all trips within the city are taken by bicycle. This bicycle-friendly transport system has greatly reduced air pollution and traffic congestion while providing daily exercise for the city's residents.[44]

Not only are new technologies becoming available, but some of these technologies can be combined to create entirely new outcomes. Gas-electric hybrid cars with a second storage battery and a plug-in capacity, combined with investment in wind farms feeding cheap electricity into the grid, could mean that much of our daily driving could be done with electricity, with the cost of off-peak wind-generated electricity at the equivalent of 50¢-a-gallon gasoline. Domestic wind energy can be substituted for imported oil.[45]

The challenge is to build a new economy and to do it at wartime speed before we miss so many of nature's deadlines that the economic system begins to unravel. This introductory chapter leads into five chapters outlining the leading environmental challenges facing our global civilization. Following these are seven chapters that outline Plan B, both describing where we want to go and offering a roadmap of how to get there.

Participating in the construction of this enduring new economy is exhilarating. So is the quality of life it will bring. We will be able to breathe clean air. Our cities will be less congested, less noisy, and less polluted. The prospect of living in a world where population has stabilized, forests are expanding, and carbon emissions are falling is an exciting one.

I

A Civilization in Trouble

2

Beyond the Oil Peak

When the price of oil climbed above $50 a barrel in late 2004, public attention began to focus on the adequacy of world oil supplies—and specifically on when production would peak and begin to decline. Analysts are far from a consensus on this issue, but several prominent ones now believe that the oil peak is imminent.[1]

Oil has shaped our twenty-first century civilization, affecting every facet of the economy from the mechanization of agriculture to jet air travel. When production turns downward, it will be a seismic economic event, creating a world unlike any we have known during our lifetimes. Indeed, when historians write about this period in history, they may well distinguish between before peak oil (BPO) and after peak oil (APO).

The peaking of oil production is approaching at a time when the world is facing many challenges, such as rising temperatures, falling water tables, and numerous other damaging environmental trends. Adjusting to a shrinking oil supply is part of the economic restructuring needed to put the economy on a path that will sustain progress.

The Coming Decline of Oil

The oil prospect can be analyzed in several different ways. Oil companies, oil consulting firms, and national governments rely heavily on computer models to project future oil production and prices. The results from these models vary widely according to the quality of data and the assumptions fed into the models. Here we review several different analytical methods.

One approach—use of the reserves/production relationship to gain a sense of future production trends—was pioneered several decades ago by the legendary King Hubbert, a geologist with the U.S. Geological Survey. Given the nature of oil production, Hubbert theorized that the time lag between the peaking of new discoveries and the peaking of production was predictable. Noting that the discovery of new reserves in the United States had peaked around 1930, he predicted that U.S. oil production would peak in 1970. He hit it right on the head. As a result of this example and other more recent country experiences, his basic model is now used by many oil analysts.[2]

A second approach, separating the world's principal oil-producing countries into two groups—those where production is falling and those where it is still rising—is illuminating. Of the 23 leading oil producers, output appears to have peaked in 15 and to still be rising in eight. The post-peak countries range from the United States (the only country other than Saudi Arabia to ever pump more than 9 million barrels of oil per day) and Venezuela (where oil production peaked in 1970) to the two North Sea oil producers, the United Kingdom and Norway, where production peaked in 1999 and 2000 respectively. U.S. oil production, which peaked at 9.6 million barrels a day in 1970, dropped to 5.4 million barrels a day in 2004—a decline of 44 percent. Venezuela's production has dropped 31 percent since 1970.[3]

The eight pre-peak countries are dominated by the world's leading oil producers, Saudi Arabia and Russia, producing roughly 11 million and 9 million barrels of oil a day in the fall of 2005. Other countries with substantial potential for increasing production are Canada, largely because of its tar sands, and Kazakhstan, which is still developing its oil resources. The other four pre-peak countries are Algeria, Angola, China, and Mexico.[4]

The biggest question mark among these eight countries is Saudi Arabia. Its production technically peaked in 1980 at

9.9 million barrels a day and output is now nearly 1 million barrels a day below that. It is included as a country with rising production only on the basis of statements by Saudi officials that the country could produce far more. However, some analysts doubt whether the Saudis can raise output much beyond its current production. Some of its older oil fields are largely depleted, and it remains to be seen whether pumping from new fields will be sufficient to more than offset the loss from the old ones.[5]

This analysis comes down to whether production will actually increase enough in the eight pre-peak countries to offset the declines under way in the 15 countries where production has already peaked. In volume of output, the two groups have essentially the same total production capacity. If production begins to fall in any one of the eight, however, this may well tilt the global balance to decline.[6]

A third way to consider oil production prospects is to look at the actions of the major oil companies themselves. While some CEOs sound very bullish about the growth of future production, their actions suggest a less confident outlook.

One bit of evidence of this is the decision by leading oil companies to invest heavily in buying up their own stocks. ExxonMobil, for example, with the largest quarterly profit of any company on record—$8.4 billion in the last quarter of 2004—invested nearly $10 billion in buying back its own stock. ChevronTexaco used $2.5 billion of its profits to buy back stock. With little new oil to be discovered and world oil demand growing fast, companies appear to be realizing that their reserves will become even more valuable in the future.[7]

Closely related to this behavior is the lack of any substantial increases in exploration and development in 2005 even though oil prices are well above $50 a barrel. This suggests that the companies agree with petroleum geologists who say that 95 percent of all the oil in the world has already been discovered. "The whole world has now been seismically searched and picked over," says independent geologist Colin Campbell. "Geological knowledge has improved enormously in the past 30 years and it is almost inconceivable now that major fields remain to be found." This also implies that it may take a lot of costly exploration and drilling to find that remaining 5 percent.[8]

This shrinkage of reserves is strikingly evident in the ratio between new oil discoveries and production of the major oil companies. Among those reporting that their 2004 oil production greatly exceeded new discoveries were Royal Dutch/Shell, ChevronTexaco, and Conoco-Phillips. The bottom line is that the oil reserves of major companies are shrinking yearly. On a global scale, geologist Walter Youngquist, author of *GeoDestinies: The Inevitable Control of Earth Resources Over Nations and Individuals*, notes that in 2004 the world produced 30.5 billion barrels of oil but discovered only 7.5 billion barrels of new oil.[9]

The influence on oil production in the years immediately ahead that is most difficult to measure is the emergence of what I call a "depletion psychology." Once oil companies or oil-exporting countries realize that output is about to peak, they will begin to think seriously about how to stretch out their remaining reserves. As it becomes clear that even a moderate cut in production may double world oil prices, the long-term value of their oil will become much clearer.

The geological evidence suggests that world oil production will be peaking sooner rather than later. Matt Simmons, head of the oil investment bank Simmons and Company International and an industry leader, says in reference to new oil fields: "We've run out of good projects. This is not a money issue...if these oil companies had fantastic projects, they'd be out there [developing new fields]." Kenneth Deffeyes, a highly respected geologist and former oil industry employee now at Princeton University, says in his 2005 book, *Beyond Oil*, "It is my opinion that the peak will occur in late 2005 or in the first few months of 2006." Walter Youngquist and A.M. Samsan Bakhtiari of the Iranian National Oil Company both project that oil will peak in 2007.[10]

Sadad al-Husseini, recently retired as head of exploration and production at Aramco, the Saudi national oil company, discussed the world oil prospect with Peter Maass for the *New York Times*. His basic point was that new oil output coming online had to be sufficient to cover both annual growth in world demand of at least 2 million barrels a day and the annual decline in production from existing fields of over 4 million barrels a day. "That's like a whole new Saudi Arabia every couple of years," Husseini said. "It's not sustainable."[11]

Where are companies looking for more oil? Aside from conventional petroleum, the kind that can easily be pumped to the surface, vast amounts of oil are stored in tar sands and can be produced from oil shale. The Athabasca tar sand deposits in Alberta, Canada, may total 1.8 trillion barrels. Of this total, however, it is thought that not more than 300 billion barrels is recoverable. Venezuela also has a large deposit of extra heavy oil, estimated at 1.2 trillion barrels. Perhaps a third of it can be readily recovered. If Venezuela's heavy oil is developed on a large enough scale, its oil production could one day exceed its 1970 historical peak. Oil shale concentrated in Colorado, Wyoming, and Utah in the United States also holds large quantities of kerogen, an organic material that can be converted into oil and gas.[12]

How much oil can be economically produced from oil shale? In the late 1970s the United States launched a major effort to develop oil shale on the western slope of the Rocky Mountains in Colorado. When oil prices dropped in 1982, the oil shale industry collapsed. Exxon quickly pulled out of its $5-billion Colorado project, and the remaining companies soon followed suit. Since this process requires several barrels of water for each barrel of oil produced, water shortages in the region may limit its revival.[13]

The one project that is moving ahead is the tar sands project in Canada's Alberta Province. This initiative, which began in the early 1980s, is now producing a million barrels of oil per day, enough to supply 5 percent of current U.S. oil use. This tar sand oil is not cheap, however, and it wreaks environmental havoc on a vast scale. Heating and extracting the oil from the sands relies on the extensive use of natural gas, production of which has peaked in North America.[14]

Thus although these reserves of oil in tar sands and shale may be vast, gearing up for production is a costly, time-consuming process. At best, the development of tar sands and oil shale is likely only to slow the decline in world oil production.[15]

The Oil Intensity of Food

Modern agriculture depends heavily on the use of gasoline and diesel fuel in tractors for plowing, planting, cultivating, and harvesting. Irrigation pumps use diesel fuel, natural gas, and coal-fired electricity. Fertilizer production is also energy-intensive: the mining, manufacture, and international transport of phosphates and potash all depend on oil. Natural gas, however, is used to synthesize the basic ammonia building block in nitrogen fertilizers.[16]

In the United States, for which reliable historical data are available, the combined use of gasoline and diesel fuel in agriculture has fallen from its historical high of 7.7 billion gallons in 1973 to 4.6 billion in 2002, a decline of 40 percent. For a broad sense of the fuel efficiency trend in U.S. agriculture, the gallons of fuel used per ton of grain produced dropped from 33 in 1973 to 13 in 2002, an impressive decrease of 59 percent.[17]

One reason for this was a shift to minimum and no-till cultural practices on roughly two fifths of U.S. cropland. No-till cultural practices are now used on roughly 95 million hectares worldwide, nearly all of them concentrated in the United States, Brazil, Argentina, and Canada. The United States—with 25 million hectares of minimum or no-till—leads the field, closely followed by Brazil.[18]

While U.S. agricultural use of gasoline and diesel has been declining, in many developing countries it is rising as the shift from draft animals to tractors continues. A generation ago, for example, cropland in China was tilled largely by animals. Today much of the plowing is done with tractors.[19]

Fertilizer accounts for 20 percent of U.S. farm energy use. Worldwide, the figure may be slightly higher. On average, the world produces 13 tons of grain for each ton of fertilizer used. But this varies widely among countries. For example, in China a ton of fertilizer yields 9 tons of grain, in India it yields 11 tons, and in the United States, 18 tons.[20]

U.S. fertilizer efficiency is high because U.S. farmers routinely test their soils to precisely determine crop nutrient needs and because the United States is also the leading producer of soybeans, a leguminous crop that fixes nitrogen in the soil. Soybeans, which rival corn for area planted in the United States, are commonly grown in rotation with corn and, to a lesser degree, with winter wheat. Since corn has a voracious appetite for nitrogen, alternating corn and soybeans in a two-year rotation substantially reduces the nitrogen fertilizer needed for the corn.[21]

Urbanization increases demand for fertilizer. As rural people migrate to cities, it becomes more difficult to recycle the nutrients in human waste back into the soil. Beyond this, the growing international food trade can separate producer and consumer by thousands of miles, further disrupting the nutrient cycle. The United States, for example, exports some 80 million tons of grain per year—grain that contains large quantities of basic plant nutrients: nitrogen, phosphorus, and potassium. The ongoing export of these nutrients would slowly drain the inherent fertility from U.S. cropland if the nutrients were not replaced in chemical form.[22]

Factory farms, like cities, tend to separate producer and consumer, making it difficult to recycle nutrients. Indeed, the nutrients in animal waste that are an asset to farmers become a liability in large feeding operations, often with costly disposal. As oil, and thus fertilizer, become more costly, the economics of factory farms may become less attractive.

Irrigation, another major energy claimant, is taking more and more energy worldwide. In the United States, close to 19 percent of agricultural energy use is for pumping water. In the other two large food producers—China and India—the number is undoubtedly much higher, since irrigation figures so prominently in both countries.[23]

Since 1950 the world's irrigated area has tripled, climbing from 94 million hectares to 277 million hectares in 2002. In addition, the shift from large dams with gravity-fed canal systems that dominated the last century's third quarter to drilled wells that tap underground water resources has also boosted irrigation fuel use.[24]

Some trends, such as the shift to no tillage, are making agriculture less oil-intensive. But rising fertilizer use, the spread of farm mechanization, and falling water tables are making food production more oil-dependent. This helps explain why farmers are becoming involved in the production of biofuels, both ethanol to replace gasoline and biodiesel to replace diesel. (Renewed interest in these fuels is discussed later in this chapter.)

Although attention commonly focuses on energy use on the farm, this accounts for only one fifth of total food system energy use in the United States. Transport, processing, packaging, marketing, and kitchen preparation of food account for nearly four fifths of food system energy use. Indeed, my colleague Danielle Murray notes that the U.S. food economy uses as much energy as France does in its entire economy.[25]

The 14 percent of energy used in the food system to move

goods from farmer to consumer is roughly equal to two thirds of the energy used to produce the food. And an estimated 16 percent of food system energy use is devoted to processing—canning, freezing, and drying food—everything from frozen orange juice concentrate to canned peas.[26]

Food staples, such as wheat, have traditionally moved over long distances by ship, traveling from the United States to Europe, for example. What is new is the shipment of fresh fruits and vegetables over vast distances by air. Few economic activities are more energy-intensive.[27]

Food miles—the distance food travels from producer to consumer—have risen with cheap oil. Among the longest hauls are the flights during the northern hemisphere winter that carry fresh produce, such as blueberries from New Zealand to the United Kingdom. At my local supermarket in downtown Washington, D.C., the fresh grapes in winter typically come by plane from Chile, traveling almost 5,000 miles. Occasionally they come from South Africa, in which case the distance from grape arbor to dining room table is 8,000 miles, nearly a third of the way around the earth.[28]

One of the most routine long-distance movements of fresh produce is from California to the heavily populated U.S. East Coast. Most of this produce moves by refrigerated trucks. In assessing the future of long-distance produce transport, one oil analyst observed that the days of the 3,000-mile Caesar salad may be numbered.[29]

Packaging is also surprisingly energy-intensive, accounting for 7 percent of food system energy use. It is not uncommon for the energy invested in packaging to exceed that of the food it contains. And worse, nearly all the packaging in a modern supermarket is designed to be discarded after one use.[30]

The most energy-intensive segment of the food chain is the kitchen. Much more energy is used to refrigerate and prepare food in the home than is used to produce it in the first place. The big energy user in the food system is the kitchen refrigerator, not the farm tractor.[31]

While the use of oil dominates the production end of the food system, electricity (usually produced from coal or gas) dominates the consumption end. The oil-intensive modern food system that evolved when oil was cheap will not survive as it is now structured with higher energy prices. Among the principal adjustments will be more local food production and movement down the food chain as consumers react to rising food prices by buying fewer high-cost livestock products.

The Falling Wheat-Oil Exchange Rate

While we focus on the oil used to produce food, the amount of oil that food will buy is falling precipitously. The shift in terms of trade between wheat and oil is both dramatic and ongoing. From 1950 to 1973, the prices of both wheat and oil were remarkably stable, as was the relationship between the two. At any time during the 23-year span, a bushel of wheat could be traded for a barrel of oil in the world market. (See Table 2–1.)[32]

Since 1973, however, the relative values of wheat and oil have shifted dramatically. In 2005, it took 13 bushels of wheat to buy a barrel of oil. The two countries most affected by this dramatic shift are the leading exporters of these two commodities: the United States and Saudi Arabia.[33]

Table 2–1. *The Wheat/Oil Exchange Rate, 1950–2005*

Year	Bushel of Wheat	Barrel of Oil	Bushels Per Barrel
	(dollars)		(ratio)
1950	1.89	1.71	1
1955	1.81	2.11	1
1960	1.58	1.85	1
1965	1.62	1.79	1
1970	1.49	1.79	1
1975	4.06	11.45	3
1980	4.70	35.71	8
1985	3.70	27.37	7
1990	3.69	22.99	6
1995	4.82	17.20	4
2000	3.10	28.23	9
2005*	3.90	52.00	13

*2005 figures are author's estimates based on January–August data.
Source: See endnote 32.

The United States, both the largest importer of oil and the largest exporter of grain, is paying dearly for this shift in the wheat-oil exchange rate. The 13-fold shift since 1973 is contributing to the largest U.S. trade deficit in history and a record external debt. In contrast, Saudi Arabia—the world's leading oil exporter and a leading grain importer—is benefiting handsomely.[34]

While the exchange rate between grain and oil was deteriorating, U.S. oil imports were climbing. During the early 1970s, before the OPEC oil price hikes, the United States largely could pay its oil import bill with grain exports. But in 2004, grain exports covered only 13 percent of the staggering U.S. oil import bill of $132 billion.[35]

The first big adjustment between oil and wheat came when OPEC tripled the price of oil at the end of 1973. During 1974–78, it took roughly three bushels of wheat to buy a barrel of oil. Then after the second OPEC oil price hike, which boosted oil from $13 per barrel in 1978 to $30 in 1980, it took eight bushels of wheat to buy a barrel of oil.[36]

This steep rise in the buying power of oil led to one of the most abrupt transfers of wealth in history. The coffers of Saudi Arabia, Kuwait, Iraq, and Iran began to overflow with dollars while those of oil-importing countries were being emptied.

No one knows exactly what will happen to the wheat-oil exchange rate in the years ahead, but as the number of grain-based ethanol distilleries producing automotive fuel grows, the profitability of converting grain into fuel may stabilize the wheat-oil exchange rate.

The United States is pressing the Saudis to produce more oil. Yet the answer is not for the Saudis to produce more, even if they can, but for the United States to consume less. Unless the United States assumes a leadership role, Saudi Arabia will continue to dictate not only the exchange rate between oil and grain but also U.S. gasoline prices.

Food and Fuel Compete for Land

Historically, the world's farmers produced food, feed, and fiber. Today they are starting to produce fuel as well. Since nearly everything we eat can be converted into automotive fuel, the high price of oil is becoming the support price for farm products. It is also determining the price of food. On any given day there are now two groups of buyers in world commodity markets: one representing food processors and another representing biofuel producers. The line between the food and fuel economies has suddenly blurred as service stations compete with supermarkets for the same commodities.

First triggered by the oil shocks of the 1970s, production of biofuels—principally ethanol from sugarcane in Brazil and corn in the United States—grew rapidly for some years but then stagnated during the 1990s. After 2000, as oil prices edged upward, it began to again gain momentum. (See Figure 2–1.) Europe, meanwhile, led by Germany and France, was starting to extract biodiesel from oilseeds.[37]

Production of biofuels in 2005 equaled nearly 2 percent of world gasoline use. From 2000 to 2005, ethanol production worldwide increased from 4.6 billion to 12.2 billion gallons, a jump of 165 percent. Biodiesel, starting from a small base of 251 million gallons in 2000, climbed to an estimated 790 million gallons in 2005, more than tripling.[38]

Governments support biofuel production because of concerns about climate change and a possible shrinkage in the flow of imported oil. Since substituting biofuels for gasoline reduces carbon emissions, governments see this as a way to meet their carbon reduction goals. Biofuels also have a domestic econom-

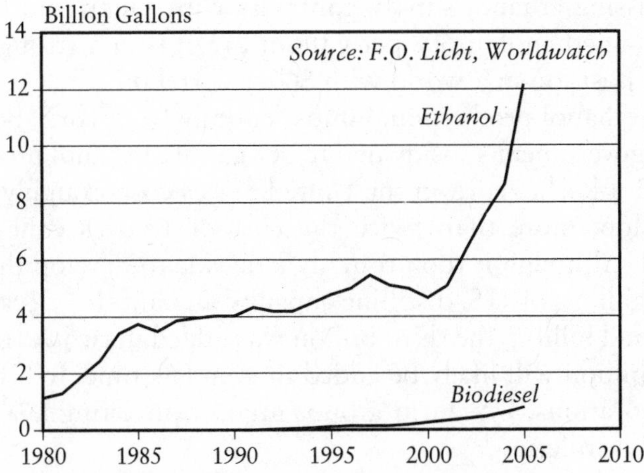

Figure 2–1. *World Ethanol and Biodiesel Production, 1980–2005*

ic appeal partly because locally produced fuel creates jobs and keeps money within the country.

Brazil, using sugarcane as the feedstock for ethanol, is producing some 4 billion gallons a year, satisfying 40 percent of its automotive fuel needs. The United States, using corn as the feedstock, produced 3.4 billion gallons of ethanol in 2004, supplying just under 2 percent of the fuel used by its vast automotive fleet. Forecasts for 2005 show U.S. ethanol output overtaking that of Brazil, at least temporarily. Europe ranks third in fuel ethanol output, the lion's share from France, the United Kingdom, and Spain. Europe's distillers use mostly sugar beets, wheat, and barley.[39]

Interest in biofuels has escalated sharply since oil prices reached $40 per barrel in mid-2004. Brazil, the world's largest sugarcane producer, is emerging as the world leader in farm fuel production. In 2004, half of its sugarcane crop was used for sugar and half for ethanol. Expanding the sugarcane area from 5.3 million hectares in 2005 to some 8 million hectares would enable it to become self-sufficient in automotive fuel within a matter of years while maintaining its sugar production and exports.[40]

Even though Brazil has phased out ethanol subsidies, by mid-2005 the private sector had committed $5.1 billion to investment in sugar mills and distilleries over the next five years. Thinking beyond its currently modest exports of ethanol, Brazil is discussing ethanol supply contracts with Japan and China. Producing ethanol at 60¢ per gallon, Brazil is in a strong competitive position in a world with $60-a-barrel oil.[41]

U.S. ethanol production, almost entirely from corn, benefits from a government subsidy of 51¢ per gallon. Ethanol produced from $3-a-bushel corn in the United States costs roughly $1.40 per gallon, more than twice the cost of Brazil's cane-based ethanol. Although it took roughly a decade to develop the first billion gallons of U.S. distilling capacity and another decade for the second billion, the third billion was added in two years. The fourth billion will likely be added in even less time. In addition to corporations, U.S. farm groups are also investing heavily in ethanol distilleries.[42]

India, the world's second largest producer of sugarcane, has 10 ethanol plants in operation and expects to have 20 additional plants up and running by the end of 2005. China is projected to bring on-line four plants producing up to 360 million gallons of additional fuel ethanol by the end of 2005, mostly from corn and wheat.[43]

Colombia and the Central American countries represent the other biofuel hot spot. Colombia is off to a fast start, opening one new ethanol distillery each month from August 2005 through the end of the year. The challenge is to coordinate growth in distillery construction with growth in the land in sugarcane.[44]

For biofuels used in diesel engines, Europe is the leader. Germany, producing 326 million gallons of biodiesel in 2004, is now covering 3 percent of its diesel fuel needs. Relying almost entirely on rapeseed (the principal source of cooking oil in Europe), it plans to expand output by half within the next few years.[45]

France, where biodiesel production totaled 150 million gallons in 2004, plans to double its output by 2007. Like Germany, it uses rapeseed as its feedstock. In both countries the impetus for biodiesel production comes from the European Union's goal of meeting 5.75 percent of automotive fuel needs with biofuels by 2010. Biofuels in Europe are exempted from the hefty taxes levied on gasoline and diesel.[46]

In the United States, a latecomer to biodiesel production, output is growing rapidly since the 2003 adoption of a $1-per-gallon subsidy that took effect in January 2005. Iowa, a leading soybean producer and a hotbed of soy-fuel enthusiasm, now has three biodiesel plants in operation, another under construction, and five more in the planning stages. State officials estimate that biodiesel plants will be extracting oil from 200 million bushels of the state's 500-million-bushel annual harvest within a few years, producing 280 million gallons of biodiesel. The four fifths of the soybean left after the oil is extracted is a protein-rich livestock feed supplement, which is even more valuable than the oil itself.[47]

Other countries either producing biodiesel or planning to do so include Malaysia, Indonesia, and Brazil. Malaysia and Indonesia, the major producers of palm oil, would likely use highly productive oil palm plantations as their feedstock source. Brazil, which has ambitious plans to ramp up biodiesel production, will also likely turn to palm oil.[48]

There are two key indicators in evaluating crops for biofuel production: the fuel yield per acre and the net energy yield of the biofuels, after subtracting the energy used in both production and refining. For ethanol, the top yields per acre are 714 gallons from sugar beets in France and 662 gallons per acre for sugarcane in Brazil. (See Table 2–2.) U.S. corn comes in at 354 gallons per acre, or roughly half the beet and cane yields.[49]

With biodiesel production, oil palm plantations are a strong first, with a yield of 508 gallons per acre. Next comes coconut oil, with 230 gallons per acre, and rapeseed, at 102 gallons per acre. Soybeans, grown primarily for their protein content, yield only 56 gallons per acre.[50]

For net energy yield, ethanol from sugarcane in Brazil is in a class all by itself, yielding over 8 units of energy for each unit

Table 2–2. *Ethanol and Biodiesel Yield per Acre from Selected Crops*

Fuel	Crop	Fuel Yield (gallons)
Ethanol		
	Sugar beet (France)	714
	Sugarcane (Brazil)	662
	Cassava (Nigeria)	410
	Sweet Sorghum (India)	374
	Corn (U.S.)	354
	Wheat (France)	277
Biodiesel		
	Oil palm	508
	Coconut	230
	Rapeseed	102
	Peanut	90
	Sunflower	82
	Soybean	56*

*Author's estimate

Note: Crop yields can vary widely. Ethanol yields given are from optimal growing regions. Biodiesel yield estimates are conservative. The energy content of ethanol is about 67 percent that of gasoline. The energy content of biodiesel is about 90 percent that of petroleum diesel.

Source: See endnote 49.

invested in cane production and ethanol distillation. Once the sugary syrup is removed from the cane, the fibrous remainder, bagasse, is burned to provide the heat needed for distillation, eliminating the need for an additional external energy source. This helps explain why Brazil can produce cane-based ethanol for 60¢ per gallon.[51]

Ethanol from sugar beets in France comes in at 1.9 energy units for each unit of invested energy. Among the three principal feedstocks now used for ethanol production, U.S. corn-based ethanol, which relies largely on natural gas for distillation energy, comes in a distant third in net energy efficiency, yielding only 1.5 units of energy for each energy unit used.[52]

Another perhaps more promising option for producing ethanol is to use enzymes to break down cellulosic materials, such as switchgrass, a vigorously growing perennial grass, or fast-growing trees, such as hybrid poplars. Ethanol is now being produced from cellulose in a small demonstration plant in Canada. If switchgrass turns out to be an economic source of ethanol, as some analysts think it may, it will be a major breakthrough, since it can be grown on land that is highly erodible or otherwise not suitable for annual crops. In a competitive world market for crop-based ethanol, the future belongs to sugarcane and switchgrass.[53]

The ethanol yield per acre for switchgrass is calculated at 1,150 gallons, higher even than for sugarcane. The net energy yield, however, is roughly 4, far above the 1.5 for corn but less than the 8 for sugarcane.[54]

Aside from the prospective use of cellulose, current and planned ethanol-producing operations use food crops such as sugarcane, sugar beets, corn, wheat, and barley. The United States, for example, in 2004 used 32 million tons of corn to produce 3.4 billion gallons of ethanol. Although this is scarcely 12 percent of the huge U.S. corn crop, it is enough to feed 100 million people at average world grain consumption levels.[55]

In an oil-short world, what will be the economic and environmental effects of agriculture's emergence as a producer of transport fuels? Agriculture's role in the global economy clearly will be strengthened as it faces a vast, virtually unlimited market for automotive fuel. Tropical and subtropical countries that can produce sugarcane or palm oil will be able to fully exploit their year-round growing conditions, giving them a strong comparative advantage in the world market.

With biofuel production spreading, the world price for oil will, in effect, become a support price for farm products. If food and feed crop prices are weak and oil prices are high, commodities will go to fuel producers. For example, vegetable oils trading on European markets on any given day may end up in either supermarkets or service stations.

The risk is that economic pressures to clear land for expanding sugarcane production in the Brazilian *cerrado* and Amazon basin and for palm oil plantations in countries such as Indonesia and Malaysia will pose a major new threat to plant and animal diversity. In the absence of governmental constraints, the rising price of oil could quickly become the leading threat to biodiversity, ensuring that the wave of extinctions now under way does indeed become the sixth great extinction.

With oil prices now high enough to stimulate potentially massive investments in fuel crop production, the world farm economy—already struggling to feed 6.5 billion people—will face far greater demands. How the world manages this new incredibly complex situation will tell us a great deal about the prospect for our energy-hungry twenty-first century civilization.[56]

Cities and Suburbs After Peak Oil

Modern cities are a product of the oil age. From the first cities, which apparently took shape in Mesopotamia some 6,000 years ago, until 1900, urbanization was a slow, barely perceptible process. When the last century began, there were only a few cities with a million people. Today there are more than 400 cities that large, and 20 mega-cities have 10 million or more residents.[57]

The metabolism of cities depends on concentrating vast amounts of food and materials and then disposing of garbage and human waste. With the limited range and capacity of horse-drawn wagons, it was difficult to create large cities. Trucks running on cheap oil changed all that.

As cities grow ever larger and as nearby landfills reach capacity, garbage must be hauled longer distances to disposal sites. With oil prices rising and available landfills receding ever further from the city, the cost of garbage disposal also rises. At some point, many throwaway products may be priced out of existence.

Urban living costs will likely rise as oil production turns down and oil prices escalate. One of the intriguing questions this raises is whether urbanization will continue APO, after peak oil. Or might the process even be reversed when people seek less oil-dependent lifestyles?

Cities will be hard hit by the coming decline in oil production, but suburbs will be hit even harder. People living in poorly designed suburbs not only depend on importing everything, they are also often isolated geographically from their jobs and shops. They must drive for virtually everything they need. Living in suburban housing developments often means using a car even to get a loaf of bread or a quart of milk.

Suburbs have created a commuter culture, with the daily roundtrip commute taking, on average, close to an hour a day in the United States. While Europe's cities were largely mature before the onslaught of the automobile, those in the United States, a much younger country, were shaped by the car. While city limits are usually rather clearly defined in Europe, and while Europeans only reluctantly convert productive farmland into housing developments, Americans have few qualms about this because of a frontier mentality and because cropland was long seen as a surplus commodity.

This unsightly, aesthetically incongruous sprawl of suburbs and strip malls is not limited to the United States. It is found in Latin America, in Southeast Asia, and increasingly in China. Flying from Shanghai to Beijing provides a good view of the sprawl of buildings, including homes and factories, that is following the new roads and highways. This is in sharp contrast to the tightly built villages that shaped residential land use for millennia in China.

Shopping malls and huge discount stores, symbolized in the public mind by Wal-Mart, were all subsidized by artificially cheap oil. Isolated by high oil prices, suburbs may prove to be ecologically and economically unsustainable. Thomas Wheeler, editor of the *Alternative Press Review*, observes that "there will eventually be a great scramble to get out of the suburbs as the world oil crisis deepens and the property values of suburban homes plummet."[58]

The World After Oil Peaks

Peak oil is described as the point where oil production stops rising and begins its unavoidable long-term decline. In the face of fast-growing demand, this means rising oil prices. But even if oil production growth simply slows or plateaus, the resulting tightening in supplies will still drive the price of oil upward, albeit less rapidly.

Few countries are planning a reduction in their use of oil. Indeed, the projections of oil use by both the International Energy Agency and the U.S. Department of Energy show world oil consumption going from roughly 84 million barrels a day at present to 120 million barrels a day by 2030. According to these analyses, oil consumption in individual countries will be increasing on average by nearly half over the next 20 years. How did they come up with these "rosy" forecasts? To quote Thomas Wheeler again, are many analysts and leaders simply "oblivious to the flashing red light on the earth's fuel gauge?"[59]

Even though peak oil may be imminent, most countries are counting on much higher oil consumption in the decades ahead. Indeed, they are building automobile assembly plants, roads, highways, parking lots, and suburban housing developments as though cheap oil will last forever. New airliners are being delivered with the expectation that air travel and freight will expand indefinitely. Yet in a world of declining oil production, no country can use more oil except at the expense of others.[60]

Some segments of the global economy will be affected more than others simply because some are more oil-intensive. Among these are the automobile, food, and airline industries. Stresses within the U.S. auto industry were already evident before oil prices started climbing in mid-2004. Now General Motors and Ford, both trapped with their heavy reliance on sales of gas-hogging sport utility vehicles, have seen Standard and Poors lower their credit ratings, reducing their corporate bonds to junk bond status. In June 2005, General Motors announced that it planned to cut its U.S. workforce of 110,000 by 25,000 workers in 2007.[61]

Although it is the troubled automobile manufacturers that appear in the headlines as oil prices rise, their affiliated industries will also be affected, including auto parts and tire manufacturers.

The food sector will be affected in two ways. Food will become more costly as higher oil prices drive up production costs. As oil costs rise, diets will be altered as people move down the food chain and as they consume more local, seasonally produced food. Diets will thus become more closely attuned to local products and more seasonal in nature.

At the same time, rising oil prices will also be drawing agricultural resources into the production of fuel crops, either ethanol or biodiesel. Higher oil prices are thus setting up competition between affluent motorists and low-income food consumers for food resources, presenting the world with a complex new ethical issue.

Airlines, both passenger travel and freight, will continue to suffer as jet fuel prices climb, simply because fuel is their biggest operating expense. Although industry projections show air passenger travel growing by some 5 percent a year for the next decade, this seems highly unlikely. Cheap airfares may soon become history.[62]

Air freight may be hit even harder, perhaps leading to an absolute decline. One of the early casualties of rising oil prices could be the use of jumbo jets to transport fresh produce from the southern hemisphere to industrial countries during the northern winter. The price of fresh produce out of season may simply become prohibitive.

During the century of cheap oil, an enormous automobile infrastructure was built in industrial countries that requires large amounts of energy to maintain. The United States, for example, has 2.6 million miles of paved roads, covered mostly with asphalt, and 1.4 million miles of unpaved roads to maintain even if world oil production is falling. Higher energy prices may create a maintenance crisis.[63]

In addition to needing to use oil more efficiently, the world is also looking to other sources of energy. Although nuclear power has been getting some press attention as an alternative to fossil fuels, electricity from nuclear power plants is costly. On a level playing field with no taxpayer subsidies, nuclear power is dead. If utilities pay the full costs of nuclear waste disposal, of insurance against an accident, and of decommissioning plants that are worn out, the expense of nuclear power will take it out of the running. And with international terrorism on the rise, the vulnerability of nuclear power plants to attack combined with their use by countries as a steppingstone to the acquisition of nuclear weapons virtually eliminates nuclear fission as a future energy source.[64]

The relative abundance of coal makes it an attractive energy source in some quarters, but it is likely to soon become a victim of mounting public concern about climate change. This means a future of renewable sources of energy, including wind energy, solar cells, solar thermal panels, solar thermal power plants, geothermal energy, hydropower, wave power, and biofuels.

In the coming energy transition, there will be winners and losers. Countries that fail to plan ahead, that lag in investing in more oil-efficient technologies and new energy sources, may experience a decline in living standards. The inability of national governments to manage the energy transition could lead to a failure of confidence in leaders and to failed states.

National political leaders seem reluctant to face the coming downturn in oil and to plan for it even though it will become one of the great fault lines not only in recent economic history but in the history of civilization. Trends now taken for granted, such as urbanization and globalization, could be reversed almost overnight as oil becomes scarce and costly.

Developing countries will be hit doubly hard as still-expanding populations combine with a shrinking oil supply to steadily reduce oil use per person. Such a decline could quickly translate into a fall in living standards. If the United States, the world's largest oil consumer and importer, can sharply reduce its use of oil, it can buy the world time for a smoother transition to the post-petroleum era. What the world needs today is not more oil, but more leadership.

3

Emerging Water Shortages

Africa's Lake Chad, once a landmark for astronauts circling the earth, is now difficult for them to locate. Surrounded by Chad, Niger, and Nigeria—three countries with some of the world's fastest-growing populations—the lake has shrunk by 95 percent since the 1960s. The soaring demand for irrigation water in that area is draining dry the rivers and streams the lake depends on for its existence. As a result, Lake Chad may soon disappear entirely, its whereabouts a mystery to future generations.[1]

Every day, it seems, we read about lakes disappearing, wells going dry, or rivers failing to reach the sea. But these stories typically describe local situations. It is not until we begin to compile the numerous national studies—such as an 824-page analysis of the water situation in China, a World Bank study of the water situation in Yemen, or a detailed U.S. Department of Agriculture (USDA) assessment of the irrigation prospect in the western United States—that the extent of emerging water shortages worldwide can be grasped. Only then can we see the extent of water overuse and the decline it can bring.[2]

The world is incurring a vast water deficit—one that is largely invisible, historically recent, and growing fast. Because much of the deficit comes from aquifer overpumping, it is often not apparent. Unlike burning forests or invading sand dunes, falling water tables are often discovered only when wells go dry.

This global water deficit is recent, the result of demand tripling over the last half-century. The drilling of millions of irrigation wells has pushed water withdrawals beyond the recharge of many aquifers. The failure of governments to limit pumping to the sustainable yield of aquifers means that water tables are now falling in countries that contain more than half the world's people.[3]

Among the more visible manifestations of water scarcity are rivers running dry and lakes disappearing. A politics of water scarcity is emerging between upstream and downstream claimants both within and among countries. Water scarcity is now crossing borders via the international grain trade. Countries that are pressing against the limits of their water supply typically satisfy the growing need of cities and industry by diverting irrigation water from agriculture, and then importing grain to offset the loss of productive capacity.

The link between water and food is strong. We each drink on average nearly 4 liters of water per day in one form or another, while the water required to produce our daily food totals at least 2,000 liters—500 times as much. This helps explain why 70 percent of all water use is for one purpose—irrigation. Another 20 percent is used by industry, and 10 percent goes for residential purposes. With the demand for water growing in all three categories, competition among sectors is intensifying, with farmers almost always losing.[4]

Falling Water Tables

Scores of countries are overpumping aquifers as they struggle to satisfy their growing water needs, including each of the big three grain producers—China, India, and the United States. These three, along with a number of other countries where water tables are falling, are home to more than half the world's people. (See Table 3–1.)[5]

There are two types of aquifers: replenishable and nonreplenishable (or fossil) aquifers. Most of the aquifers in India and the shallow aquifer under the North China Plain are replen-

Table 3–1. *Countries Overpumping Aquifers in 2005*

Country	Population (million)
China	1,316
India	1,103
Iran	70
Israel	7
Jordan	6
Mexico	107
Morocco	31
Pakistan	158
Saudi Arabia	25
South Korea	48
Spain	43
Syria	19
Tunisia	10
United States	298
Yemen	21
Total	3,262

Source: See endnote 5.

ishable. When these are depleted, the maximum rate of pumping is automatically reduced to the rate of recharge.

For fossil aquifers, such as the vast U.S. Ogallala aquifer, the deep aquifer under the North China Plain, or the Saudi aquifer, depletion brings pumping to an end. Farmers who lose their irrigation water have the option of returning to lower-yield dryland farming if rainfall permits. In more arid regions, however, such as in the southwestern United States or the Middle East, the loss of irrigation water means the end of agriculture.

Falling water tables are already adversely affecting harvests in some countries, including China, the world's largest grain producer. A groundwater survey released in Beijing in August 2001 revealed that the water table under the North China Plain, which produces over half of that country's wheat and a third of its corn, is falling faster than earlier reported. Overpumping has largely depleted the shallow aquifer, forcing well drillers to turn to the region's deep fossil aquifer, which is not replenishable.[6]

The survey, conducted by the Geological Environmental Monitoring Institute (GEMI) in Beijing, reported that under Hebei Province in the heart of the North China Plain, the average level of the deep aquifer was dropping nearly 3 meters (10 feet) per year. Around some cities in the province, it was falling twice as fast. He Qingcheng, head of the GEMI groundwater monitoring team, notes that as the deep aquifer is depleted, the region is losing its last water reserve—its only safety cushion.[7]

His concerns are mirrored in a World Bank report: "Anecdotal evidence suggests that deep wells [drilled] around Beijing now have to reach 1,000 meters [more than half a mile] to tap fresh water, adding dramatically to the cost of supply." In unusually strong language for a Bank report, it foresees "catastrophic consequences for future generations" unless water use and supply can quickly be brought back into balance.[8]

The U.S. embassy in Beijing reports that wheat farmers in some areas are now pumping from a depth of 300 meters, or nearly 1,000 feet. Pumping water from this far down raises pumping costs so high that farmers are often forced to abandon irrigation and return to less productive dryland farming.[9]

Falling water tables, the conversion of cropland to nonfarm uses, and the loss of farm labor in provinces that are rapidly industrializing are combining to shrink China's grain harvest. The wheat crop, grown mostly in semiarid northern China, is particularly vulnerable to water shortages. After peaking at 123 million tons in 1997, the harvest has fallen in five of the last eight years, coming in at 95 million tons in 2005, a drop of 23 percent.[10]

The U.S. embassy also reports that the recent decline in rice production is partly a result of water shortages. After peaking at 140 million tons in 1997, the harvest dropped in four of the following eight years, falling to an estimated 127 million tons in 2005. Only corn, China's third major grain, has thus far avoided a decline. This is because corn prices are favorable and because the crop is not as irrigation-dependent as wheat and rice are.[11]

Overall, China's grain production has fallen from its historical peak of 392 million tons in 1998 to an estimated 358 million tons in 2005. For perspective, this drop of 34 million tons exceeds the annual Canadian wheat harvest. China largely covered the drop-off in production by drawing down its once vast stocks until 2004, at which point it imported 7 million tons of grain.[12]

A World Bank study indicates that China is overpumping three river basins in the north—the Hai, which flows through Beijing and Tianjin; the Yellow; and the Huai, the next river south of the Yellow. Since it takes 1,000 tons of water to produce one ton of grain, the shortfall in the Hai basin of nearly 40 billion tons of water per year (1 ton equals 1 cubic meter) means that when the aquifer is depleted, the grain harvest will drop by 40 million tons—enough to feed 120 million Chinese.[13]

Of the leading grain producers, only China has thus far experienced a substantial decline in production. Even with a worldwide grain crunch and climbing grain prices providing an incentive to boost production, it will be difficult for China to regain earlier grain production levels, given the loss of irrigation water.[14]

Serious though emerging water shortages are in China, they are even more serious in India simply because the margin between actual food consumption and survival is so precarious. In a survey of India's water situation, Fred Pearce reported in the *New Scientist* that the 21 million wells drilled in this global epicenter of well-drilling are lowering water tables in most of the country. In North Gujarat, the water table is falling by 6 meters (20 feet) per year.[15]

In Tamil Nadu, a state with more than 62 million people in southern India, wells are going dry almost everywhere. According to Kuppannan Palanisami of Tamil Nadu Agricultural University, falling water tables have dried up 95 percent of the wells owned by small farmers, reducing the irrigated area in the state by half over the last decade.[16]

As water tables fall, well drillers are using modified oil-drilling technology to reach water, going as deep as 1,000 meters in some locations. In communities where underground water sources have dried up entirely, all agriculture is rain-fed and drinking water is trucked in. Tushaar Shah, who heads the International Water Management Institute's groundwater station in Gujarat, says of India's water situation, "When the balloon bursts, untold anarchy will be the lot of rural India."[17]

At this point, the harvests of wheat and rice, India's principal food grains, are still increasing. But within the next few years, the loss of irrigation water could override technological progress and start shrinking the harvest in some areas, as it is already doing in China.[18]

In the United States, the USDA reports that in parts of Texas, Oklahoma, and Kansas—three leading grain-producing states—the underground water table has dropped by more than 30 meters (100 feet). As a result, wells have gone dry on thousands of farms in the southern Great Plains. Although this mining of underground water is taking a toll on U.S. grain production, irrigated land accounts for only one fifth of the U.S. grain harvest, compared with close to three fifths of the harvest in India and four fifths in China.[19]

Pakistan, a country with 158 million people that is growing by 3 million per year, is also mining its underground water. In the Pakistani part of the fertile Punjab plain, the drop in water tables appears to be similar to that in India. Observation wells near the twin cities of Islamabad and Rawalpindi show a fall in the water table between 1982 and 2000 that ranges from 1 to nearly 2 meters a year.[20]

In the province of Baluchistan, water tables around the capital, Quetta, are falling by 3.5 meters per year. Richard Garstang, a water expert with the World Wildlife Fund and a participant in a study of Pakistan's water situation, said in 2001 that "within 15 years Quetta will run out of water if the current consumption rate continues."[21]

The water shortage in Baluchistan is province-wide. Sardar Riaz A. Khan, former director of Pakistan's Arid Zone Research Institute in Quetta, reports that six basins have exhausted their groundwater supplies, leaving their irrigated lands barren. Khan expects that within 10–15 years virtually all the basins outside the canal-irrigated areas will have depleted their groundwater supplies, depriving the province of much of its grain harvest.[22]

Future irrigation water cutbacks as a result of aquifer depletion will undoubtedly reduce Pakistan's grain harvest. Countrywide, the harvest of wheat—the principal food staple—is continuing to grow, but more slowly than in the past.[23]

Iran, a country of 70 million people, is overpumping its aquifers by an average of 5 billion tons of water per year, the water equivalent of one third of its annual grain harvest. Under the small but agriculturally rich Chenaran Plain in northeastern Iran, the water table was falling by 2.8 meters a year in the late 1990s. New wells being drilled both for irrigation and to supply the nearby city of Mashad are responsible. Villages in eastern Iran are being abandoned as wells go dry, generating a flow of "water refugees."[24]

Saudi Arabia, a country of 25 million people, is as water-poor as it is oil-rich. Relying heavily on subsidies, it developed an extensive irrigated agriculture based largely on its deep fossil aquifer. After several years of using oil money to support wheat prices at five times the world market level, the government was forced to face fiscal reality and cut the subsidies. Its wheat harvest dropped from a high of 4.1 million tons in 1992 to 1.2 million tons in 2005, a drop of 71 percent.[25]

Craig Smith writes in the *New York Times*, "From the air, the circular wheat fields of this arid land's breadbasket look like forest green poker chips strewn across the brown desert. But they are outnumbered by the ghostly silhouettes of fields left to fade back into the sand, places where the kingdom's gamble on agriculture has sucked precious aquifers dry." Some Saudi farmers are now pumping water from wells that are 4,000 feet deep, nearly four fifths of a mile (1 mile equals 1.61 kilometers).[26]

A 1984 Saudi national survey reported fossil water reserves at 462 billion tons. Half of that, Smith reports, has probably disappeared by now. This suggests that irrigated agriculture could last for another decade or so and then will largely vanish, limited to the small area that can be irrigated with water from the shallow aquifers that are replenished by the kingdom's sparse rainfall. It is a classic example of an overshoot-and-collapse food economy.[27]

In neighboring Yemen, a nation of 21 million, the water table under most of the country is falling by roughly 2 meters a year as water use outstrips the sustainable yield of aquifers. In western Yemen's Sana'a Basin, the estimated annual water extraction of 224 million tons exceeds the annual recharge of 42 million tons by a factor of five, dropping the water table 6 meters per year. World Bank projections indicate the Sana'a Basin—site of the national capital, Sana'a, and home to 2 million people—will be pumped dry by 2010.[28]

In the search for water, the Yemeni government has drilled test wells in the basin that are 2 kilometers (1.2 miles) deep—depths normally associated with the oil industry—but they have failed to find water. Yemen must soon decide whether to bring water to Sana'a, possibly by pipeline from coastal desalting plants, if it can afford it, or to relocate the capital. Either alternative will be costly and potentially traumatic.[29]

With its population growing at 3 percent a year and with water tables falling everywhere, Yemen is fast becoming a hydrological basket case. Aside from the effect of overpumping on the capital, World Bank official Christopher Ward observes that "groundwater is being mined at such a rate that parts of the rural economy could disappear within a generation."[30]

Israel, even though it is a pioneer in raising irrigation water productivity, is depleting both of its principal aquifers—the coastal aquifer and the mountain aquifer that it shares with Palestinians. Israel's population, whose growth is fueled by both natural increase and immigration, is outgrowing its water supply. Conflicts between Israelis and Palestinians over the allocation of water in the latter area are ongoing. Because of severe water shortages, Israel has banned the irrigation of wheat.[31]

In Mexico—home to a population of 107 million that is projected to reach 140 million by 2050—the demand for water is outstripping supply. Mexico City's water problems are well known. Rural areas are also suffering. For example, in the agricultural state of Guanajuato, the water table is falling by 2 meters or more a year. At the national level, 51 percent of all the water extracted from underground is from aquifers that are being overpumped.[32]

Since the overpumping of aquifers is occurring in many countries more or less simultaneously, the depletion of aquifers and the resulting harvest cutbacks could come at roughly the same time. And the accelerating depletion of aquifers means this day may come soon, creating potentially unmanageable food scarcity.

Rivers Running Dry

While falling water tables are largely hidden, rivers that are drained dry before they reach the sea are highly visible. Two rivers where this phenomenon can be seen are the Colorado, the major river in the southwestern United States, and the Yellow, the largest river in northern China. Other large rivers that either run dry or are reduced to a mere trickle during the dry season are the Nile, the lifeline of Egypt; the Indus, which supplies most of Pakistan's irrigation water; and the Ganges in India's densely populated Gangetic basin. Many smaller rivers have disappeared entirely.[33]

As the world's demand for water has tripled over the last half-century and as the demand for hydroelectric power has grown even faster, dams and diversions of river water have drained many rivers dry. As water tables have fallen, the springs that feed rivers have gone dry, reducing river flows.[34]

Since 1950, the number of large dams, those over 15 meters high, has increased from 5,000 to 45,000. Each dam deprives a river of some of its flow. Engineers like to say that dams built to generate electricity do not take water from the river, only its energy, but this is not entirely true since reservoirs increase evaporation. The annual loss of water from a reservoir in arid or semiarid regions, where evaporation rates are high, is typically equal to 10 percent of its storage capacity.[35]

The Colorado River now rarely makes it to the sea. With the states of Colorado, Utah, Arizona, Nevada, and, most important, California depending heavily on the Colorado's water, the river is simply drained dry before it reaches the Gulf of California. This excessive demand for water is destroying the river's ecosystem, including its fisheries.[36]

A similar situation exists in Central Asia. The Amu Darya—which, along with the Syr Darya, feeds the Aral Sea—is now drained dry by Uzbek and Turkmen cotton farmers upstream. With the flow of the Amu Darya cut off, only the diminished flow of the Syr Darya keeps the Aral Sea from disappearing entirely.[37]

China's Yellow River, which flows some 4,000 kilometers through five provinces before it reaches the Yellow Sea, has been under mounting pressure for several decades. It first ran dry in 1972, and since 1985 it has often failed to reach the sea.[38]

The Nile, site of another ancient civilization, now barely makes it to the sea. Water analyst Sandra Postel, in *Pillar of Sand*, notes that before the Aswan Dam was built, some 32 billion cubic meters of water reached the Mediterranean each year.

After the dam was completed, however, increasing irrigation, evaporation, and other demands reduced its discharge to less than 2 billion cubic meters.[39]

Pakistan, like Egypt, is essentially a river-based civilization, heavily dependent on the Indus. This river, originating in the Himalayas and flowing westward to the Indian Ocean, not only provides surface water, it also recharges aquifers that supply the irrigation wells dotting the Pakistani countryside. In the face of growing water demand, it too is starting to run dry in its lower reaches. Pakistan, with a population projected to reach 305 million by 2050, is in trouble.[40]

In Southeast Asia, the flow of the Mekong is being reduced by the dams being built on its upper reaches by the Chinese. The downstream countries, including Cambodia, Laos, Thailand, and Viet Nam—countries with 168 million people—complain about the reduced flow of the Mekong, but this has done little to curb China's efforts to exploit the power and the water in the river.[41]

The same problem exists with the Tigris and Euphrates Rivers, which originate in Turkey and flow through Syria and Iraq en route to the Persian Gulf. This river system, the site of Sumer and other early civilizations, is being overused. Large dams erected in Turkey and Iraq have reduced water flow to the once "fertile crescent," helping to destroy more than 90 percent of the formerly vast wetlands that enriched the delta region.[42]

In the river systems just mentioned, virtually all the water in the basin is being used. Inevitably, if people upstream get more water, those downstream will get less.

Disappearing Lakes

As river flows are reduced or even eliminated entirely and as water tables fall from overpumping, lakes are shrinking and in some cases disappearing. As my colleague Janet Larsen notes, the lakes that are disappearing are some of the world's best known—including Lake Chad in Central Africa, the Aral Sea in Central Asia, and the Sea of Galilee (also known as Lake Tiberias).[43]

Many U.S. lakes have not fared well either. In California, Owens Lake, which covered 200 square miles when the last century began, has disappeared. After the Owens River was diverted to thirsty Los Angeles in 1913, the lake lasted little more than a decade.[44]

California's Mono Lake, geologically the oldest lake in North America and a popular feeding stop for migratory water birds, is a more recent victim of Los Angeles's seemingly unquenchable thirst. Mono Lake has experienced a 35-foot drop in its water level since 1941, when the diversion of water from its tributaries to Los Angeles began.[45]

Reuters reporter Megan Goldin writes that "walking on the Sea of Galilee is a feat a mere mortal can accomplish," due to receding shores. When I first saw the Jordan River as it enters Israel from Syria, its frailty was obvious. Indeed, in many countries it would be called a creek or a stream. And yet it has the primary responsibility for supplying water to the Sea of Galilee, which it enters at the north end and exits on the south end, continuing southward some 105 kilometers before emptying into the Dead Sea.[46]

With the Jordan's flow further diminished as it passes through Israel, the Dead Sea is shrinking even faster than the Sea of Galilee. Over the past 40 years, its water level has dropped by some 25 meters (nearly 80 feet). As a result of diversions from the Jordan River as it flows southward in Israel as well as fast-falling water tables on the Jordanian side, the Dead Sea could disappear entirely by 2050.[47]

Of all the shrinking lakes and inland seas, none has gotten as much attention as the Aral Sea. Its ports, once centers of commerce in the region, are now abandoned, looking like the ghost mining towns of the American West. Once one of the world's largest freshwater bodies, the Aral has lost four fifths of its volume since 1960. Ships that once plied its water routes are now stranded in the sand of the old seabed—with no water in sight.[48]

The seeds for the Aral Sea's demise were sown in 1960, when Soviet central planners in Moscow decided the region embracing the Syr Darya and Amu Darya basins would become a vast cotton bowl to supply the country's textile industry. As cotton planting expanded, so too did the diversion of water from the two rivers that fed the Aral Sea. As the sea shrank, the salt concentrations climbed until the fish died. The thriving fishery that once produced 50,000 tons per year disappeared, as did the jobs on the fishing boats and in the fish processing factories.[49]

With the 65-billion-cubic-meter annual influx of water from the two rivers now down to 1.5 billion cubic meters a year, the prospect for reversing the shrinkage is not good. With the sea's shoreline now up to 250 kilometers (165 miles) from the original port cities, there is a vast area of exposed seabed. Each day the wind lifts thousands of tons of sand and salt from the dry seabed, distributing the airborne particles on the surrounding grasslands and croplands and damaging them.[50]

At a 1990 Soviet Academy of Sciences conference on the future of the Aral Sea, there was an aerial tour for foreign guests. Flying over this area in a World War II–vintage single-engine biplane a few hundred feet above the dry, salt-covered seabed, I noted that it looked like the surface of the moon. There was no vegetation, no sign of any life, only total desolation.[51]

The disappearance of lakes is perhaps most pronounced in China. In western China's Qinhai province, through which the Yellow River's main stream flows, there were once 4,077 lakes. Over the last 20 years, more than 2,000 have disappeared. The situation is far worse in Hebei Province, which surrounds Beijing. With water tables falling fast throughout this region, Hebei has lost 969 of its 1,052 lakes.[52]

Lakes are disappearing in other Asian countries as well, including India, Pakistan, and Iran. For example, numerous lakes have disappeared in India's Kashmir Valley. Lake Dal, at one time covering 75 square kilometers, has shrunk to 12 square kilometers. With water tables falling in so much of India, many lakes are disappearing and others are shrinking fast.[53]

Population is also outgrowing the water supply in Mexico. Lake Chapala, the country's largest, is the primary source of water for Guadalajara, which is home to 5 million people. Expanding irrigation in the region has reduced water volume in the lake by 80 percent.[54]

Lakes are disappearing on every continent and for the same reasons: excessive diversion of water from rivers and overpumping of aquifers. No one knows exactly how many lakes have disappeared over the last half-century, but we do know that thousands of them now exist only on old maps.

Farmers Losing to Cities

Water conflicts among countries dominate the headlines. But within countries it is the jousting for water between cities and farms that preoccupies local political leaders. The economics of water use do not favor farmers in this competition, simply because it takes so much water to produce food. For example, while it takes only 14 tons of water to make a ton of steel worth $550, it takes 1,000 tons of water to grow a ton of wheat worth $150. In countries preoccupied with expanding the economy and creating jobs, the policy decision to make agriculture the residual claimant comes as no surprise.[55]

Many of the world's largest cities are located in watersheds where all available water is being used. Cities in such watersheds, such as Mexico City, Cairo, and Beijing, can increase their water consumption only by importing water from other basins or taking it from agriculture. Literally hundreds of the world's cities are now meeting their growing needs by taking irrigation water from farmers. Among the U.S. cities doing so are San Diego, Los Angeles, Las Vegas, Denver, and El Paso. A USDA study of 11 western states found that annual sales of water rights during 1996 and 1997 averaged 1.65 billion tons, enough to produce 1.65 million tons of grain.[56]

World Bank calculations for densely populated South Korea, a relatively well watered country, indicate that growth in residential and industrial water use there could reduce the supply available for agriculture from 13 billion to 7 billion tons in 2025. The Bank also projects that between 2000 and 2010, China's urban water demand will increase from 50 billion to 80 billion tons, a growth of 60 percent. Industrial water demand, meanwhile, will go from 127 billion to 206 billion tons, up 62 percent. Several hundred cities are looking to the countryside to satisfy their future water needs. In the region around Beijing, this shift has been under way since 1994, when farmers were banned from drawing on the reservoirs that supplied the city.[57]

As China attempts to accelerate the economic development of the upper Yellow River basin, emerging industries upstream get priority in the use of water. And as more water is used upstream, less reaches farmers downstream. In unusually dry years, the Yellow River fails to reach Shandong, the last province en route to the sea.[58]

Farmers in Shandong, who have traditionally received roughly half of their irrigation water from the Yellow River and half from wells, are now losing water from both sources. Irrigation water losses in a province that produces one fifth of China's corn and one seventh of its wheat help explain why China's grain harvest is declining.[59]

Literally hundreds of cities in other countries are meeting their growing water needs by taking some of the water that farmers count on. In western Turkey, for example, the city of Izmir now relies heavily on well fields from the neighboring agricultural district of Manisa.[60]

In the U.S. southern Great Plains and Southwest, where virtually all water is now spoken for, the growing water needs of cities and thousands of small towns can be satisfied only by taking water from agriculture. A monthly magazine from California, *The Water Strategist*, devotes several pages to a listing of water sales in the western United States during the preceding month. Scarcely a day goes by without another sale. Eight out of 10 sales are typically by either individual farmers or their irrigation districts to cities and municipalities.[61]

Colorado, with a fast-growing population, has one of the world's most active water markets. Growing cities and towns of all sizes in a state with high immigration are buying irrigation water rights from farmers and ranchers. In the upper Arkansas River basin, which occupies the southeastern quarter of the state, Colorado Springs and Aurora (a suburb of Denver) have already bought water rights to one third of the basin's farmland. Aurora has purchased rights to water that was once used to irrigate 9,600 hectares (23,000 acres) of cropland in the Arkansas valley.[62]

Far larger purchases are being made by cities in California. In 2003, San Diego bought annual rights to 247 million tons (200,000 acre-feet) of water from farmers in the nearby Imperial Valley—the largest rural/urban water transfer in U.S. history. This agreement covers the next 75 years. In 2004, the Metropolitan Water District, which supplies water to 18 million southern Californians in several cities, negotiated the purchase of 137 million cubic meters of water per year from farmers for the next 35 years. Without irrigation water, the highly productive land owned by these farmers is wasteland. The farmers who are selling their water rights would like to continue farming, but city officials are offering far more for the water than the farmers could possibly earn by using it to irrigate crops.[63]

In many countries, however, farmers are not compensated for a loss of irrigation water. In 2004, for example, Chinese farmers along the Juma River downstream from Beijing discovered that the river had stopped flowing. A diversion dam had been built near the capital to take river water for Yanshan Petrochemical, a state-owned industry. Although the farmers protested bitterly, it was a losing battle. For the 120,000 villagers downstream from the diversion dam, the loss of water could cripple their ability to make a living from farming.[64]

Whether it is outright government expropriation, farmers being outbid by cities, or cities simply drilling deeper wells than farmers can afford, the world's farmers are losing the water war. They are faced with not only a shrinking water supply in many situations but also a shrinking share of that shrinking supply. Slowly but surely, cities are siphoning water from the world's farmers even as they try to feed some 70 million more people each year.[65]

Scarcity Crossing National Borders

Historically, water scarcity was a local issue. It was up to national governments to balance water supply and demand. Now this is changing as scarcity crosses national boundaries via the international grain trade. Since producing one ton of grain takes 1,000 tons of water (1,000 cubic meters), as noted earlier, importing grain is the most efficient way to import water. Countries are, in effect, using grain to balance their water books. Similarly, trading in grain futures is in a sense trading in water futures.[66]

After China and India, there is a second tier of countries with large water deficits—Algeria, Egypt, Iran, Mexico, and Pakistan. Three of these—Algeria, Egypt, and Mexico—already import much of their grain. However, in a parallel move with China, water-short Pakistan abruptly turned to the world market in 2004 for imports of 1.5 million tons of wheat. Its need for imports is likely to climb in the years ahead.[67]

The Middle East and North Africa—from Morocco in the west through Iran in the east—has become the world's fastest-growing grain import market. The demand for grain is driven both by rapid population growth and by rising affluence, much of the latter derived from the export of oil. With virtually every country in the region pressing against its water limits, the growing urban demand for water can be satisfied only by taking irrigation water from agriculture.[68]

Egypt, with some 74 million people, has become a major importer of wheat in recent years, vying with Japan—traditionally the leading wheat importer—for the top spot. It now imports 40 percent of its total grain supply, a number that edges steadily upward as its population outgrows the grain harvest produced with the Nile's water.[69]

Algeria, with 33 million people, imports more than half of its grain, which means that the water embodied in the imported grain exceeds the use of water for all purposes from domestic sources. Because of its heavy dependence on imports, Algeria is particularly vulnerable to disruptions, such as grain export embargoes.[70]

Overall, the water required to produce the grain and other farm products imported into the Middle East and North Africa last year equaled the annual flow of the Nile River at Aswan. In effect, the region's water deficit can be thought of as another Nile flowing into the region in the form of imported grain.[71]

It is often said that future wars in the Middle East will more likely be fought over water than oil, but the competition for water is taking place in world grain markets. The countries that are financially the strongest, not necessarily those that are militarily the strongest, will fare best in this competition.

Knowing where grain import needs will be concentrated tomorrow requires looking at where water deficits are developing today. Thus far, the countries importing much of their grain have been smaller ones. Now we are looking at fast-growing water deficits in both China and India, each with more than a billion people.[72]

Each year the gap between world water consumption and the sustainable water supply widens. Each year the drop in the water table is greater than the year before. Both aquifer depletion and the diversion of water to cities will contribute to the growing irrigation water deficit and hence to a growing grain deficit in many water-short countries.

A Food Bubble Economy

As noted earlier, overpumping is a way of satisfying growing food demand that virtually guarantees a future drop in food production when aquifers are depleted. Many countries are in essence creating a "food bubble economy"—one in which food production is artificially inflated by the unsustainable mining of groundwater.

The effects of overdrafting were not obvious when farmers began pumping on a large scale a few decades ago. The great attraction of pumping groundwater in contrast to large-scale surface water systems is that farmers can apply the water to crops precisely when it is needed, thereby maximizing water use efficiency. Groundwater is also available during the dry season, enabling many farmers in mild climatic regions to double crop.

To illustrate, yields of foodgrains in the Punjab on land irrigated from wells was 5.5 tons per hectare, while yields on land irrigated from canals averaged 3.2 tons per hectare. Similar data for the southern state of Andhra Pradesh also showed a strong advantage in pumped irrigation, with foodgrain yields averaging 5.7 tons per hectare compared with 3.4 tons on land irrigated with canal water.[73]

In the United States, 37 percent of all irrigation water comes from underground; the other 63 percent comes from surface sources. Yet three of the top grain-producing states—Texas, Kansas, and Nebraska—each get 70–90 percent of their irrigation water from the Ogallala aquifer, which is essentially a fossil aquifer with little recharge. The high productivity of groundwater-based irrigation means that the food production losses will be disproportionately large when the groundwater runs out.[74]

At what point does water scarcity translate into food scarcity? In which countries will the irrigation water losses from aquifer depletion translate into a drop in grain production? David Seckler and his colleagues at the International Water Management Institute, the world's premier water research group, summarized this issue well: "Many of the most populous countries of the world—China, India, Pakistan, Mexico, and nearly all the countries of the Middle East and North Africa—have literally been having a free ride over the past two or three decades by depleting their groundwater resources. The penalty for mismanagement of this valuable resource is now coming due and it is no exaggeration to say that the results could be catastrophic for these countries and, given their importance, for the world as a whole."[75]

Since expanding irrigation helped triple the world grain harvest from 1950 to 2000, it comes as no surprise that water losses can shrink harvests. With water for irrigation, many countries are in a classic overshoot-and-decline mode. If countries that are overpumping do not move quickly to reduce water use and stabilize water tables, then an eventual drop in food production is almost inevitable.[76]

4

Rising Temperatures and Rising Seas

In 2004, Sir David King, the U.K. government's Chief Scientific Advisor, reported some revealing new research confirming the relationship between carbon dioxide (CO_2) levels and temperature over the last 740,000 years. Analysis of an Antarctic ice core drilled to a depth of three kilometers by British scientists showed that the atmospheric concentrations of CO_2 consistently fluctuated between 200 parts per million (ppm) during the ice ages and 270 ppm during the warm intervals. This shift from ice age to warm period occurred many times and always within this CO_2 range.[1]

When the Industrial Revolution began, the atmospheric CO_2 level was roughly 270 ppm. The 377 ppm registered for 2004 is not only far above any level over the last 740,000 years, it may be nearing a level not seen for 55 million years. At that time the earth was a tropical planet. There was no polar ice; sea level was 80 meters (260 feet) higher than it is today.[2]

The destructive effects of higher temperatures are visible on many fronts. Crop-withering heat waves have lowered grain harvests in key food-producing regions in recent years. In 2002, record-high temperatures and associated drought reduced grain harvests in India, the United States, and Canada, dropping the world harvest 90 million tons, or 5 percent below consumption. The record-setting 2003 European heat wave contributed to a world harvest shortfall of 90 million tons. Intense heat and drought in the U.S. Corn Belt in 2005 contributed to a world shortfall of 34 million tons.[3]

Such intense heat waves also take a direct human toll. In 1995, 700 residents of Chicago died in a heat wave. In May 2002, in a heat wave in India that reached 50 degrees Celsius (122 degrees Fahrenheit), more than 1,000 people died in the state of Andhra Pradesh alone.[4]

In 2003, the searing heat wave that broke temperature records across Europe claimed 49,000 lives in eight countries. Italy alone lost more than 18,000 people, while 14,800 died in France. More than 15 times as many people died in Europe in this heat wave as died during the terrorist attacks on the World Trade Center and the Pentagon on September 11, 2001.[5]

Among the various manifestations of rising temperatures, ice melting and its effect on sea level are drawing scientists' attention. As sea level rises, low-lying island countries like Tuvalu and the Maldives and coastal cities like London, New York, and Shanghai will be among the first to feel the consequences.[6]

The insurance industry is painfully aware of the relationship between higher temperatures and storm intensity. As weather-related damage claims have soared, the last few years have brought a drop in earnings and a flurry of lowered credit ratings for both insurance companies and the reinsurance companies that insure them. Companies using historical records as a basis for calculating insurance rates for future storm damage are realizing that the past is no longer a reliable guide to the future.[7]

This is a problem not only for the insurance industry, but for all of us. We are altering the earth's climate, setting in motion trends we do not always understand with consequences we cannot anticipate.

Rising Temperature and Its Effects

Scientists at the National Aeronautics and Space Administration's Goddard Institute for Space Studies gather data from a global network of some 800 climate-monitoring stations to measure changes in the earth's average temperature. Their records go back 125 years, to 1880.[8]

Since 1970, the earth's average temperature has risen by 0.8 degrees Celsius, or nearly 1.4 degrees Fahrenheit. During this span, the rise in temperature each decade was greater than in the preceding one. (See Figure 4–1.) Meteorologists note that the 22 warmest years on record have come since 1980. And the six warmest years since recordkeeping began in 1880 have come in the last eight years. Three of these six—2002, 2003, and 2005—were years in which major food-producing regions saw their crops wither in the face of record temperatures.[9]

The amount of CO_2 in the atmosphere has risen substantially since the Industrial Revolution, with most of the rise coming since recordkeeping began in 1959. Since then it has risen every year, making this one of the world's most predictable environmental trends. As shown in Figure 4–2, CO_2 levels turned sharply upward around 1960. Roughly a decade later, around 1970, the temperature too began to climb.[10]

Against this backdrop of record increases, the projections of the Intergovernmental Panel on Climate Change (IPCC) that the earth's average temperature will rise 1.4–5.8 degrees Celsius (2.5–10.4 degrees Fahrenheit) during this century seem all too possible. Recent data on the temperature rise in some northern regions—such as Alaska, western Canada, and Siberia—cou-

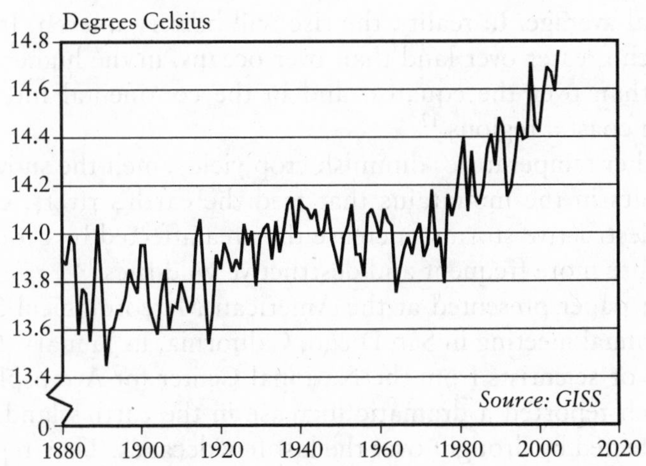

Figure 4–1. *Average l bal Temperature, 1880–2005*

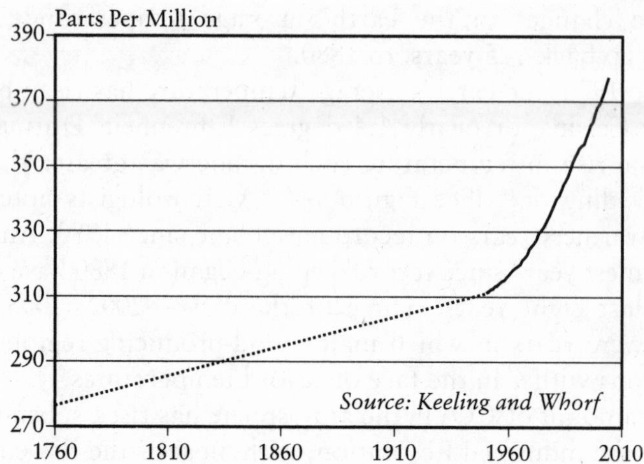

Figure 4–2. *Global Atmospheric Concentrations of Carbon Dioxide, 1760–2004*

pled with the accelerated melting of the Greenland ice sheet, the melting of glaciers in mountain ranges throughout the world, and the likelihood at this writing that the global temperature for 2005 will set yet another record high all suggest that the global temperature rise will be close to the upper end of the IPCC projected range. Such an increase—of 5.8 degrees Celsius by 2100, a rise comparable to that between the last Ice Age and today—will create a world far different from the one we know.[11]

At a practical level, the IPCC projected rise in temperature is a global average. In reality, the rise will be very uneven. It will be much greater over land than over oceans, in the higher latitudes than over the equator, and in the continental interiors than in coastal regions.[12]

Higher temperatures diminish crop yields, melt the snow/ice reservoirs in the mountains that feed the earth's rivers, cause more destructive storms, increase the area affected by drought, and cause more frequent and destructive wild fires.[13]

In a paper presented at the American Meteorological Society's annual meeting in San Diego, California, in January 2005, a team of scientists from the National Center for Atmospheric Research reported a dramatic increase in the earth's land surface affected by drought over the last few decades. They reported that the land experiencing very dry conditions expanded from less than 15 percent of the earth's total land area in the 1970s to roughly 30 percent by 2002. They attributed part of the change to a rise in temperature and part to reduced precipitation, with high temperatures becoming progressively more important during the latter part of the period. Lead author Aiguo Dai reported that most of the drying was concentrated in Europe and Asia, Canada, western and southern Africa, and eastern Australia.[14]

Researchers with the U.S. Department of Agriculture's Forest Service reported that even a 1.6-degree-Celsius rise in summer temperature could double the area of wildfires in the 11 western states. The study, published in the August 2004 issue of *Conservation Biology*, drew on 85 years of fire and temperature records.[15]

The National Wildlife Federation (NWF) reports that if temperatures continue to rise, by 2040 one out of five of the Pacific Northwest's rivers will be too hot for salmon, steelhead, and trout to survive. Paula Del Giudice, Director of the Federation's Northwest Natural Resource Center, notes that "global warming will add an enormous amount of pressure onto what's left of the region's prime cold-water fish habitat."[16]

Ecosystems everywhere will be affected by higher temperatures, sometimes in ways we cannot easily predict. The Pew Center for Global Climate Change sponsored a mega-study analyzing some 40 scientific reports that link temperature with changes in ecosystems. Among the many changes reported are spring arriving nearly two weeks earlier in the United States, tree swallows nesting nine days earlier than they did 40 years ago, and a northward shift of red fox habitat that has it encroaching on the Arctic fox's range. Inuits were surprised by the appearance of robins, a bird they had never seen before. Indeed, there is no word in Inuit for "robin."[17]

Hector Galbraith of the University of Colorado-Boulder, a co-author of the Pew study, indicated that "the effects of this change are occurring much more rapidly than . . . expected." He also said "that ecosystems are much more sensitive to climate change than believed a decade ago." A study sponsored by Conservation International has predicted that continued climate change could drive more than a quarter of all land animals and plants to extinction.[18]

Douglas Inkley, NWF Senior Science Advisor and senior author of a report to the Wildlife Society, notes, "We face the prospect that the world of wildlife that we now know—and many of the places we have invested decades of work in conserving as refugees and habitats for wildlife—will cease to exist as we know them, unless we change this forecast."[19]

The Crop Yield Effect

One of the economic trends most sensitive to higher temperatures is crop yields. Crops in many countries are grown at or near their thermal optimum, making them vulnerable to any rise in temperature. Even a relatively minor increase during the growing season of 1 or 2 degrees Celsius can shrink the grain harvest in major food-producing regions, such as the North China Plain, the Gangetic Plain of India, or the U.S. Corn Belt.[20]

Higher temperatures can reduce or even halt photosynthesis, prevent pollination, and lead to crop dehydration. Although the elevated concentrations of atmospheric carbon dioxide that raise temperature can also raise crop yields, the detrimental effect of higher temperatures on yields overrides the CO_2 fertilization effect for the major crops.

In a study of local ecosystem sustainability, Mohan Wali and his colleagues at Ohio State University noted that as temperature rises, photosynthetic activity in plants increases until the temperature reaches 20 degrees Celsius (68 degrees Fahrenheit). The rate of photosynthesis then plateaus until the temperature hits 35 degrees Celsius (95 degrees Fahrenheit), whereupon it begins to decline, until at 40 degrees Celsius (104 degrees Fahrenheit), photosynthesis ceases entirely.[21]

The most vulnerable part of a plant's life cycle is the pollination period. Of the world's three food staples—rice, wheat, and corn—corn is particularly vulnerable. In order for corn to reproduce, pollen must fall from the tassel to the strands of silk that emerge from the end of each ear of corn. Each of these silk strands is attached to a kernel site on the cob. If the kernel is to develop, a grain of pollen must fall on the silk strand and then journey to the kernel site. When temperatures are uncommonly high, the silk strands quickly dry out and turn brown, unable to play their role in the fertilization process.

The effects of temperature on rice pollination have been

studied in detail in the Philippines. Scientists there report that the pollination of rice falls from 100 percent at 93 degrees Fahrenheit (34 degrees Celsius) to near zero at 104 degrees Fahrenheit, leading to crop failure.[22]

High temperatures can also dehydrate plants. While it may take a team of scientists to understand how temperature affects rice pollination, anyone can tell when a cornfield is suffering from heat stress. When a corn plant curls its leaves to reduce exposure to the sun, photosynthesis is reduced. And when the stomata on the underside of the leaves close to reduce moisture loss, CO_2 intake is reduced, thereby restricting photosynthesis. At elevated temperatures, the corn plant, which under ideal conditions is so extraordinarily productive, goes into thermal shock.

Within the last few years, crop ecologists in several countries have been focusing on the precise relationship between temperature and crop yields. In an age of rising temperatures, their findings are disturbing. One of the most comprehensive of these studies was conducted at the International Rice Research Institute (IRRI) in the Philippines, the world's premier rice research organization. The team of eminent crop scientists there noted that from 1979 to 2003, the annual mean temperature at the research site rose by roughly 0.75 degrees Celsius.[23]

Using crop yield data from the experimental field plots for irrigated rice under optimal management practices for the years 1992–2003, the team's finding confirmed the rule of thumb emerging among crop ecologists—that a 1-degree-Celsius rise in temperature above the norm lowers wheat, rice, and corn yields by 10 percent. The IRRI finding was consistent with those of other recent research projects. The scientists concluded that "temperature increases due to global warming will make it increasingly difficult to feed Earth's growing population."[24]

While this study analyzing rice yields was under way, an empirical historical analysis of the effect of temperature on corn and soybean yields was being conducted in the United States. It concluded that higher temperatures had an even greater effect on yields of these crops. Using data for 1982–98 from 618 counties for corn and 444 counties for soybeans, David Lobell and Gregory Asner concluded that for each 1-degree Celsius rise in temperature, yields declined by 17 percent.

Given the projected temperature increases in the U.S. Corn Belt, where a large share of the world's corn and soybeans is produced, these findings should be of concern to those responsible for world food security.[25]

Two scientists in India, K.S. Kavi Kumar and Jyoti Parikh, assessed the effect of higher temperatures on wheat and rice yields. Basing their model on data from 10 sites, they concluded that in north India a 1-degree Celsius rise in mean temperature did not meaningfully reduce wheat yields, but a 2-degree rise lowered yields at almost all the sites. When they looked at temperature change alone, a 2-degree Celsius rise led to a decline in irrigated wheat yields ranging from 37 percent to 58 percent. When they combined the negative effects of higher temperature with the positive effects of CO_2 fertilization, the decline in yields among the various sites ranged from 8 percent to 38 percent. For a country projected to add 500 million people by midcentury, this is a troubling prospect.[26]

Reservoirs in the Sky

Snow/ice masses in mountains are nature's freshwater reservoirs—nature's way of storing water to feed rivers during the summer dry season. Now they are being threatened by the rise in temperature. Even a 1-degree rise in temperature in mountainous regions can markedly reduce the share of precipitation falling as snow and can boost that coming down as rain. This in turn increases flooding during the rainy season and reduces the snowmelt to feed rivers during the dry season.

In some agricultural regions, these "reservoirs in the sky" are the leading source of irrigation and drinking water. In the southwestern United States, for instance, the Colorado River—the primary source of irrigation water—depends on snowfields in the Rockies for much of its flow. California, in addition to depending heavily on the Colorado, also relies on snowmelt from the Sierra Nevada in the eastern part of the state. Both the Sierra Nevada and the coastal range supply irrigation water to California's Central Valley, the world's fruit and vegetable basket.

Preliminary results of an analysis of rising temperature effects on three major river systems in the western United States—the Columbia, the Sacramento, and the Colorado—indicate that the winter snow pack in the mountains feeding them will be dramatically reduced and that winter rainfall and flooding will increase accordingly.[27]

With a business-as-usual energy policy, global climate models project a 70-percent reduction in the amount of snow pack for the western United States by mid-century. A detailed study of the Yakima River Valley, a vast fruit-growing region in Washington state, by the U.S. Department of Energy's Pacific Northwest National Laboratory shows progressively heavier harvest losses as the snow pack shrinks, reducing irrigation water flows. A 2-degree-Celsius rise in temperature would reduce farm income in the valley by $92 million; a rise of 4 degrees Celsius would cut farm income by $163 million, nearly a quarter of the current harvest.[28]

In Central Asia, the agriculture in several countries—Uzbekistan, Turkmenistan, Kyrgyzstan, Kazakhstan, Tajikistan, and Afghanistan—depends heavily on snowmelt from the Hindu Kush, Pamir, and Tien Shan mountain ranges for irrigation water. Nearby Iran gets much of its water from the snowmelt in the 5,700-meter Alborz Mountains between Tehran and the Caspian Sea.[29]

Largest of all, where farmers are concerned, is the vast snow/ice mass in the Himalayas. Every major river in Asia, where half the world's people live, originates in the Himalayas, including the Indus, the Ganges, the Mekong, the Yangtze, and the Yellow. If warmer temperatures increase rainfall and reduce snowfall in the Himalayas, there will be more flooding during the rainy season and less snowmelt to feed rivers during the dry season.[30]

Reduced snow pack to feed the Yellow River flow will shrink China's wheat harvest, the largest in the world. Alterations in the flow of the Yangtze River will directly affect China's rice harvest—also the world's largest. And India's wheat harvest, second only to China's, will be affected by the flows of both the Indus and the Ganges. Anything that lowers the summer flow of the Mekong will affect the rice harvest of Viet Nam, a leading exporter.[31]

The shrinking of glaciers in the Himalayas could affect the water supply for hundreds of millions of people. In countries like India and China, the water stored during the rainy season as snow and ice for release in the dry season would be reduced or, in some cases, disappear entirely. The result would be more destructive floods alternating annually with more severe early summer water shortages.[32]

There are many more mountain ranges where snow/ice regimes are shifting, including the Alps and the Andes. The snow/ice masses in the world's leading mountain ranges and the water they store as ice and snow is taken for granted simply because it has been there since before agriculture began. Now that is changing. If we continue raising the earth's temperature, we risk losing these reservoirs in the sky on which cities and farmers depend.[33]

Melting Ice and Rising Seas

In its landmark third report, released in early 2001, the IPCC projected that sea level would rise during this century by 0.09–0.88 meters (4–35 inches) as a result of thermal expansion and ice melting. Numerous new studies during the four years since then indicate that the earth's ice cover is melting even faster than IPCC scientists projected.[34]

A 2002 study by two scientists from the University of Colorado's Institute of Arctic and Alpine Research showed that the melting of large glaciers on the west coast of Alaska and in northern Canada is accelerating. Earlier data had indicated that this melting was raising sea level by 0.14 millimeters per year, but new data for the 1990s show that the more rapid melting is now raising sea level by 0.32 millimeters a year—more than twice as fast.[35]

This study is reinforced by a U.S. Geological Survey (USGS) study that indicated glaciers are now shrinking in all 11 of Alaska's glaciated mountain ranges. An earlier USGS study reported that the number of glaciers in Glacier National Park in the United States had dwindled from 150 in 1850 to fewer than 50 today. The remaining glaciers are projected to disappear within 30 years, leaving future generations of visitors to puzzle over the park's name.[36]

Another team of USGS scientists, which used satellite data to measure changes in the area covered by glaciers worldwide, described an accelerated melting of glaciers in several mountainous regions, including the South American Andes, the Swiss Alps, and the French and Spanish Pyrenees.[37]

The melting of glaciers is gaining momentum throughout the Andes. Glaciologist Lonnie Thompson of Ohio State University reports that the Qori Kalis glacier, on the west side of the Quelccaya ice cap in the Peruvian Andes, shrank three times as fast each year from 1998 to 2000 as it did between 1995 and 1998. And the earlier rate, in turn, was nearly double the annual rate of retreat from 1993 to 1995. Thompson projects that the Quelccaya ice cap will disappear entirely between 2010 and 2020. In nearby Ecuador, the Antisana glacier, which supplies half of the water for Quito, has retreated nearly 100 meters in the last eight years.[38]

Bernard Francou, research director for the French government's Institute of Research and Development, believes that 80 percent of South American glaciers will disappear within the next 15 years. For countries like Bolivia, Peru, and Ecuador, which rely on glaciers for water for household and irrigation use, this is not good news.[39]

The European Alps are also suffering a meltdown. Scientists at Zurich University report that glaciers in Switzerland shrank by 1 percent from 1973 to 1985 but that the area covered shrank 18 percent between 1985 and 2000. They observed that "the changes could also impact tourism, a crucial pillar of the Swiss economy, as the country's scenic glacial valleys become barren and rocky." As the glaciers disappear and the snowline retreats upward, the winter ski season will shrink.[40]

Lonnie Thompson's studies of Kilimanjaro show that between 1989 and 2000, Africa's tallest mountain lost 33 percent of its ice field. He projected that its snowcap could disappear entirely by 2015. In March 2005, the *Guardian* in London reported: "Africa's tallest mountain, with its white peak, is one of the most instantly recognizable sites in the world. But as this aerial photograph shows, Kilimanjaro's trademark snowy cap, at 5,895 meters (19,340 feet), is now all but gone—15 years before scientists predicted it."[41]

The vast snow/ice mass in the Himalayas is also retreating. The Union Internationale des Associations d'Alpinisme reports that the glacier that ended at the base camp from which Edmund Hillary and Tenzing Norgay launched their history-making ascent of Everest in 1953 has retreated about 5 kilometers (3 miles). Geologist Jeffrey Kargel, who studies the

Himalayas, is not surprised by this. "That fits in with the general picture of what's happening in Nepal, India, Bhutan and, to a smaller extent, Tibet," he says.[42]

A recently completed study by a team of more than 50 U.S. and Chinese scientists over 26 years measured the accelerated melting of the glaciers in Western China. The study reported that the volume of China's 46,298 glaciers has shrunk by 5.5 percent in the past 24 years. The melting of glaciers in this region, as in most other areas of the world, has accelerated sharply since the early 1990s.[43]

Yao Tandong, a leading Chinese glaciologist and contributor to the study, predicted that two thirds of China's glaciers could be gone by 2060. Melting of the vast Himalayan ice fields, which contain more ice than any region outside of the poles, could dramatically raise sea level. Yao Tandong summarized the situation: "The full-scale glacier shrinkage in the plateau region will eventually lead to an ecological catastrophe."[44]

Another recent study, *Impacts of a Warming Arctic*, concluded that the Arctic is warming almost twice as fast as the rest of the planet. Conducted by the Arctic Climate Impact Assessment (ACIA) team, an international group of 300 scientists, the study found that in the regions surrounding the Arctic, including Alaska, western Canada, and Eastern Russia, winter temperatures have already climbed by 3–4 degrees Celsius (4–7 degrees Fahrenheit) over the last half-century. Robert Corell, chair of ACIA observes, "The impacts of global warming are affecting people now in the Arctic." This region, he says, "is experiencing some of the most rapid and severe climate change on Earth."[45]

In testimony before the U.S. Senate Commerce Committee, Sheila Watt-Cloutier, an Inuit speaking on behalf of the 155,000 Inuits who live in Alaska, Canada, Greenland, and the Russian Federation, described the Inuits' struggle to survive in the fast-changing Arctic climate as "a snapshot of what is happening to the planet." She called the warming of the Arctic "a defining event in the history of this planet." And she went on to say "the Earth is literally melting."[46]

The ACIA report described how the retreat of the sea ice has "devastating consequences for polar bears" whose very survival may be at stake. Also threatened are the ice-living seals, a basic food source for the Inuits.[47]

Higher Arctic temperatures are also thawing what had been perpetually frozen soils of the region. As the tundra thaws, it destabilizes and damages buildings, pipelines, and roads. The melting of the tundra has effects far beyond local structural problems. A report in *Science* says, "No one knows exactly how much carbon is locked up in boreal and alpine permafrost, but estimates range from 350 to 450 gigatons [billion tons]—perhaps a quarter to a third of all soil carbon. The big question is what will happen if even a fraction of this massive carbon store is liberated." This compares with the 7 billion tons of carbon that is emitted from burning fossil fuels each year.[48]

The scientists chronicling the warming of the Arctic are perhaps most concerned about the effect on Greenland. If all the ice in the Arctic Sea melts, it will not affect sea level because that ice is already in the water. But if the warming of the Arctic melts the Greenland ice sheet, which is a mile and a half thick in some places, sea level would rise by 7 meters (23 feet). Such a melting of the Greenland ice sheet would be measured in centuries, not years. Nonetheless, recent maps show rapid melting around the ice sheet's outer edges, particularly on the eastern coast.[49]

Scientists are concerned about the melting of the Greenland ice sheet not only because of its obvious effect on sea level, but also because it might disrupt ocean circulation, particularly the flow of the Gulf Stream. Under current conditions, the Gulf Stream that brings warm surface water northward from the South Atlantic supports Western Europe's mild climate. As the high-salinity warm water moves northward, it cools as a result of heat loss and evaporation, becoming more dense and salty. This eventually causes it to sink and then flow southward as deep water. An influx of fresh water from melting of the Greenland ice sheet or of Arctic sea ice could disrupt this circulation pattern, resulting in somewhat lower temperatures in the northeastern United States and eastern Canada and a sharp temperature drop in Europe. Historical evidence suggests that such shifts have sometimes come quickly—in a matter of years or decades.[50]

As the Arctic sea ice melts, it opens the possibility of using the Arctic Sea as a shipping route between the Atlantic and Pacific Oceans. The search for the Northwest Passage, a dream of early explorers who otherwise had to sail around the Cape of

Good Hope, could become a nightmare for our early twenty-first century society. Shipping companies are already looking at potential shortcuts. The trip from Europe to Asia via the Panama Canal typically covers some 12,600 nautical miles, according to an article in Canada's *Globe and Mail*, while the trip via the Northwest Passage would be shortened to 7,900 nautical miles. The risk is that the environmental damage from any accidents, such as an oil spill in the Arctic Sea, could last for decades if not longer in this frigid environment.[51]

At the other end of the earth, the 2-kilometer thick Antarctic ice sheet covers a continent about twice the size of Australia and contains 70 percent of the world's fresh water. Ice shelves that extend from the continent into the surrounding seas are beginning to break up at an alarming pace.[52]

The ice shelves surrounding Antarctica are formed by the flow of glaciers off the continent to lower levels in the surrounding sea. This flow of ice, fed by the continuous formation of new ice on land and culminating in the breakup of the shelves on the outer fringe and the calving of icebergs, is not new. What is new is the pace of this process. When Larsen A, a huge ice shelf on the east coast of the Antarctic Peninsula, broke up in 1995, it was a signal that all was not well in the region. Then in 2000, a huge iceberg nearly the size of Connecticut—11,000 square kilometers, or 4,250 square miles—broke off the Ross Ice Shelf.[53]

After Larsen A broke up, it was only a matter of time, given the rise in temperature in the region, before Larsen B would do the same. In November 2001, an alert went out to the scientific community from a researcher at the Instituto Antártico Argentino, who noted the unusually warm spring temperature and the 20-percent acceleration in the flow of the ice shelf. So when the northern part of the Larsen B ice shelf collapsed into the sea in March 2002, it was not a total surprise. At about the same time, a huge chunk of ice broke off the Thwaite Glacier. Covering 5,500 square kilometers, this iceberg was the size of Rhode Island.[54]

Even veteran ice watchers are amazed at how quickly the disintegration is occurring. "The speed of it is staggering," said Dr. David Vaughan, a glaciologist at the British Antarctic Survey, which has been monitoring the Larsen Ice Shelf closely. Along the Antarctic Peninsula, in the vicinity of the Larsen Ice Shelf, the average temperature has risen 2.5 degrees Celsius over the last five decades. Higher temperatures lead to ice melting on the surface of the ice shelves. Scientists theorize that as the melted water on the surface penetrates fractures, it weakens the ice, making it vulnerable to further fracturing.[55]

When ice shelves already in the water break off from the continental ice mass, this does not have much direct effect on sea level per se. But without the ice shelves to impede the flow of glacial ice, typically moving 400–900 meters a year, the flow of ice from the continent could accelerate, leading to a thinning of the ice sheet on the edges of the Antarctic continent. If this were to happen, sea level would rise. Dr. Neal Young of the Antarctic Cooperative Research Centre at the University of Tasmania in Australia notes that after Larsen A broke off, the upstream rate of glacial flow at least doubled.[56]

The accelerated melting of ice, which is consistent with the accelerating rise in temperature that has occurred since 1980, is of great concern in low-lying regions of coastal countries and low-lying island countries. Perhaps the most easily measured effect of rising sea level is the inundation of coastal areas. Donald F. Boesch, with the University of Maryland's Center for Environmental Sciences, estimates that for each 1-meter rise in sea level, the shoreline will retreat by an average 1,500 meters, or nearly a mile.[57]

In 2000, the World Bank published a map showing that a 1-meter rise in sea level would inundate half of Bangladesh's riceland. With a rise in sea level of up to 1 meter forecast for this century, tens of millions of Bangladeshis would be forced to migrate. In a country with 142 million people—already one of the most densely populated on earth—this would be a traumatic experience. Rice-growing river floodplains in other Asian countries would also be affected, including India, Thailand, Viet Nam, Indonesia, and China. With a 1-meter rise in sea level, more than a third of Shanghai, a city of 13 million people, would be under water.[58]

Such a rise would cost the United States 36,000 square kilometers (14,000 square miles) of land, most of it in the middle Atlantic and Mississippi Gulf states. With a 50-year storm surge, large portions of Lower Manhattan and the

National Mall in the center of Washington, D.C., would be flooded with seawater.[59]

While public attention focuses on the effect of ice melting on sea level rise, the thermal expansion of the oceans as a result of rising temperature is also raising sea level. At present, scientists estimate the relative contributions of ice melting and thermal expansion to sea level rise to be about the same. Together, the two are raising sea level at a measurable rate. It has become an indicator to watch—a trend that could force a human migration of unimaginable dimensions. It also raises questions about responsibility to future generations that humanity has never before faced.[60]

More Destructive Storms

Rising seas are not the only threat that comes with elevated global temperatures. Higher surface water temperatures in the tropical oceans mean more energy radiating into the atmosphere to drive tropical storm systems, leading to more frequent and more destructive storms. The combination of rising seas, more powerful storms, and stronger storm surges can be devastating.[61]

In the fall of 1998, Hurricane Mitch—one of the most powerful storms ever to come out of the Atlantic, with winds approaching 200 miles per hour—hit the east coast of Central America. As atmospheric conditions stalled the normal northward progression of the storm, some 2 meters of rain were dumped on parts of Honduras and Nicaragua within a few days. The deluge collapsed homes, factories, and schools, leaving them in ruins. It destroyed roads and bridges. Seventy percent of the crops and much of the topsoil in Honduras were washed away—topsoil that had accumulated over long stretches of geological time. Huge mudslides destroyed villages, sometimes burying local populations.[62]

The storm left 11,000 dead. Thousands more were never found. The basic infrastructure—the roads and bridges in Honduras and Nicaragua—was largely destroyed. President Flores of Honduras summed it up this way: "Overall, what was destroyed over several days took us 50 years to build." The damage from this storm, exceeding the annual gross national product of the two countries, set their economic development back by 20 years.[63]

In 2004, Japan experienced a record 10 typhoons (hurricanes) that collectively caused $10 billion worth of losses. During the same season, the state of Florida was hit by 4 of the 10 most costly hurricanes in U.S. history. These 4 hurricanes together generated insurance claims of $22 billion.[64]

A year later, these storms were dwarfed when Hurricane Katrina came onshore in the U.S. Gulf region with a storm surge of more than 20 feet that totally destroyed many coastal towns. The storm also flooded New Orleans, leaving much of it uninhabitable. Altogether the storm generated hundreds of thousands of refugees from Alabama, Mississippi, and Louisiana. This powerful storm, fueled by higher temperatures of surface waters in the Gulf, left in its wake a bill estimated early on at $200 billion. Since it will take years for the region to fully recover, the cost could climb even higher.[65]

Against this backdrop, insurance companies and reinsurance companies find it difficult to calculate a safe level of premiums, since the historical record traditionally used to calculate insurance fees is no longer a guide to the future. For example, the number of major flood disasters worldwide has grown during each of the last several decades, increasing from 6 major floods in the 1950s and 1960s to 8 in the 1970s, 18 in the 1980s, and 26 in the 1990s.[66]

The insurers are convinced that with higher temperatures and more energy driving storm systems, future losses will be even greater. They are concerned about whether the industry can remain solvent under this onslaught of growing damages. So, too, is Moody's Investors Service, which in 2002 downgraded the creditworthiness of several of the world's leading reinsurance companies. Since then, one of these firms—Munich Re—reported that 2004 was a record year of claims for the insurance industry worldwide even after adjusting for inflation.[67]

Thomas Loster, a Munich Re climate expert, said at the end of 2004: "As in 2002 and 2003, the overall balance of natural catastrophes is again clearly dominated by weather-related disasters, many of them exceptional and extreme.... We need to stop this dangerous experiment humankind is conducting on the Earth's atmosphere." The insurance industry is particularly concerned about new climate-related risks that may be emerging, such as Hurricane Catarina, which developed in 2004 in the South Atlantic, where water temperatures are not usually high enough to generate a hurricane. Whether Catarina, which came onshore in southern Brazil, is an anomalous event or the beginning of a disturbing new trend remains to be seen.[68]

Munich Re has published a list of storms with insured losses of $1 billion or more. The first such natural disaster came in 1983, when Hurricane Alicia struck the United States, racking up $1.5 billion in insured losses. Of the 49 natural catastrophes with $1 billion or more of insured losses recorded through the end of 2004, 3 were earthquakes, including the devastating 2004 Asian tsunami; the other 46 were weather-related—storms, floods, hurricanes, or wildfires. During the 1980s, there were 3 such events; during the 1990s, there were 26; and during the first half of the current decade, 2000 through 2004, there were 17.[69]

Prior to Hurricane Katrina, the two largest events in terms of total damage were Hurricane Andrew in 1992, which took down 60,000 homes and racked up $30 billion worth of damage, and the flooding of China's Yangtze river basin in 1998, which also cost an estimated $30 billion, a sum comparable to the value of China's rice harvest. Part of the growing damage toll is due to greater urban and industrial development in coastal areas and river floodplains. But part is due to more frequent, more destructive storms.[70]

The regions most vulnerable to more powerful storms currently are the Atlantic and Gulf Coasts of the United States and the Caribbean countries. In Asia, it is East and Southeast Asia, including the Philippines, Taiwan, Japan, China, and Viet Nam, that are likely to bear the brunt of the powerful storms crossing the Pacific. Further west, in the Bay of Bengal, Bangladesh and the east coast of India are particularly vulnerable.

Western Europe, traditionally experiencing a heavily damaging winter storm perhaps once in a century, had its first winter storm to exceed $1 billion in 1987—one that wreaked $3.7 billion in destruction, $3.1 billion of which was covered by insurance. Since then, it has had eight winter storms with insured losses ranging from $1.3 billion to $5.9 billion.[71]

Andrew Dlugolecki, a senior officer at the CGMU Insurance Group, the largest insurance company in the United Kingdom, notes that damage from atmospherically related events has increased by roughly 10 percent a year. "If such an increase were to continue indefinitely," he notes, "by 2065 storm damage would exceed the gross world product. The world obviously would face bankruptcy long before then." In the real world, few growth trends continue at a fixed rate for several decades, but Dlugolecki's basic point is that climate change can be destructive, disruptive, and very costly.[72]

Subsidizing Climate Change

At a time of mounting public concern about climate change driven by the burning of fossil fuels, the world fossil fuel industry is still being subsidized by taxpayers at more than $210 billion per year. Fossil fuel subsidies belong to another age, a time when development of the oil and coal industries was seen as a key to economic progress—not as a threat to our twenty-first century civilization. Once in place, subsidies lead to special interest lobbies that fight tooth and nail against eliminating them, even those that were not appropriate in the first place.[73]

In the United States, oil and gas companies are now perhaps the most powerful lobbyists in Washington. Between 1990 and 2004, they amassed $181 million in campaign contributions in an effort to protect special tax deductions worth billions. In testimony before the House Ways and Means Committee in 1999, Donald Lubick, U.S. Treasury Assistant Secretary for Tax Policy, said in reference to oil and gas companies: "This is an industry that probably has a larger tax incentive relative to its size than any other industry in the country." That such profitable investments are possible is a measure of the corruption of the U.S. political system, particularly the capacity of those with money to shape the economy to their advantage.[74]

Subsidies permeate and distort every corner of the global economy. Germany's coal mining subsidy was initially justified in part as a job protection measure, for example. At its peak, the government was subsidizing the industry to the tune of nearly $90,000 per year for each worker. In purely economic terms, it would have made more sense to close the mines and pay miners not to work.[75]

Many subsidies are largely hidden from taxpayers. This is especially true of the fossil fuel industry, whose subsidies include such things as a depletion allowance for oil pumping in the United States. Even more dramatic are the routine U.S. military expenditures to protect access to Middle Eastern oil, which were calculated by analysts at the Rand Corporation before the most recent Iraq war to fall between $30 billion and $60 billion a year, while the oil imported from the region was worth only $20 billion.[76]

A 2001 study by Redefining Progress shows U.S. taxpayers subsidizing automobile use at $257 billion a year, or roughly $2,000 per taxpayer. In addition to subsidizing carbon emissions, this also means that taxpayers who do not own automobiles, including those too poor to afford them, are subsidizing those who do.[77]

One of the bright spots about this subsidization of fossil fuels is that it provides a reservoir of tax deductions that can be diverted to climate-benign, renewable sources of energy, such as wind, solar, and geothermal energy. Shifting these subsidies from fossil fuels to the development of renewable sources would be a win-win situation, as described in Chapter 12. To subsidize the use of fossil fuels is to subsidize crop-withering heat waves, melting ice, rising seas, and more destructive storms. Perhaps it is time for the world's taxpayers to ask if this is how they want their hard-earned money to be spent.

5

Natural Systems Under Stress

In 1938, Walter Lowdermilk, a senior official in the Soil Conservation Service of the U.S. Department of Agriculture (USDA), traveled abroad to look at lands that had been cultivated for thousands of years, seeking to learn how these older civilizations had coped with soil erosion. He found that some had managed their land well, maintaining its fertility over long stretches of history, and were thriving. Others had failed to do so and left only remnants of their illustrious pasts.[1]

In a section of his report entitled "The Hundred Dead Cities," he described a site in northern Syria, near Aleppo, where ancient buildings were still standing in stark isolated relief, but they were on bare rock. During the seventh century, the thriving region had been invaded, initially by a Persian army and later by nomads out of the Arabian Desert. In the process, soil and water conservation practices used for centuries were abandoned. Lowdermilk noted, "Here erosion had done its worst....if the soils had remained, even though the cities were destroyed and the populations dispersed, the area might be re-peopled again and the cities rebuilt, but now that the soils are gone, all is gone."[2]

Now fast forward to a trip in 2002 by a U.N. team to assess the food situation in Lesotho, a small country of 2 million people imbedded within South Africa. Their finding was straightforward: "Agriculture in Lesotho faces a catastrophic future; crop production is declining and could cease altogether over large tracts of the country if steps are not taken to reverse soil erosion, degradation, and the decline in soil fertility." Michael Grunwald, reports in the *Washington Post* that nearly half of the children under five in Lesotho are stunted physically. "Many," he says, "are too weak to walk to school."[3]

Whether the land is in northern Syria, Lesotho, or elsewhere, the health of the people living on it cannot be separated from the health of the land itself. A large share of the world's 852 million hungry people live on land with soils worn thin by erosion.[4]

Merciless human demands are putting stresses on forests, rangelands, and fisheries that they cannot withstand. We are also destroying many of the plant and animal species with which we share the planet. Worldwide, species are now disappearing at 1,000 times the rate at which new species evolve: we have put the extinction clock on fast-forward.[5]

Shrinking Forests: The Costs

In early December 2004, Philippine President Gloria Macapagal Arroyo "ordered the military and police to crack down on illegal logging, after flash floods and landslides, triggered by rampant deforestation, killed nearly 340 people," according to news reports. Fifteen years earlier, in 1989, the government of Thailand announced a nationwide ban on tree cutting following severe flooding and the heavy loss of life in landslides. And in August 1998, following several weeks of record flooding in the Yangtze river basin and a staggering $30 billion worth of damage, the Chinese government banned all tree cutting in the basin, home to 400 million people. Each of these governments had belatedly learned a costly lesson, namely that services provided by forests, such as flood control, may be far more valuable to society than the lumber in those forests.[6]

At the beginning of the twentieth century, the earth's forested area was estimated at 5 billion hectares. Since then it has shrunk to 3.9 billion hectares—with the remaining forests rather evenly divided between tropical and subtropical forests in developing countries and temperate/boreal forests in industrial countries.[7]

World forest loss is concentrated in developing countries. Since 1990, the loss in these nations has averaged 13 million hectares a year, an area roughly the size of Kansas. Overall, the developing world is losing 6 percent of its forests per decade. The industrial world is actually gaining an estimated 3.6 million hectares of forestland each year, principally from abandoned cropland that is returning to forests on its own, as in Russia, and from the spread of commercial forestry plantations.[8]

Unfortunately, even these official data from the U.N.'s Food and Agriculture Organization (FAO) do not reflect the gravity of the situation. For example, tropical forests that are clearcut or burned off rarely recover. They simply become wasteland or at best scrub forest, yet they are still included in official forestry numbers. Plantations, too, count as forest area, yet they also are a far cry from the old-growth forest they sometimes replace.

The World Resources Institute (WRI) reports that of the forests that do remain standing, "the vast majority are no more than small or highly disturbed pieces of the fully functioning ecosystems they once were." Only 40 percent of the world's remaining forest cover can be classified as frontier forest, which WRI defines as "large, intact, natural forest systems relatively undisturbed and big enough to maintain all of their biodiversity, including viable populations of the wide-ranging species associated with each type."[9]

Pressures on forests continue to mount. Use of firewood, paper, and lumber is expanding. Of the 3.34 billion cubic meters of wood harvested worldwide in 2003, over half was used for fuel. In developing countries, fuelwood accounts for nearly three fourths of the total.[10]

Deforestation to supply fuelwood is extensive in the Sahelian zone of Africa and the Indian subcontinent. As urban firewood demand surpasses the sustainable yield of nearby forests, the woods slowly retreat from the city in an ever larger circle, a process clearly visible from satellite photographs taken over time. As the circles enlarge, the transport costs of firewood increase, triggering the development of a charcoal industry, a more concentrated form of energy with lower transport costs. March Turnbull writes in *Africa Geographic Online*: "Every large Sahelian town is surrounded by a sterile moonscape. Dakar and Khartoum now reach out further than 500 kilometers for charcoal, sometimes into neighboring countries."[11]

Logging for lumber also takes a heavy toll, as is most evident in Southeast Asia and Africa. In almost all cases, logging is done by foreign corporations more interested in maximizing a one-time harvest than in managing for a sustainable yield in perpetuity. Once a country's forests are gone, companies move on, leaving only devastation behind. Nigeria and the Philippines have both lost their once-thriving tropical hardwood export industries and are now net importers of forest products.[12]

Forest losses from clearing land for farming and ranching, usually by burning, are concentrated in the Brazilian Amazon, the Congo basin, and Borneo. After having lost 97 percent of its Atlantic rainforest, Brazil is now destroying the Amazon rainforest. This huge forest, roughly the size of Europe, was largely intact until 1970. Since then, 20 percent of it has been lost.[13]

The fast-rising demand for palm oil led to an 8-percent annual growth rate in the palm plantation area in Malaysian Borneo (Sarawak and Sabah) between 1998 and 2003. In Kalimantan, the Indonesian part of Borneo, growth in palm oil plantings is higher, at over 11 percent. Now that palm oil is emerging as a leading biodiesel fuel, growth in its cultivation will likely climb even faster. The near-limitless demand for biodiesel now threatens the remaining tropical forests in Borneo and elsewhere.[14]

Haiti, a country of 8 million people, was once largely covered with forests. Now there are forests standing on scarcely 2 percent of its land largely because trees are cut for firewood. In September 2004, tropical storm Jeanne left 1,500 dead and over 1,000 more missing and presumed dead. With the trees gone, the soil had washed away and there was little left to hold the downpour. Once a tropical paradise, Haiti is now a case study of a country committing ecological and economic suicide. As its forests have shrunk and its soils have eroded, Haiti has been caught in an ecological/economic downward spiral from which it has not been able to escape. It is a country sustained by international life-support systems of food aid and economic assistance.[15]

Haiti is a classic case of overshoot and collapse. First the trees go, then the soil, and finally the society itself. Without food from abroad, Haiti's population might now be declining as a result of hunger. Haiti is a microcosm of what much of the earth will be like if deforestation continues.

Scores of countries are suffering from disastrous flooding as a result of deforestation. In 2000, Mozambique was partially inundated as the Limpopo overflowed its banks, taking thousands of lives and destroying homes and crops on an unprecedented scale. The Limpopo river basin, which has lost 99 percent of its original tree cover, will likely face many more such floods.[16]

The biologically rich rainforest of Madagascar is also disappearing fast. As the trees are cut either to produce charcoal or to clear land in order to grow food for the island's increasing population, the sequence of events is all too familiar. Environmentalists warn that Madagascar could soon become a landscape of scrub growth and sand.[17]

While deforestation accelerates the flow of water back to the ocean, it also can reduce the recycling of rainfall inland. Some 20 years ago, two Brazilian scientists, Eneas Salati and Peter Vose, pointed out in *Science* that when rainfall coming from clouds moving in from the Atlantic fell on healthy Amazon rainforest, one fourth of the water ran off and three fourths evaporated into the atmosphere to be carried further inland to provide more rainfall. When land is cleared for grazing or farming, however, the amount that runs off and returns to the sea increases dramatically while that which is recycled inland falls alarmingly.[18]

Ecologist Philip Fearnside, who has spent his career studying the Amazon, observes that the agriculturally prominent south-central part of Brazil depends on water that is recycled inland via the Amazon rainforest. If the Amazon is converted into a cattle pasture, he notes, there will be less rainfall to support agriculture.[19]

A similar situation may be developing in Africa, where deforestation and land clearing are proceeding rapidly as firewood use mounts and as logging firms clear large tracts of virgin forests. In Malawi, a country of 13 million in East Africa, forest cover has shrunk from 47 percent of the country's land area to some 28 percent in a matter of years. The cutting of trees to produce charcoal and to cure tobacco is leading to a sequence of events paralleling that in Haiti.[20]

As the trees disappear, rainfall runoff increases and the land is deprived of the water from evapotranspiration. Consulting hydrogeologist Jim Anscombe notes: "Driven by energy from the sun, the trees pump water from the water table, through the roots, trunk and leaves, up into the process of transpiration. Collectively the forest pumps millions of liters of water daily to the atmosphere." Given the local climate conditions, this evapotranspiration translates into summer rainfall, helping to sustain crops. When the forests disappear, this rainfall declines and crop yields follow.[21]

More and more countries are beginning to recognize the risks associated with deforestation. Among the countries that now have total or partial bans on logging in primary forests are China, New Zealand, the Philippines, Sri Lanka, Thailand, and Viet Nam. Unfortunately, all too often a ban in one country simply shifts the deforestation to others or drives illegal logging. For example, the 1998 ban in China following the Yangtze flooding led to sharp increases in logging in Myanmar (formerly Burma) and Russia, much of it illegal.[22]

Losing Soil

The thin layer of topsoil that covers the planet's land surface is the foundation of civilization. This soil, measured in inches over much of the earth, was formed over long stretches of geological time as new soil formation exceeded the natural rate of erosion. As soil accumulated over the eons, it provided a medium in which plants could grow. In turn, plants protect the soil from erosion. Human activity is disrupting this relationship.

Sometime within the last century, soil erosion began to exceed new soil formation in large areas. Perhaps a third or more of all cropland is losing topsoil faster than new soil is forming, thereby reducing the land's inherent productivity. Today the foundation of civilization is crumbling. The seeds of collapse of some early civilizations, such as the Mayans, may have originated in soil erosion that undermined the food supply.[23]

The accelerating soil erosion over the last century can be seen in the dust bowls that form as vegetation is destroyed and wind erosion soars out of control. Among those that stand out are the Dust Bowl in the U.S. Great Plains during the 1930s, the dust bowls in the Soviet Virgin Lands in the 1960s, the huge one that is forming today in northwest China, and the one taking shape in the Sahelian region of Africa. Each of these is associated with a familiar pattern of overgrazing, deforestation, and agricultural expansion onto marginal land, followed by retrenchment as the soil begins to disappear.[24]

Twentieth-century population growth pushed agriculture onto highly vulnerable land in many countries. The overplowing of the U.S. Great Plains during the late nineteenth and early twentieth centuries, for example, led to the 1930s Dust Bowl. This was a tragic era in U.S. history, one that forced hundreds of thousands of farm families to leave the Great Plains. Many migrated to California in search of a new life, a move immortalized in John Steinbeck's *The Grapes of Wrath*.[25]

Three decades later, history repeated itself in the Soviet Union. The Virgin Lands Project between 1954 and 1960 centered on plowing an area of grassland for wheat that was larger than the wheatland in Canada and Australia combined. Initially this resulted in an impressive expansion in Soviet grain production, but the success was short-lived as a dust bowl developed there as well.[26]

Kazakhstan, at the center of this Virgin Lands Project, saw its grainland area peak at just over 25 million hectares around 1980, then shrink to 14 million hectares today. Even on the remaining land, however, the average wheat yield is scarcely 1 ton per hectare, a far cry from the nearly 8 tons per hectare that farmers get in France, Western Europe's leading wheat producer.[27]

A similar situation exists in Mongolia, where over the last 20 years half the wheatland has been abandoned and wheat yields have also fallen by half, shrinking the harvest by three fourths. Mongolia—a country almost three times the size of France with a population of 2.6 million—is now forced to import nearly 60 percent of its wheat.[28]

Dust storms originating in the new dust bowls are now faithfully recorded in satellite images. On January 9, 2005, the National Aeronautics and Space Administration (NASA) released images of a vast dust storm moving westward out of central Africa. This vast cloud of tan-colored dust stretched over some 5,300 kilometers (roughly 3,300 miles). NASA noted that if the storm were relocated to the United States, it would cover the country and extend into the oceans on both coasts.[29]

Andrew Goudie, Professor of Geography at Oxford University, reports that Saharan dust storms—once rare—are now commonplace. He estimates they have increased 10-fold during the last half-century. Among the countries in the region most affected by topsoil loss from wind erosion are Niger, Chad, Mauritania, northern Nigeria, and Burkino Faso. In Mauritania, in Africa's far west, the number of dust storms jumped from 2 a year in the early 1960s to 80 a year today.[30]

The Bodélé Depression in Chad is the source of an estimated 1.3 billion tons of wind-borne soil a year, up 10-fold from 1947 when measurements began. The 2–3 billion tons of fine soil particles that leave Africa each year in dust storms are slowly draining the continent of its fertility and, hence, its biological productivity. In addition, dust storms leaving Africa travel westward across the Atlantic, depositing so much dust in the Caribbean that they cloud the water and damage coral reefs there.[31]

In China, plowing excesses became common in several provinces as agriculture pushed northward and westward into the pastoral zone between 1987 and 1996. In Inner Mongolia (Nei Monggol), for example, the cultivated area increased by 1.1 million hectares, or 22 percent, during this period. Other provinces that expanded their cultivated area by 3 percent or more during this nine-year span include Heilongjiang, Hunan, Tibet (Xizang), Qinghai, and Xinjiang. Severe wind erosion of soil on this newly plowed land made it clear that its only sustainable use was controlled grazing. As a result, Chinese agriculture is now engaged in a strategic withdrawal in these provinces, pulling back to land that can sustain crop production.[32]

Water erosion also takes a toll on soils. This can be seen in the silting of reservoirs and in muddy, silt-laden rivers flowing into the sea. Pakistan's two large reservoirs, Mangla and Tarbela, which store Indus River water for the country's vast irrigation network, are losing roughly 1 percent of their storage capacity each year as they fill with silt from deforested watersheds.[33]

Ethiopia, a mountainous country with highly erodible soils on steeply sloping land, is losing an estimated 1 billion tons of topsoil a year, washed away by rain. This is one reason Ethiopia always seems to be on the verge of famine, never able to accumulate enough grain reserves to provide a meaningful measure of food security.[34]

Deteriorating Rangelands

One tenth of the earth's land surface is cropland, but an area twice this size is rangeland—land that is too dry, too steeply sloping, or not fertile enough to sustain crop production. This area—one fifth of the earth's land surface, most of it semi-arid—supports the world's 3.2 billion cattle, sheep, and goats. These livestock are ruminants, animals with complex digestive systems that enable them to digest roughage, converting it into beef, mutton, and milk.[35]

An estimated 180 million people worldwide make their living as pastoralists tending cattle, sheep, and goats. Many countries in Africa depend heavily on their livestock economies for food and employment. The same is true for large populations in the Middle East, Central Asia, Mongolia, northwest China, and much of India. India, with the world's largest cattle herd, depends on cattle not only for milk but also for draft power and fuel.[36]

In other parts of the world, rangelands are exploited by large-scale commercial ranching. Australia, whose land mass is dominated by rangeland, has a flock of 95 million sheep, five times its human population. Grass-based livestock economies also predominate in Argentina, Brazil, Mexico, and Uruguay. And in the Great Plains of North America, semiarid lands that are not suited to growing wheat are devoted to grazing cattle.[37]

Although public attention often focuses on the role of feedlots in beef production, the share of the world's cattle in feedlots at any one time is a tiny fraction of the vast numbers feeding on grass. Even in the United States, which has most of the world's feedlots, the typical steer is in a feedlot for only a matter of months.

Beef and mutton tend to dominate meat consumption where grazing land is abundant relative to population size. Among the leading beef consumers are the people of Argentina, Brazil, the United States, and Australia. Mutton looms large in diets in New Zealand and Kazakhstan.[38]

These same ruminants that are uniquely efficient at converting roughage into food also supply leather and wool. The world's leather goods and woolen industries, the livelihood of millions, depend on rangelands for raw materials.

Worldwide, almost half of all grasslands are lightly to moderately degraded and 5 percent are severely degraded. The problem is highly visible throughout Africa, the Middle East, Central Asia, and India, where livestock numbers track the growth in human numbers. In 1950, 238 million Africans relied on 273 million livestock. By 2004, there were 887 million people and 725 million livestock. Demands of the livestock industry, a cornerstone of the African economy virtually everywhere, often exceed grassland carrying capacity by half or more.[39]

Iran—one of the Middle East's most populous countries, with 70 million people—illustrates the pressures facing that region. With more than 9 million cattle and 80 million sheep and goats—the source of wool for its fabled rug-making industry—Iran's rangelands are deteriorating from overstocking. In a country where sheep and goats outnumber humans, mutton consumption is widespread. With rangelands now pushed beyond their limits, however, the current livestock population is not sustainable.[40]

China faces similarly difficult challenges. After the economic reforms in 1978 that shifted the responsibility for farming from state-organized production teams to farm families, the government lost control of livestock numbers. As a result, China's cattle, sheep, and goat population spiraled upward. While the United States, a country with comparable grazing capacity, has 95 million cattle, China has 107 million. And while the United States has 7 million sheep and goats, China has 339 million. Concentrated in China's western and northern provinces, sheep and goats are destroying the land's protective vegetation. The wind then does the rest, removing the soil and converting productive rangeland into desert.[41]

Fodder needs of livestock in nearly all developing countries now exceed the sustainable yield of rangelands and other forage resources. In India, the demand for fodder greatly outpaces the supply, leaving millions of emaciated, unproductive cattle.[42]

Land degradation from overgrazing is taking a heavy economic toll in lost livestock productivity. In the early stages of overgrazing, the costs show up as lower land productivity. But as the process continues, it destroys vegetation, leading to erosion and the eventual creation of wasteland and desert. At some point, growth in the livestock population begins to shrink the biologically productive area and thus the earth's capacity to sustain civilization.[43]

Advancing Deserts

Desertification, the process of converting productive land to wasteland through overuse and mismanagement, is unfortunately all too common. Anything that removes protective grass or trees leaves soil vulnerable to wind and water erosion. In the early stages of desertification, the finer particles of soil are removed by the wind, creating the dust storms described earlier. Once the fine particles are removed, then the coarser particles—the sand—are also carried by the wind in localized sand storms.

Large-scale desertification is concentrated in Asia and Africa—two regions that together contain nearly 4.8 billion of the world's 6.5 billion people. Populations in countries across the top of Africa are being squeezed by the northward advance of the Sahara.[44]

In the vast east-to-west swath of semiarid Africa between the Sahara Desert and the forested regions to the south lies the Sahel, a region where farming and herding overlap. In countries from Senegal and Mauritania in the west to Sudan, Ethiopia, and Somalia in the east, the demands of growing human and livestock numbers are converting more and more land into desert.[45]

Nigeria, Africa's most populous country, is losing 351,000 hectares of rangeland and cropland to desertification each year. While Nigeria's human population was growing from 33 million in 1950 to 132 million in 2005, a fourfold expansion, its livestock population grew from roughly 6 million to 66 million, an 11-fold increase. With the forage needs of Nigeria's 15 million cattle and 51 million sheep and goats exceeding the sustainable yield of the country's grasslands, the northern part of the country is slowly turning to desert. If Nigeria continues toward 258 million people by 2050, the deterioration will only accelerate.[46]

Iran is also losing its battle with the desert. Mohammad Jarian, who heads Iran's Anti-Desertification Organization, reported in 2002 that sand storms had buried 124 villages in the southeastern province of Sistan-Baluchistan, forcing their abandonment. Drifting sands had covered grazing areas, starving livestock and depriving villagers of their livelihood.[47]

Neighboring Afghanistan is faced with a similar situation. The Registan Desert is migrating westward, encroaching on agricultural areas. A U.N. Environment Programme (UNEP) team reports that "up to 100 villages have been submerged by windblown dust and sand." In the country's northwest, sand dunes are moving onto agricultural land in the upper reaches of the Amu Darya basin, their path cleared by the loss of stabilizing vegetation from firewood gathering and overgrazing. The UNEP team observed sand dunes 15 meters high blocking roads, forcing residents to establish new routes.[48]

China is being affected by desertification more than any other major country. Wang Tao, Director of the Cold and Arid Regions Environmental and Engineering Research Institute, describes the country's accelerating desertification. He reports that from 1950 to 1975 an average of 1,560 square kilometers of land were lost to desert each year. Between 1975 and 1987, this climbed to 2,100 square kilometers a year. From then until the century's end, it jumped to 3,600 square kilometers of land going to desert annually.[49]

China is now at war. It is not invading armies that are claiming its territory, but expanding deserts. Old deserts are advancing and new ones are forming like guerrilla forces striking unexpectedly, forcing Beijing to fight on several fronts. Wang Tao reports that over the last half-century, some 24,000 villages in northern and western China have been entirely or partly abandoned as a result of being overrun by drifting sand.[50]

People in China are all too familiar with the dust storms that originate in the northwestern area and western Mongolia, but the rest of the world typically learns about this fast-growing ecological catastrophe from the massive dust storms that travel outside the region. On April 18, 2001, the western United States—from the Arizona border north to Canada—was blanketed with dust. It came from a huge dust storm that originated in northwestern China and Mongolia on April 5. Measuring 1,800 kilometers across when it left China, the storm carried millions of tons of topsoil, a vital resource that will take centuries to replace through natural processes.[51]

Almost exactly one year later, on April 12, 2002, South Korea was engulfed by a huge dust storm from China that left people in Seoul literally gasping for breath. Schools were closed, airline flights were cancelled, and clinics were overrun with patients having difficulty breathing. Retail sales fell. Koreans have come to dread the arrival of what they now call "the fifth season," the dust storms of late winter and early spring.[52]

These two dust storms, among the 10 or so major dust storms that occur each year in China, are one of the externally visible indicators of the ecological catastrophe unfolding in northern and western China. Overgrazing is the principal culprit.[53]

A U.S. Embassy report entitled "Desert Mergers and Acquisitions" describes satellite images showing two deserts in north-central China expanding and merging to form a single, larger desert overlapping Inner Mongolia and Gansu provinces. To the west in Xinjiang Province, two even larger deserts—the Taklimakan and Kumtag—are also heading for a merger. Highways running through the shrinking regions between them are regularly inundated by sand dunes.[54]

In Latin America, deserts are expanding in both Brazil and Mexico. In Brazil, where some 58 million hectares of land are affected, economic losses from desertification are estimated at $300 million per year, much of it concentrated in the country's northeast. Mexico, with a much larger share of arid and semi-arid land, is even more vulnerable. The degradation of cropland now prompts some 700,000 Mexicans to leave the land each year in search of jobs in nearby cities or in the United States.[55]

In scores of countries, the overgrazing, overplowing, and overcutting that are driving the desertification process are intensifying as the growth in human and livestock numbers continues. Stopping the desertification process from claiming more productive land may now rest on stopping the growth in human and livestock numbers.

Collapsing Fisheries

After World War II, accelerating population growth and steadily rising incomes drove the demand for seafood upward at a record pace. At the same time, advances in fishing technologies, including huge refrigerated processing ships that enabled trawlers to exploit distant oceans, enabled fishers to respond to the growing world demand.

In response, the oceanic fish catch climbed from 19 million tons in 1950 to its historic high of 93 million tons in 1997. This fivefold growth—more than double that of population during this period—raised the wild seafood supply per person worldwide from 7 kilograms in 1950 to a peak of 17 kilograms in 1988. Since then, it has fallen to 14 kilograms.[56]

As population grows and as modern food marketing systems give more people access to these products, seafood consumption is growing. Indeed, the human appetite for seafood is outgrowing the sustainable yield of oceanic fisheries. Today 75 percent of fisheries are being fished at or beyond their sustainable capacity. As a result, many are in decline and some have collapsed. In some fisheries, the breeding stocks have been mostly destroyed.[57]

A 2003 landmark study by a Canadian-German science team, published in *Nature*, concluded that 90 percent of the large fish in the oceans had disappeared over the last 50 years. Ransom Myers, a fisheries biologist at Canada's Dalhousie University and lead scientist in this study, says: "From giant blue marlin to mighty blue fin tuna, from tropical groupers to Antarctic cod, industrial fishing has scoured the global ocean. There is no blue frontier left."[58]

Myers goes on to say, "Since 1950, with the onset of industrialized fisheries, we have rapidly reduced the resource base to less than 10 percent—not just in some areas, not just for some stocks, but for entire communities of these large fish species from the tropics to the poles."[59]

Fisheries are collapsing throughout the world. The 500-year-old cod fishery of Canada failed in the early 1990s, putting some 40,000 fishers and fish processors out of work. Fisheries off the coast of New England were not far behind. And in Europe, cod fisheries are in decline, approaching a free fall. Like the Canadian cod fishery, the European ones may have been depleted to the point of no return. Countries that fail to meet nature's deadlines for halting overfishing face fishery decline and collapse.[60]

Atlantic stocks of the heavily fished bluefin tuna, where a large specimen headed for Tokyo's sushi restaurants can bring $50,000, have been cut by a staggering 94 percent. It will take years for such long-lived species to recover, even if fishing were to stop altogether. The harvest of the Caspian Sea sturgeon, source of the world's most prized caviar, has fallen from a record 27,700 tons in 1977 to 461 tons in 2000. Overfishing, much of it illegal, is responsible for the dramatic drop.[61]

Overfishing is not the only threat to the world's seafood supply. Some 90 percent of fish residing in the ocean rely on coastal wetlands, mangrove swamps, or rivers as spawning areas. Well over half the original area of mangrove forests in tropical and subtropical countries has been lost. The disappearance of coastal wetlands in industrial countries is even greater. In Italy, whose coastal wetlands are the nurseries for many Mediterranean fisheries, the loss is a whopping 95 percent.[62]

Damage to coral reefs, breeding grounds for fish in tropical and subtropical waters, is also taking a toll. Between 2000 and 2004, the share of destroyed reefs worldwide expanded from 11 percent to 20 percent. As the reefs deteriorate, so do the fisheries that depend on them.[63]

While oceanic fisheries face numerous threats, it is overfishing that directly threatens their survival. Oceanic harvests expanded as new technologies evolved, ranging from sonar for tracking schools of fish to vast driftnets that are collectively long enough to circle the earth many times over.

Commercial fishing is now largely an economics of today versus tomorrow. Governments are seeking to protect tomorrow's catches by forcing fishers to keep their ships idle; fishing communities are torn between the need for income today versus the future. Ironically, one reason for excess fleet capacity is long-standing government subsidized loans for investing in new boats and fishing gear.[64]

Fishing subsidies were based on an unfounded belief that past trends in oceanic harvests could be projected into the future—that past growth meant future growth. The advice of marine biologists, who had long warned that marine harvests would someday reach a limit, was largely ignored.[65]

Even among countries accustomed to working together, such as those in the European Union (EU), the challenge of negotiating catch limits at sustainable levels can be difficult. In April 1997, after prolonged negotiations, agreement was reached in Brussels to reduce the fishing capacity of EU fleets by 30 percent for endangered species, such as cod, herring, and sole in the North Sea, and by 20 percent for overfished stocks, such as cod in the Baltic Sea, the bluefin tuna, and swordfish off the Iberian peninsula. The good news was that the EU finally reached agreement on reducing the catch. The bad news was that these cuts were not sufficient to arrest the decline of the region's fisheries.[66]

In January 2001, the EU went further, announcing a complete ban on fishing for cod, haddock, and whiting during the 12-week spring spawning period. With the annual cod catch falling from 300,000 tons during the mid-1980s to 50,000 tons in 2000, this step was a desperate effort to save the fishery. EU officials are all too aware that Canada's vast Newfoundland cod fishery has not recovered since collapsing in 1992, despite the total ban on fishing imposed then. In December 2002, the European Union adopted a still stronger fisheries management plan.[67]

When some fisheries collapse, it puts more pressure on those that remain. Local shortages quickly become global shortages. With restrictions on the catch in overfished EU waters, the heavily subsidized EU fishing fleet has turned to the west coast of Africa, buying licenses to fish off the coasts of Senegal, Mauritania, Morocco, Guinea-Bissau, and Cape Verde. They are competing there with fleets from Japan, South Korea, Taiwan, Russia, and China. For impoverished countries like Mauritania and Guinea-Bissau, income from fishing licenses can account for up to half of government revenue.[68]

Unfortunately for the Africans, their fisheries too are collapsing. In Senegal, where local fishers with small boats once could quickly fill their crafts with fish, on many days now they cannot catch enough fish to cover even their fuel costs. As one Senegalese tribal elder said, "Poverty came to Senegal with these fishing agreements."[69]

If the oceans cannot sustain a catch of more than 95 million tons and if world population continues to grow as projected, the oceanic fish catch per person will likely be declining for the foreseeable future. The generation that came of age during World War II saw the fish catch per person double during their lifetimes. Their grandchildren, the children of today, will experience a steady decline in seafood consumption.[70]

The bottom line is that the growing worldwide demand for seafood can no longer be satisfied by expanding the oceanic fish catch. If it is to be satisfied, it will be by expanding fish farming. But once fish are put in ponds or cages they have to be fed, further intensifying the pressure on land resources.

Disappearing Plants and Animals

The archeological record shows five great extinctions since life began, each representing an evolutionary setback, a wholesale impoverishment of life on earth. The last of these mass extinctions occurred some 65 million years ago, quite possibly because an asteroid collided with our planet, spewing vast amounts of dust and debris into the atmosphere. The resultant abrupt cooling obliterated the dinosaurs and at least one fifth of all other extant life forms.[71]

We are now in the early stage of the sixth great extinction. Unlike previous extinction events, which were caused by natural phenomena, this one is of human origin. For the first time in the earth's long history, one species has evolved, if that is the right word, to where it can eradicate much of life.

As various life forms disappear, they diminish the services provided by nature, such as pollination, seed dispersal, insect control, and nutrient cycling. This loss of species is weakening the web of life, and if it continues it could tear huge gaps in its fabric, leading to irreversible changes in the earth's ecosystem.

Species of all kinds are threatened by habitat destruction, principally through the loss of tropical rainforests. As we burn off the Amazon rainforest, we are in effect burning one of the great repositories of genetic information. Our descendents may one day view the wholesale burning of this genetic library much as we view the burning of the library in Alexandria in 48 BC.

Habitat alteration from rising temperatures, chemical pollution, or the introduction of exotic species can also decimate both plant and animal species. As human population grows, the number of species with which we share the planet shrinks. We cannot separate our fate from that of all life on the earth. If the rich diversity of life that we inherited is continually impoverished, eventually we will be impoverished as well.[72]

The share of birds, mammals, and fish that are vulnerable or in immediate danger of extinction is now measured in double digits: 12 percent of the world's nearly 10,000 bird species; 23 percent of the world's 4,776 mammal species; and 46 percent of the fish species analyzed.[73]

Among mammals, the 240 known species of primates other than humans are most at risk. The World Conservation Union–IUCN reports that nearly half of these species are threatened with extinction. Some 95 of the world's primate species live in Brazil, where habitat destruction poses a particular threat. Hunting, too, is a threat, particularly in West and Central Africa, where the deteriorating food situation and newly constructed logging roads are combining to create a lively market for "bushmeat."[74]

The bonobos of West Africa, great apes that are smaller than the chimpanzees of East Africa, may be our closest living relative both genetically and in social behavior. But this is not saving them from the bushmeat trade or the destruction of their habitat by loggers. Concentrated in the dense forest of the Democratic Republic of the Congo, their numbers fell from an estimated 100,000 in 1980 to only 3,000 today. In one human generation, 97 percent of the bonobos have disappeared.[75]

Birds, because of their high visibility, are a useful indicator of the diversity of life. Of the 9,775 known bird species, roughly 70 percent are declining in number. Of these, an estimated 1,212 species are in imminent danger of extinction. Habitat loss and degradation affect 86 percent of all threatened bird species. For example, 61 bird species have become locally extinct with the extensive loss of lowland rainforest in Singapore. Some once-abundant species may have already dwindled to the point of no return. The great bustard, once widespread in Pakistan and surrounding countries, is being hunted to extinction. Ten of the world's 17 species of penguins are threatened or endangered, potential victims of global warming. Stanford University biologist Çagan Sekercioglu, who led a separate study on the status of the world's birds said, "We are changing the world so much that even birds cannot adapt."[76]

A particularly disturbing recent event is the precipitous decline in the populations of Britain's most popular songbirds. Within the last 30 years the populations of well-known species such as the willow warbler, the song thrush, and the spotted flycatcher have fallen 50–80 percent; no one seems to know why, although there is speculation that habitat destruction and pesticides may be playing a role. Without knowing the source of the decline, it is difficult to take actions that will arrest the plunge in numbers.[77]

The threat to fish may be the greatest of all. The principal causes are overfishing, water pollution, and the excessive extraction of water from rivers and other freshwater ecosystems. An estimated 37 percent of the fish species that once inhabited the lakes and streams of North America are either extinct or in jeopardy. Ten North American freshwater fish species have disappeared during the last decade. In semiarid regions of Mexico, 68 percent of native and endemic fish species have disappeared. The situation may be even worse in Europe, where some 80 species of freshwater fish out of a total of 193 are threatened, endangered, or of special concern. Two thirds of the 94 fish species in South Africa need special protection to avoid extinction.[78]

The leatherback turtle, one of the most ancient animal species, and one that can reach a weight of 360 kilograms (800 pounds), also is fast disappearing. Its numbers dropped from 115,000 in 1982 to 34,500 in 1996. At the Playa Grande nesting colony on Costa Rica's west coast, the number of nesting females dropped from 1,367 in 1989 to 117 in 1999. Writing in *Nature*, James Spotila and colleagues warn that "if these turtles are to be saved, immediate action is needed to minimize mortality through fishing and to maximize hatchling production."[79]

A World Resources Institute report on coral reefs in the Caribbean notes that 35 percent of Caribbean reefs are threatened by sewage discharge, water-based sediment, and pollution from fertilizer and that 15 percent are threatened by pollution from cruise ship discharges. In economic terms, the Caribbean coral reefs supply goods and services worth at least $3.1 billion per year.[80]

The spectacular coral reefs of the Red Sea, some of the most strikingly beautiful reefs anywhere, are facing extinction due to destructive fishing practices, dredging, sedimentation, and sewage discharge. Anything that reduces sunlight penetration in the sea impairs the growth of corals, leading to die-off. Coral reefs play an important role as nurseries for many forms of sea life, including numerous commercial species of fish.[81]

One of the fastest-growing threats to the diversity of plant and animal life today is the extraordinary agricultural expansion now under way in Brazil as land is cleared to plant soybeans and, more recently, to produce sugarcane for ethanol. Farmers and ranchers are opening up vast areas in the Amazon basin and in the *cerrado*, a Europe-sized savanna-like region south of the Amazon basin. Although there are mechanisms in place that are designed to protect the rich biological diversity of the Amazon, such as the requirement that landowners clear no more than one fifth of their land, the government lacks enforcement capacity.[82]

Like the Amazon, the *cerrado* is also biologically rich, with thousands of endemic plant and animal species. It contains many large mammals, including the maned wolf, giant armadillo, giant anteater, deer, and several large cats—jaguar, puma, ocelot, and jaguarundi. The *cerrado* contains 837 species of birds, including the rhea, a cousin of the ostrich, which grows six feet tall. More than 1,000 species of butterflies have been identified. Conservation International reports that the *cerrado* also contains some 10,000 plant species—at least 4,400 of which are endemic, not found anywhere else.[83]

One of the newer worldwide threats to species, and one that is commonly underestimated, is the introduction of alien species, which can alter local habitats and communities, driving native species to extinction. For example, non-native species may be responsible for 30 percent of the threatened bird species on the IUCN *Red List*. For plants, alien species are implicated in 15 percent of all the listings.[84]

Efforts to save wildlife traditionally have centered on the creation of parks or wildlife reserves. Unfortunately, this approach may now be less effective, for if we cannot stabilize climate, there is not an ecosystem on earth that we can save. Everything will change. As the number of species with which we share the planet diminishes, so too does the prospect for our civilization.

In the new world we are entering, protecting the diversity of life on earth is no longer simply a matter of setting aside tracts of land, fencing them, and calling them parks and preserves. Success in this effort depends on stabilizing both climate and population.

On the plus side, we now have more information on the state of the earth and the life on it than ever before. While knowledge is not a substitute for action, it is a prerequisite for saving the earth's natural systems—and the civilization that they support.

6

Early Signs of Decline

In recent years U.N. demographers have stunned the world by announcing that life expectancy among the 750 million people living in sub-Saharan Africa has dropped from 61 to 48 years. This precipitous drop was primarily the result of governments' failure to check the spread of the HIV virus. While industrial countries held HIV infection rates among adults under 1 percent, in some African countries they climbed above 30 percent.[1]

For the first time in the modern era, life expectancy, a seminal indicator of development, has dropped for a large segment of humanity. For the people of sub-Saharan Africa, a failure of leadership is quite literally reversing the march of progress. Is this failure of the political system an anomaly? Or is it an early sign that the scale of emerging problems can overwhelm our political institutions?

During the decades following World War II, life expectancy climbed throughout the world with advances in public health, vaccines, antibiotics, and food production. But as the twentieth century drew to a close, the HIV epidemic brought this trend to an end in many countries.[2]

Today the variation in life expectancy among countries is

wider than at any time in history, ranging from a low of 33 in Swaziland and 37 in Botswana to a high of 82 in Japan and 81 in Iceland. Not surprisingly, life expectancy usually correlates with income levels except where the distribution of income is heavily skewed. In the United States, where income is concentrated among the wealthy and where some 24 million Americans are without health insurance, life expectancy is shorter than in countries like Sweden, Germany, or Japan. Indeed, U.S. life expectancy of 77 years now lags behind the 78 years of Costa Rica, a developing country.[3]

The stresses in our early twenty-first century civilization take many forms. Economically we see them in the widening income gap between the world's rich and poor. Socially they take the form of the widening gap in education and health care and a swelling flow of environmental refugees as productive land turns to desert and as wells go dry. Politically we see them manifest in conflict over basic resources such as cropland, grazing land, and water. And perhaps most fundamentally, we see the stresses the world is facing in the growing number of failed and failing states.

Our Socially Divided World

The social and economic gap between the world's richest 1 billion people and its poorest 1 billion has no historical precedent. Not only is this gap wide, it is widening. The poorest billion are trapped at a subsistence level and the richest billion are becoming wealthier with each passing year. The economic gap can be seen in the contrasts in nutrition, education, disease patterns, family size, and life expectancy.

World Health Organization (WHO) data indicate that roughly 1.2 billion people are undernourished, underweight, and often hungry. At the same time, roughly 1.2 billion people are overnourished and overweight, most of them suffering from excessive caloric intake and exercise deprivation. So while 1 billion people worry whether they will eat, another billion should worry about eating too much.[4]

Disease patterns also reflect the widening gap. The billion poorest suffer mostly from infectious diseases—malaria, tuberculosis, dysentery, and AIDS. Malnutrition leaves infants and small children even more vulnerable to such infectious diseases.

Unsafe drinking water takes a heavier toll on those with hunger-weakened immune systems, resulting in millions of fatalities each year. In contrast, among the billion at the top of the global economic scale, it is diseases related to aging and lifestyle excesses, including obesity, smoking, diets rich in fat and sugar, and exercise deprivation, that cause most deaths.[5]

Education levels reflect the deep divide between the rich and the poor. In some industrial countries—for example, Canada and Japan—more than half of all young people now graduate from college with either two- or four-year degrees. By contrast, in developing countries 115 million youngsters of elementary school age are not in school at all. Although five centuries have passed since Gutenberg invented the printing press, nearly 800 million adults are illiterate. Unable to read, they are also excluded from the use of computers and the Internet. Without adult literacy programs, their prospects of escaping poverty are not good.[6]

Close to 1 billion people live in countries where population size is essentially stable. But another billion or so live in countries where population is projected to double by 2050. The world's illiterates are concentrated in a handful of the more populous countries, most of them in Asia and Africa. Prominent among these are India, China, Pakistan, Bangladesh, Nigeria, Egypt, Indonesia, and Ethiopia, plus Brazil and Mexico in Latin America. From 1990 to 2000, China and Indonesia made large gains in reducing illiteracy. Other countries also making meaningful progress were Mexico, Nigeria, and Brazil. However, in four other populous countries—Bangladesh, Egypt, Pakistan, and India—the number of illiterates increased.[7]

Illiteracy and poverty tend to reinforce each other because illiterate women typically have much larger families than literate women do and because each year of schooling raises earning power by 10–20 percent. In Brazil, for instance, illiterate women have more than six children each on average; literate women have only two. Additionally, illiterate women are trapped by large families and minimal earning power.[8]

To be poor often means to be sick. As with illiteracy, poverty and ill health are closely linked. Health is closely related to access to safe water, something that 1.1 billion people lack. Waterborne diseases claim more than 3 million lives each year, mostly as a result of dysentery and cholera. These and other waterborne diseases take their heaviest toll among children. Infant mortality in affluent societies averages 8 per 1,000 live births; in the 50 poorest countries, it averages 97 per 1,000—nearly 13 times as high.[9]

The poor and uneducated often do not understand the mechanisms of infectious disease transfer and thus fail to take steps to protect themselves. In addition, those with immune systems weakened by hunger are more vulnerable to common infectious diseases. Poverty also means children are often not vaccinated for routine infectious diseases, even though the cost may be just pennies per child.[10]

The connection between poverty and disease is strong, but it has been broken for most of humanity by economic development. The challenge now is to break this link for that remaining minority who do not have access to safe water, vaccines, education, and basic health care.

Hunger is the most visible face of poverty. The U.N. Food and Agriculture Organization estimates that 852 million of the world's people are chronically hungry. They are not getting enough food to achieve full physical and mental development and to maintain adequate levels of physical activity.[11]

The majority of the underfed and underweight are concentrated in the Indian subcontinent and sub-Saharan Africa—regions that contain 1.4 billion and 750 million people, respectively. Twenty-five years ago, the nutritional status of Asia's population giants, India and China, was similar, but since then China has eliminated most of its hunger, whereas India has made limited progress. During this last quarter-century, China has accelerated the shift to smaller families. While gains in food production in India during this period were absorbed largely by population growth, those in China went mostly to raising individual consumption.[12]

Malnutrition takes its heaviest toll among the young, who are most vulnerable during their rapid physical and mental development. In both India and Bangladesh, almost half of all children under five are underweight and malnourished. In Ethiopia, 47 percent of children are undernourished, while in Nigeria the figure is 31 percent—and these are two of Africa's most populous countries.[13]

Although it is not surprising that those who are underfed

and underweight are concentrated in developing countries, it is perhaps surprising that most of them live in rural communities. More often than not, the undernourished are either landless or they live on plots of land so small that they are effectively landless. Those who live on the well-watered plains are usually better nourished. It is those who live on marginal land—land that is steeply sloping or semiarid—who are hungry.[14]

The penalties of being undernourished begin at birth. Gary Gardner and Brian Halweil of Worldwatch Institute cite a U.N. report that estimates 20 million underweight infants are born each year to mothers who also are malnourished. The study indicates that these children suffer lasting effects in the form of "impaired immune systems, neurological damage, and retarded physical growth." David Barker of Britain's University of Southampton observes soberly "that 60 percent of all newborns in India would be in intensive care had they been born in California."[15]

Health Challenge Growing

Health challenges are becoming more numerous as new infectious diseases such as SARS, the West Nile virus, and avian flu emerge. In addition, the accumulation of chemical pollutants in the environment is starting to take a toll. While some infectious diseases, such as malaria and cholera, have been around a long time and are diseases with which health authorities are quite familiar, the health effects of many environmental pollutants are only now being determined.

Among the leading infectious diseases, malaria claims more than 1 million lives each year, 89 percent of them in Africa. The number who are infected, and often suffer from it most of their lives, is many times greater. Economist Jeffrey Sachs, head of Columbia University's Earth Institute, estimates that reduced worker productivity and other costs associated with malaria are cutting economic growth by a full percentage point in countries with heavily infected populations.[16]

Although diseases such as malaria and cholera exact a heavy toll, there is no precedent for the number of lives affected by the HIV epidemic. To find anything similar to such a potentially devastating loss of life, we have to go back to the smallpox decimation of Native American communities in the sixteenth century or to the bubonic plague that took roughly a fourth of Europe's population during the fourteenth century. HIV should be seen for what it is—an epidemic of epic proportions that, if not checked soon, could take more lives during this century than were claimed by all the wars of the last century.[17]

Since the human immunodeficiency virus was identified in 1981, this infection has spread worldwide. By 1990, an estimated 10 million people were infected with the virus. By the end of 2004, the number who had been infected climbed to 78 million. Of this total, 38 million have died; 39 million are living with the virus. Twenty-five million HIV-positive people today live in sub-Saharan Africa, but only 500,000 or so are being treated with anti-retroviral drugs. Seven million live in South and Southeast Asia, with over 5 million of them in India alone.[18]

Infection rates are climbing. In the absence of effective treatment, the parts of sub-Saharan Africa with the highest infection rates face a staggering loss of life. Adding the heavy mortality from the epidemic to the normal mortality of older adults means that countries like Botswana and Zimbabwe will lose half of their adult populations within a decade.[19]

The HIV epidemic is not an isolated phenomenon. It is affecting every facet of life and every sector of the economy. Food production per person, already lagging in most countries in sub-Saharan Africa, is now falling fast as the number of field workers shrinks. As food production falls, hunger intensifies among the dependent groups of children and the elderly. The downward spiral in family welfare typically begins when the first adult falls victim to the illness—a development that is doubly disruptive because for each person who is sick and unable to work, another adult must care for that person.[20]

The massive loss of young adults to AIDS is already beginning to cut into economic activity. Rising worker health insurance costs in industry are shrinking or even eliminating company profit margins, forcing some firms into the red. In addition, companies are facing increased sick leave, decreased productivity, and the burden of recruiting and training replacements when employees die.[21]

Education is also affected. The ranks of teachers are being decimated by the virus. In 2001, for instance, Zambia lost 815 primary school teachers to AIDS, the equivalent of 45 percent of new teachers trained that year. With students, when one or both parents die, more children are forced to stay home simply because there is not enough money to buy books and to pay school fees. Universities are also feeling the effects. At the University of Durbin in South Africa, for example, 25 percent of the student body is HIV-positive.[22]

The effects on health care are equally devastating. In many hospitals in eastern and southern Africa, a majority of the beds are now occupied by AIDS victims, leaving less space for those with other illnesses. Already overworked doctors and nurses are often stretched to the breaking point. With health care systems now unable to provide even basic care, the toll of traditional disease is also rising. Life expectancy is dropping not only because of AIDS, but also because of the deterioration in health care.[23]

The epidemic is leaving millions of orphans in its wake. Sub-Saharan Africa is expected to have 18.4 million "AIDS orphans" by 2010—children who have lost at least one parent to the disease. There is no precedent for millions of street children in Africa. The extended family, once capable of absorbing orphaned children, is now itself being decimated by the loss of adults, leaving children, often small ones, to take care of themselves. For some girls, the only option is what has come to be known as "survival sex." Michael Grunwald of the *Washington Post* writes from Swaziland, "In the countryside, teenage Swazi girls are selling sex—and spreading HIV—for $5 an encounter, exactly what it costs to hire oxen for a day of plowing."[24]

The HIV epidemic in Africa is now a development problem, a matter of whether a society can continue to function as needed to support its people. It is a food security problem. It is a national security problem. It is an educational system problem. And it is a foreign investment problem. Stephen Lewis, the U.N. Special Envoy for HIV/AIDS in Africa, says that the epidemic can be curbed and the infection trends can be reversed, but it will take help from the international community. The failure to fully fund the Global Fund to Fight AIDS, Tuberculosis and Malaria, he says, is "mass murder" by complacency.[25]

Writing in the *New York Times*, Alex de Waal, an adviser to the U.N. Economic Commission for Africa and to UNICEF, sums up the effects of the epidemic well: "Just as HIV destroys the body's immune system, the epidemic of HIV and AIDS has disabled the body politic. As a result of HIV, the worst hit African countries have undergone a social breakdown that is now reaching a new level: African societies' capacity to resist famine is fast eroding. Hunger and disease have begun reinforcing each other. As daunting as the prospect is, we will have to fight them together, or we will succeed against neither."[26]

While the HIV epidemic is currently concentrated in Africa, air and water pollutants are damaging the health of people everywhere. A joint study by the University of California and the Boston Medical Center shows some 200 human diseases, ranging from cerebral palsy to testicular atrophy, linked to pollutants. Other diseases that can be caused by pollutants include an astounding 37 forms of cancer, plus heart disease, kidney disease, high blood pressure, diabetes, dermatitis, bronchitis, hyperactivity, deafness, sperm damage, and Alzheimer's and Parkinson's diseases.[27]

In July 2005, the Environmental Working Group in collaboration with Commonweal released an analysis of umbilical cord blood from 10 randomly selected newborns in U.S. hospitals. They detected a total of 287 chemicals in these tests. "Of the 287 chemicals we detected… we know that 180 cause cancer in humans or animals, 217 are toxic to the brain and nervous system, and 208 cause birth defects or abnormal development in animal tests." Everyone on the planet shares this "body burden" of toxic chemicals, but infants are at greater risk because they are in the highly vulnerable formative stage of early development.[28]

WHO reports an estimated 3 million deaths worldwide each year from air pollutants—three times the number of traffic fatalities. A study in *Lancet* concluded that air pollution claims 40,000 lives per year in France, Austria, and Switzerland, half of them attributable to vehicle emissions. In the United States, air pollution each year claims 70,000 lives, nearly double the 40,000 traffic fatalities.[29]

A U.K. research team reports a surprising rise in Alzheimer's and Parkinson's diseases and in motor neuron disease broadly in 10 industrial countries—six in Europe plus the United States, Japan, Canada, and Australia. In England and Wales, deaths from these brain diseases increased from 3,000 per year in the late 1970s to 10,000 in the late 1990s. Over an 18-year period, death rates from these dementias, mainly Alzheimer's, more than tripled for men and nearly doubled for women. This increase in dementia is linked to a rise in the concentration of pesticides, industrial effluents, car exhaust, and other pollutants in the environment.[30]

Horror stories of the health effects of uncontrolled industrial pollution in Russia are commonplace. For example, in the industrial town of Karabash in the foothills of the Ural Mountains, children routinely suffer from lead, arsenic, and cadmium poisoning. This translates into congenital defects, neurological disorders, and cancer. Pollutants also impair immune systems.[31]

Scientists are becoming increasingly concerned about the various effects of mercury, a potent neurotoxin, which now permeates the environment in virtually all countries with coal-burning power plants and many of those with gold mines. Gold miners release an estimated 200,000 pounds of mercury into the Amazon ecosystem each year, and coal-burning power plants release over 100,000 pounds of mercury into the air in the United States. The U.S. Environmental Protection Agency (EPA) reports that "mercury from power plants settles over water ways, polluting rivers and lakes, and contaminating fish."[32]

In 2004, 48 of the 50 states in the United States (all but Alaska and Wyoming) issued a total of 3,221 fish advisories warning against eating fish from local lakes and streams because of their mercury content. EPA research indicates that one out of every six women of childbearing age in the United States has enough mercury in her blood to harm a developing fetus. This means that 360,000 of the 4 million babies born in the country each year may face neurological damage from mercury exposure before birth. In a 2005 study by the Mt. Sinai Center for Children's Health and the Environment, a team of doctors calculated that lower I.Q. levels as a result of mercury exposure in the womb cost the United States $8.7 billion a year in lost earnings potential.[33]

No one knows exactly how many chemicals are manufactured today, but with the advent of synthetic chemicals the number of chemicals in use has climbed to over 100,000. A random blood test of Americans will show measurable amounts of easily 200 chemicals that did not exist a century ago.[34]

Most of these new chemicals have not been tested for toxicity. Those that are known to be toxic are included in a list of 667 chemicals whose discharge by industry into the environment must be reported to the EPA. The Toxic Release Inventory (TRI), now accessible on the Internet, also provides information on a community-by-community basis, arming local groups with data needed to evaluate the potential threats to their health and that of the environment. Since the TRI was inaugurated in 1988, reported toxic chemical emissions have declined steadily.[35]

Although we have been hearing about the carcinogenic effects of pesticides since Rachel Carson launched the environmental era with her book *Silent Spring*, we are not yet adequately dealing with this threat. Since then we have learned a great deal about the health effects from chemicals released into the environment, particularly the endocrine disruptors described by Theo Colborn and her colleagues in *Our Stolen Future*. This family of chemicals disrupts the reproductive and developmental processes not only in humans but in many other species as well.[36]

Throwaway Economy in Trouble

Another distinctly unhealthy economic trend has been the emergence over the last half-century of a throwaway economy. First conceived following World War II as a way of providing consumers with products, it soon came to be seen also as a vehicle for creating jobs and sustaining economic growth. The more goods produced and discarded, the reasoning went, the more jobs there would be.

What sold throwaways was their convenience. For example, rather than washing cloth towels or napkins, consumers welcomed disposable paper versions. Thus we have substituted facial tissues for handkerchiefs, disposable paper towels for hand towels, disposable table napkins for cloth ones, and throwaway beverage containers for refillable ones. Even the shopping bags we use to carry home throwaway products become part of the garbage flow.

This one-way economy depends on cheap energy. It is also facilitated by what are known in the United States as municipal solid waste management systems. Helen Spiegelman and Bill Sheehan of the Product Policy Institute write that these "have become a perverse public subsidy for the Throwaway Society. More and better waste management at public expense is giving unlimited license to proliferate discards. Today these systems collect 3.4 pounds of product waste a day for each American man, woman, and child—twice as much as in 1960 and ten times as much as 100 years ago. It is time to revamp the system so that it no longer supports the throwaway habit."[37]

The throwaway economy is on a collision course with the earth's geological limits. Aside from running out of landfills near cities, the world is also fast running out of the cheap oil that is used to manufacture and transport throwaway products. Perhaps more fundamentally, there is not enough readily accessible lead, tin, copper, iron ore, or bauxite to sustain the throwaway economy beyond another two or three generations. Assuming an annual 2-percent growth in extraction, U.S. Geological Survey data on current economically recoverable reserves show the world has 18 years of reserves remaining for lead, 20 years for tin, 25 years for copper, 64 years for iron ore, and 69 years for bauxite.[38]

The cost of hauling garbage from cities is rising as nearby landfills fill up and the price of oil climbs. One of the first major cities to exhaust its locally available landfills was New York. When the Fresh Kills landfill, the local destination for New York's garbage, was permanently closed in March 2001, the city found itself hauling garbage to landfill sites in New Jersey, Pennsylvania, and even Virginia—with some of the sites being 300 miles away.[39]

Given the 12,000 tons of garbage produced each day in New York and assuming a load of 20 tons of garbage for each of the tractor-trailers used for the long-distance hauling, some 600 rigs are needed to move garbage from New York City daily. These tractor-trailers form a convoy nearly nine miles long—impeding traffic, polluting the air, and raising carbon emissions. This daily convoy led Deputy Mayor Joseph J. Lhota, who supervised the Fresh Kills shutdown, to observe that getting rid of the city's trash is now "like a military-style operation on a daily basis."[40]

Fiscally strapped local communities in other states are willing to take New York's garbage—if they are paid enough. Some see it as an economic bonanza. State governments, however, are saddled with increased road maintenance costs, traffic congestion, increased air pollution, noise, potential water pollution from landfill leakage, and complaints from nearby communities.

Virginia Governor Jim Gilmore wrote to Mayor Rudy Giuliani in 2001 complaining about the use of Virginia as a dumping ground. "I understand the problem New York faces," he noted, "but the home state of Washington, Jefferson and Madison has no intention of becoming New York's dumping ground."[41]

Garbage travails are not limited to New York City. Toronto, Canada's largest city, closed its last remaining landfill on December 31, 2002, and now ships all its 1.1-million-ton-per-year garbage to Wayne County, Michigan. Ironically, the state of New Jersey, the recipient of some of New York's waste, is now shipping up to 1,000 tons of demolition debris 600 miles—also to Wayne County in Michigan.[42]

The challenge is to replace the throwaway economy with a reduce-reuse-recycle economy. For cities like New York, the challenge should be less what to do with the garbage than how to avoid producing it in the first place.

Population and Resource Conflicts

As land and water become scarce, we can expect competition for these vital resources to intensify within societies, particularly between the wealthy and those who are poor and dispossessed. The shrinkage of life-supporting resources per person that comes with population growth is threatening to drop the living standards of millions of people below the survival level. This could lead to unmanageable social tensions that will translate into broad-based conflicts.[43]

Access to land is a prime source of social tension. Expanding world population has cut the grainland per person in half, from 0.23 hectares in 1950 to 0.10 hectares in 2004. One tenth of a hectare is half of a building lot in an affluent U.S. suburb. This ongoing shrinkage of grainland per person makes it more difficult for the world's farmers to feed adequately the 70 million or more people added each year.[44]

The shrinkage in cropland per person not only threatens livelihoods; in largely subsistence societies, it threatens survival itself. Tensions within communities begin to build as landholdings shrink to an area smaller than that needed for survival. The Sahelian zone of Africa, with one of the world's fastest-growing populations, is also an area of spreading conflict.[45]

In troubled Sudan, 2 million people have died and over 4 million have been displaced in the long-standing conflict of more than 20 years between the Muslim north and the Christian south. The conflict in the Darfur region in western Sudan that began in 2003 illustrates the mounting tensions between two Muslim groups—Arab camel herders and black African subsistence farmers. Government troops are backing Arab militias, who are engaging in the wholesale slaughter of black Africans in an effort to drive them off their land, sending them into refugee camps in neighboring Chad. To date, some 140,000 people have been killed in the conflict and another 250,000 have died in the refugee camps of hunger and disease.[46]

In Nigeria, where 132 million people are crammed into an area not much larger than Texas, overgrazing and overplowing are converting grassland and cropland into desert, putting farmers and herders in a war for survival. As Somini Sengupta reported in the *New York Times* in June 2004, "in recent years, as the desert has spread, trees have been felled and the populations of both herders and farmers have soared, the competition for land has only intensified."[47]

Unfortunately, the division between herders and farmers is also often the division between Muslims and Christians. The competition for land, amplified by religious differences and combined with a large number of frustrated young men with guns, has created what the *New York Times* described as a "combustible mix" that has "fueled a recent orgy of violence across this fertile central Nigerian state [Kebbi]. Churches and mosques were razed. Neighbor turned against neighbor. Reprisal attacks spread until finally, in mid-May, the government imposed emergency rule."[48]

Similar divisions exist between herders and farmers in northern Mali, the *New York Times* noted, where "swords and sticks have been chucked for Kalashnikovs, as desertification and population growth have stiffened the competition between the largely black African farmers and the ethnic Tuareg and Fulani herders. Tempers are raw on both sides. The dispute, after all, is over livelihood and even more, about a way of life."[49]

Rwanda has become a classic case study in how mounting population pressure can translate into political tension and conflict. James Gasana, who was Rwanda's Minister of Agriculture and Environment in 1990–92, offers some insights. As the chair of a national agricultural commission in 1990, he had warned that without "profound transformations in its agriculture, [Rwanda] will not be capable of feeding adequately its population under the present growth rate." Although the country's demographers projected major future gains in population, Gasana said in 1990 that he did not see how Rwanda would reach 10 million inhabitants without social disorder "unless important progress in agriculture, as well as other sectors of the economy, were achieved."[50]

Gasana's warning of possible social disorder was prophetic. He further described how siblings inherited land from their parents and how, with an average of seven children per family, plots that were already small were fragmented further. Many farmers tried to find new land, moving onto steeply sloping mountains. By 1989, almost half of Rwanda's cultivated land was on slopes of 10 to 35 degrees, land that is universally considered uncultivable.[51]

In 1950, Rwanda's population was 2.4 million. By 1993, it was 7.5 million, making it the most densely populated country in Africa. As population grew, so did the demand for firewood. By 1991, the demand was more than double the sustainable yield of local forests. As trees disappeared, straw and other crop residues were used for cooking fuel. With less organic matter in the soil, land fertility declined.[52]

As the health of the land deteriorated, so did that of the people dependent on it. Eventually there was simply not enough food to go around. A quiet desperation developed. Like a drought-afflicted countryside, it could be ignited with a single match. That match ignited with the crash of a plane on April 6, 1994, shot down as it approached the capital of Kigali, killing President Juvenal Habyarimana. The crash unleashed an organized attack by Hutus, leading to an estimated 800,000 deaths of Tutsis and moderate Hutus in 100 days. In some villages, whole families were slaughtered lest there be survivors to claim the family plot of land.[53]

Many other African countries, largely rural in nature, are on a demographic track similar to Rwanda's. Tanzania's population of 38 million in 2005 is projected to increase to 67 million by 2050. Eritrea, where the average family has six children, is projected to grow from 4 million to 11 million by 2050. In the Democratic Republic of the Congo, the population is projected to triple, going from 58 million to 177 million.[54]

Africa is not alone. In India, tension between Hindus and Muslims is never far below the surface. As each successive generation further subdivides already small plots, pressure on the land is intense. The pressure on water resources is even greater.

With India's population projected to grow from 1.1 billion in 2005 to 1.6 billion in 2050, a collision between rising human numbers and shrinking water supplies seems inevitable. The risk is that India could face social conflicts that would dwarf those in Rwanda. As Gasana notes, the relationship between population and natural systems is a national security issue, one that can spawn conflicts along geographic, tribal, ethnic, or religious lines.[55]

Disagreements over the allocation of water among countries that share river systems is a common source of international political conflict, especially where populations are outgrowing the flow of the river. Nowhere is this potential conflict more stark than among Egypt, Sudan, and Ethiopia in the Nile River valley. Agriculture in Egypt, where it rarely rains, is wholly dependent on water from the Nile. Egypt now gets the lion's share of the Nile's water, but its current population of 74 million is projected to reach 126 million by 2050, thus greatly expanding the demand for grain and for water. Sudan, whose 36 million people also depend heavily on food produced with Nile water, is expected to have 67 million by 2050. And the number of Ethiopians, in the country that controls 85 percent of the river's headwaters, is projected to expand from 77 million to 170 million.[56]

Since there is already little water left in the Nile when it reaches the Mediterranean, if either Sudan or Ethiopia takes more water, Egypt will get less, making it increasingly difficult to feed an additional 52 million people. Although there is an existing water rights agreement among the three countries, Ethiopia receives only a minuscule share of water. Given its aspirations for a better life, and with the headwaters of the Nile being one of its few natural resources, Ethiopia will undoubtedly want to take more. With income per person averaging only $860 a year in Ethiopia compared with nearly $4,300 in Egypt, it is hard to argue that Ethiopia should not get more of the Nile water.[57]

To the north, Turkey, Syria, and Iraq share the waters of the Tigris and Euphrates river system. Turkey, controlling the headwaters, is developing a massive project on the Tigris to increase the water available for irrigation and power. Syria and Iraq, which are both projected to double their respective populations of 19 million and 29 million, are concerned because they too will need more water.[58]

In the Aral Sea basin in Central Asia, there is an uneasy arrangement among five countries over the sharing of the two rivers, the Amu Darya and the Syr Darya, that drain into the sea. The demand for water in Kazakhstan, Kyrgyzstan, Tajikistan, Turkmenistan, and Uzbekistan already exceeds the flow of the two rivers by 25 percent. (See Chapter 3.) Turkmenistan, which is upstream on the Amu Darya, is planning to develop another half-million hectares of irrigated agriculture. Racked by insurgencies, the region lacks the cooperation needed to manage its scarce water resources. On top of this, Afghanistan, which controls the headwaters of the Amu Darya, plans to use some of the water for its development. Geographer Sarah O'Hara of the University of Nottingham, who studies the region's water problems, says, "We talk about the developing world and the developed world, but this is the deteriorating world."[59]

Environmental Refugees on the Rise

As natural systems deteriorate, people are forced to migrate, sometimes to other countries. In mid-October 2003, Italian authorities discovered a boat carrying refugees from Africa bound for Italy. Adrift for more than two weeks and without fuel, food, and water, many of the passengers had died. At first the dead were tossed overboard. But after a point, the remaining survivors lacked the strength to hoist the bodies over the side. The dead and the living sharing the boat resembled what a rescuer described as "a scene from Dante's *Inferno*."[60]

The refugees were believed to be Somalis who had embarked from Libya, but they would not reveal their country of origin. We do not know whether they were political, economic, or environmental refugees. Failed states like Somalia produce all three. We do know that Somalia is an ecological basket case, with overpopulation, overgrazing, and desertification already destroying its pastoral economy.[61]

For Central American countries, including Honduras, Guatemala, Nicaragua, and El Salvador, Mexico is often the gateway to the United States. In 2003, Mexican authorities arrested and deported some 147,000 illegal immigrants, up from roughly 120,000 the previous year.[62]

In the city of Tapachula on the Guatemala-Mexico border, young men in search of jobs wait along the tracks for a slow-moving freight train moving through the city en route to the north. Some make it onto the train. Others do not. The Jesús el Buen Pastor refuge is home to 25 amputees who lost their grip and fell under the train while trying to board. For these young men, says Olga Sánchez Martínez, the director of the refuge, this is the "end of their American dream." A local priest, Flor María Rigoni, calls the migrants attempting to board the trains "the kamikazes of poverty."[63]

Environmental refugees also flow to the United States from Haiti, a widely recognized ecological disaster. In a rural economy where the land is stripped of vegetation and the soil is washing into the sea, the people are not far behind. Many drown in rough waters when attempting to make the trip to Florida in small craft not designed for the high seas.[64]

Today, bodies washing ashore in Italy, Spain, and Turkey are a daily occurrence, the result of desperate acts by desperate people. And each day Mexicans risk their lives in the Arizona desert trying to reach jobs in the United States. Some 400 to 600 Mexicans leave rural areas every day, abandoning plots of land too small or too eroded to make a living. They either head for Mexican cities or try to cross illegally into the United States. Many of those who try to cross the Arizona desert perish in its punishing heat—scores of bodies are found along the Arizona border each year.[65]

Although the modern world has extensive experience with political and economic refugees, we are now seeing a swelling flow of refugees driven from their homes by environmental pressures. This harkens back to the Dust Bowl era some 70 years ago, when nearly 3 million Americans were displaced.[66]

The United States is again contending with environmental refugees but now for different reasons. In Alaska, where the temperature rise in recent decades of 2–4 degrees Celsius (4–7 degrees Fahrenheit) is perhaps as great as anywhere in the world, thousands of indigenous peoples will almost certainly be forced to evacuate their villages as a result of ice melting and flooding. Newtok, a village of 340 Yupik Eskimos on Alaska's west coast, is being overrun by a swelling torrent of ice melt water from the Ninglick River. An engineering study estimated the cost of relocating the village at a minimum of $50 million—or $150,000 per villager. If the Newtok Indians do not move, they risk drowning in the floodwater. Although relocating villages is not a simple matter, there are 23 other Alaskan villages waiting to be relocated.[67]

With the vast majority of the nearly 3 billion people to be added to the world by 2050 living in countries where water tables are already falling, water refugees are likely to become commonplace. They will be most common in arid and semiarid regions where populations are outgrowing the water supply and sinking into hydrological poverty. Villages in northwestern India are being abandoned as aquifers are depleted and people can no longer find water. Millions of villagers in northern and western China and in parts of Mexico may have to move because of a lack of water.[68]

Advancing deserts are also displacing people, squeezing expanding populations into an ever smaller geographic area. Whereas the U.S. Dust Bowl displaced a few million people, the abandonment or partial depopulation of 24,000 villages in China's dust bowl provinces is displacing tens of millions.[69]

In Iran, villages abandoned because of spreading deserts or a lack of water already number in the thousands. In the vicinity of Damavand, a small town within an hour's drive of Tehran, 88 villages have been abandoned. And as the desert takes over in Nigeria, farmers and herders are forced to move, squeezed into a shrinking area of productive land. Desertification refugees typically end up in cities, many in squatter settlements. Many more migrate abroad.[70]

Another upcoming source of refugees, potentially a huge one, is rising seas. The largest potential displacement would come in low-lying Bangladesh, where even a 1-meter rise in sea level would not only inundate half of the country's riceland but would also force the relocation of easily 40 million people. In a densely populated country with 142 million people, internal relocation would not be easy. But where else can they go? How many countries would accept even a million Bangladeshi refugees displaced by rising sea level? Other Asian countries with rice-growing river deltas and floodplains, including China, India, Indonesia, Pakistan, the Philippines, South Korea, Thailand, and Viet Nam, could also suffer a mass exodus from rising seas.[71]

The refugee flows from falling water tables and expanding deserts are just beginning. How large these flows and those from rising seas will become remains to be seen. But the numbers could be huge, offering yet another reason for stabilizing climate and population.

Failed States and Terrorism

After a half-century of forming new states from former colonies and from the breakup of the Soviet Union, the international community is now focusing on the disintegration of states. The term "failed states" is now part of our working vocabulary, describing countries where there is no longer a central government. As one study observes, "Failed states have made a remarkable odyssey from the periphery to the very center of global politics."[72]

Recognizing this increasingly common phenomenon, various groups concerned with economic development and international affairs have begun to identify failing or failed states and the indicators associated with their failure. The World Bank, for example, has constructed a list of 30 "low-income countries under stress." Motivated by a similar concern, the United Kingdom's Department for International Development has identified 46 "fragile" states. The U.S. Central Intelligence Agency has constructed a list of 20 failing states. Most recently, the Fund for Peace and the Carnegie Endowment for International Peace have worked together to identify a list of 60 states, ranking them according to "their vulnerability to violent internal conflict."[73]

This analysis, published in *Foreign Policy*, is based on 12 social, economic, political, and military indicators. It puts Côte d'Ivoire at the top of the list of failed states, followed by the Democratic Republic of the Congo, Sudan, Iraq, Somalia, Sierra Leone, Chad, Yemen, Liberia, and Haiti. Next in line are three countries that have been much in the news in recent years: Afghanistan, Rwanda, and North Korea.[74]

Five oil-exporting countries make the top 60 list, including the two largest exporters and producers—Saudi Arabia (forty-fifth on the list) and Russia (fifty-ninth)—plus Venezuela (twenty-first), Indonesia (forty-sixth), and Nigeria (fifty-fourth). Two countries with nuclear arsenals are also on the list: Pakistan and Russia.[75]

The three top indicators used in constructing the *Foreign Policy* scorecard are uneven development, the loss of governmental legitimacy, and demographic pressure. Uneven development typically means that a small segment of the population is accumulating wealth while much of the society may be suffering a decline in living conditions. This unevenness, often associated with political corruption, creates unrest and can lead to civil conflict.[76]

Governments that fail to effectively manage emerging issues and provide basic services are seen as useless. This often causes segments of the population to shift their allegiance to warlords, tribal chieftains, or religious leaders. A loss of political legitimacy is an early sign of state decline.[77]

The third top indicator is demographic pressure. All the countries in the top 20 on the *Foreign Policy* list have fast-growing populations. In many that have experienced rapid population growth for several decades, governments are suffering from demographic fatigue, unable to cope with the steady shrinkage in per capita cropland and fresh water supplies or to build schools fast enough for the swelling ranks of children.[78]

Foreign investment drying up and a resultant rise in unemployment are also part of the decline syndrome. An earlier study by Population Action International showed that one of the key indicators of political instability in a society is the number of unemployed young men, a number that is high in countries at the top of the *Foreign Policy* article list.[79]

Another characteristic of failing states is a deterioration of the physical infrastructure—roads and power, water, and sewage systems. Care for natural systems is also neglected as people struggle to survive. Forests, grasslands, and croplands deteriorate, creating a downward economic spiral.[80]

Among the most conspicuous indications of state failure is a breakdown in law and order and a related loss of personal security. In Haiti, armed gangs rule the streets. Kidnapping for ransom of local people who are lucky enough to be among the 30 percent of the labor force that is employed is commonplace. In Afghanistan it is the local warlords, not the central government, that control the country outside of Kabul. Somalia, which now exists only on maps, is ruled by tribal leaders, each claiming a piece of what was once a country.[81]

Some of these countries are involved in long-standing civil conflicts. The Democratic Republic of the Congo, occupying a large part of the Congo River basin in the heart of Africa, has been the site of an ongoing civil conflict for six years, a conflict that has claimed 3.8 million lives and driven millions more from their homes. According to the International Rescue Committee, for each violent death in this conflict there are 62 nonviolent deaths related to it, including deaths from hunger, respiratory illnesses, diarrhea, and other diseases.[82]

Some potential sources of instability are taking the world into uncharted territory. In sub-Saharan Africa, where HIV infection rates sometimes exceed 30 percent of all adults, there will be millions of orphans in the years ahead, as noted earlier. With the number of orphans overwhelming society's capacity to care for them, many will become street children. Growing up without parental guidance and appropriate role models, and with their behavior shaped by the desperation of survival, these orphans will become a new threat to stability and progress.[83]

Failing states are of growing international concern because they are a source of terrorists, drugs, weapons, and refugees. Not only was Afghanistan a training ground for terrorists, but it quickly became, under the Allied occupation, the world's leading supplier of heroin. Refugees from Rwanda, including thousands of armed soldiers, contributed to the destabilization of the Congo. As *The Economist* notes, "Like a severely disturbed individual, a failed state is a danger not just to itself, but to those around it and beyond."[84]

In many countries, the United Nations or other internationally organized peacekeeping forces are trying to keep the peace, often unsuccessfully. Among the countries with U.N. peacekeeping forces are the Democratic Republic of the Congo, Sierra Leone, and Liberia. Other countries with multinational peacekeeping forces include Afghanistan, Haiti, and Sudan. All too often these are token forces, not nearly large enough to assure stability.[85]

Countries like Haiti and Afghanistan are surviving today because they are on international life-support systems. Economic assistance—including, it is worth noting, food aid—is helping to sustain them. But there is not now enough assistance to overcome the reinforcing trends of deterioration and replace them with state stability and sustained economic progress.[86]

II

THE RESPONSE—PLAN B

7

Eradicating Poverty, Stabilizing Population

The new century began on an inspiring note when the countries that belong to the United Nations adopted the goal of cutting the number of people living in poverty in half by 2015. And as of 2005, the world is ahead of schedule for reaching this goal. There are two big reasons for this: China and India. China's economic growth of 9 percent a year over the last quarter-century and India's acceleration to close to 6 percent a year over the last decade are together lifting hundreds of millions out of poverty.[1]

In China, the number living in poverty dropped from 648 million in 1981 to 218 million in 2001, the greatest reduction in poverty in history. India is also making impressive progress on the economic front. Under the dynamic new leadership of Prime Minister Manmohan Singh, who took office in 2004, and his skilled team, poverty is being attacked directly by upgrading infrastructure at the village level. Targeted investments are aimed at the poorest of the poor. If the international community actively reinforces this effort in reform-minded India, hundreds of millions more could be lifted out of poverty.[2]

It is time for the international community to make sure that India has the resources needed to maintain the momentum it has built. With India now on the move economically, the world can then begin to concentrate intensively on the remaining poverty concentrated in sub-Saharan Africa and a scattering of smaller countries in Latin America and Central Asia.

Several countries in Southeast Asia are making impressive gains as well, including Thailand, Viet Nam, and Indonesia. Barring any major economic setbacks, these gains in Asia virtually ensure that the U.N. Millennium Development Goal for reducing poverty by 2015 will be reached.[3]

That is the good news. The bad news is that sub-Saharan Africa—with 750 million people—is sliding deeper into poverty. Hunger, illiteracy, and disease are on the march, offsetting some of the gains in China and India. Africa, selected as a focus of discussion at the G-8 meeting in July 2005, needs special attention.[4]

In an increasingly integrated world, eradicating poverty and stabilizing population are national security issues. Slowing population growth helps eradicate poverty and its distressing symptoms, and, conversely, eradicating poverty helps slow population growth. With time running out, the urgency of moving simultaneously on both fronts is clear.

In addition to the goal of cutting the number of people living in poverty in half by 2015, the other U.N. Millennium Development Goals include cutting the number who are hungry in half, achieving universal primary school education, providing access to safe drinking water for all, and reversing the spread of infectious diseases, especially HIV and malaria. Closely related to these are the goals of reducing maternal mortality by three fourths and under-five child mortality by two thirds.[5]

While goals for cutting poverty in half by 2015 appear to be running slightly ahead of schedule, those for halving the number of hungry are not. The number of children with a primary school education appears to be increasing substantially, however, largely on the strength of progress in India. And mortality of children under five fell from 15 million in 1980 to 11 million in 2003 and is expected to continue falling.[6]

Universal Basic Education

One way of narrowing the gap between rich and poor is by ensuring universal education. This means ensuring that 115 million children who do not attend school are able to. Children without any formal education are starting life with a severe handicap, one that almost ensures they will remain in abject poverty and that the gap between the poor and the rich will continue to widen. In an increasingly integrated world, this widening gap becomes a source of instability. Nobel Prize–winning economist Amartya Sen focuses the point nicely: "Illiteracy and innumeracy are a greater threat to humanity than terrorism."[7]

Recognizing the central role of education in human progress, the United Nations set universal primary education by 2015 as one of its Millennium Development Goals. The World Bank has taken the lead with its Education for All plan, where any country with a well-designed plan to achieve universal primary education is eligible for financial support. The three principal requirements are that a country submit a sensible plan to reach universal basic education, commit a meaningful share of its own resources to the plan, and have transparent budgeting and accounting practices. If fully implemented, all children in poor countries would get a primary school education by 2015.[8]

The benefits of education are many, particularly for women. The achievement level of children correlates closely with the educational level of their mothers. Children of educated mothers are better nourished not necessarily because the family income is higher but because their mother's better understanding of nutrition leads to a better choice of foods and healthier methods of preparation. Educating women is the key to breaking the poverty cycle.[9]

The education of girls leads to smaller families. In every society for which data are available, fertility falls as female educational levels rise. And mothers with at least five years of school lose fewer infants during childbirth or early illnesses than their less educated peers do. Among other things, these women can read the instructions on medications and they have a better understanding of how to take care of themselves during pregnancy. Economist Gene Sperling concluded in a 2001 study of 72 countries that "the expansion of female secondary education may be the single best lever for achieving substantial reductions in fertility."[10]

Basic education increases agricultural productivity. Agricultural extension services that cannot use printed materials to disseminate information on improved agricultural practices are severely handicapped. So too are farmers who cannot read the instructions on a bag of fertilizer. The inability to read instructions on a pesticide container can be life-threatening.

At a time when HIV is spreading throughout the world, schools provide the institutional means to educate young people about the risks of infection. The time to inform and educate children about the virus and about the lifestyles that foster its spread is when they are young, not when they are already infected. Young people can also be mobilized to conduct educational campaigns among their peers.

One great need in developing countries, particularly those where the ranks of teachers are being decimated by AIDS, is more teacher training. Providing scholarships for promising students from poor families to attend training institutes in exchange for a commitment to teach for a fixed period of time, say five years, could be a highly profitable investment. It would help ensure that the human resources are available to reach the universal primary education goal, and it would also open the door for an upwelling of talent from the poorest segments of society.

Gene Sperling believes that every plan should provide for getting to the hardest-to-reach segments of society, especially poor girls in rural areas. He notes that Ethiopia has pioneered this with Girls Advisory Committees. Representatives of these groups go to the parents who are seeking early marriage for their daughters and encourage them to keep their children in school. Some countries, Brazil and Bangladesh among them, actually provide small scholarships for girls where needed, thus helping girls from poor families get a basic education.[11]

As the world becomes ever more integrated economically, its nearly 800 million illiterate adults are severely handicapped. This deficit can perhaps best be dealt with by launching adult literacy programs, relying heavily on volunteers. The international community could offer seed money to provide educational materials and outside advisors where needed. Bangladesh and Iran, both of which have successful adult literacy programs, can serve as models.[12]

The World Bank estimates that external funding of roughly $12 billion a year would be needed to achieve universal primary education in the more than 80 countries that are unlikely to reach this goal by 2015. At a time when education gives children access not only to books but also to personal computers and the vast information resources of the Internet, having children who never go to school is no longer acceptable.[13]

Few incentives to get children in school are as effective as a school lunch program, especially in the poorest countries. Since 1946, every child in public school in the United States has had access to a school lunch program, ensuring one good meal each day. There is no denying the benefits of this national program that has continued uninterrupted for so many years. George McGovern and Robert Dole, both former members of the U.S. Senate agricultural committee and former candidates for President, want to provide school lunch programs in all the world's poorest countries.[14]

Children who are ill or hungry miss many days of school. And even when they can attend, they do not learn as well. Jeffrey Sachs notes, "Sick children often face a lifetime of diminished productivity because of interruptions in schooling together with cognitive and physical impairment." But when school lunch programs are launched in low-income countries, school enrollment jumps. The children's attention span increases. Their academic performance goes up. Fewer days are missed, and children spend more years in school.[15]

Girls benefit especially. Drawn to school by the lunch, they stay in school longer, marry later, and have fewer children. This is a win-win-win situation. Adopting a school lunch program in the 44 lowest-income countries would cost an estimated $6 billion per year beyond what the United Nations is now spending in its efforts to reduce hunger.[16]

Greater efforts are also needed to improve nutrition before children even get to school age, so they can benefit from school lunches later. George McGovern notes that "a women, infants and children (WIC) program, which offers nutritious food supplements to needy pregnant and nursing mothers," should also be available in the poor countries. Based on 25 years of experience, it is clear that the U.S. WIC program has been enormously successful in improving nutrition, health, and the development of preschool children from low-income families. If this were expanded to reach pregnant women, nursing mothers, and small children in the 44 poorest countries, it would help eradicate hunger among millions of small children at a stage in their lives when it could make a huge difference.[17]

These efforts, though costly, are not expensive compared with the annual losses in productivity from hunger. McGovern and Dole think that this initiative can help "dry up the swamplands of hunger and despair that serve as potential recruiting grounds for terrorists." In a world where vast wealth is accumulating among the rich, it makes little sense for children to be going to school hungry.[18]

Stabilizing Population

Some 42 countries now have populations that are either essentially stable or declining slowly. In countries with the lowest fertility rates, including Japan, Russia, Germany, and Italy, populations will likely decline over the next half-century.[19]

A larger group of countries has reduced fertility to the replacement level or just below. They are headed for population stability after large groups of young people move through their reproductive years. Included in this group are China, the world's most populous country, and the United States, the third most populous one. A third group of countries is projected to more than double their populations by 2050, including Ethiopia, the Democratic Republic of the Congo, and the Sudan.[20]

U.N. projections show world population growth under three different assumptions about fertility levels. The medium projection, the one most commonly used, has world population reaching 9.1 billion by 2050. The high one reaches 10.6 billion. The low projection, which assumes that the world will quickly move below replacement-level fertility to 1.6 children per couple, has population peaking at 7.8 billion in 2041 and then declining. If the goal is to eradicate poverty, hunger, and illiteracy, we have little choice but to strive for the lower projection.[21]

Slowing world population growth means that all women who want to plan their families should have access to the family planning services they need. Unfortunately, at present 201 million couples cannot obtain the services they need to limit the size of their families. Filling the family planning gap may be the most urgent item on the global agenda. The benefits are enormous and the costs are minimal.[22]

The good news is that countries that want to help couples to reduce the size of their families quickly can do so. My colleague Janet Larsen writes that in just one decade Iran dropped its population growth rate from one of the world's fastest to one of the lowest in the developing world. When Ayatollah Khomeini assumed leadership in Iran in 1979, he immediately dismantled the family planning programs that the Shah had put in place in 1967 and instead advocated large families. At war with Iraq between 1980 and 1988, Khomeini wanted large families to increase soldiers for Islam. His goal was an army of 20 million. In response to his pleas, fertility levels climbed, pushing Iran's population growth up to a peak of 4.2 percent in the early 1980s, a level approaching the biological maximum. As this enormous growth began to burden the economy and the environment, the country's leaders realized that overcrowding, environmental degradation, and unemployment were undermining Iran's future.[23]

In 1989 the government did an about-face and Iran restored its family planning program. In May 1993, a national family planning law was passed. The resources of several government ministries, including education, culture, and health, were mobilized to encourage smaller families. Iran Broadcasting was given responsibility for raising awareness of population issues and of the availability of family planning services. Some 15,000 "health houses" or clinics were established to provide rural populations with health and family planning services.[24]

Religious leaders were directly involved in what amounted to a crusade for smaller families. Iran introduced a full panoply of contraceptive measures, including male sterilization—a first among Muslim countries. All forms of birth control, including contraceptives such as the pill and sterilization, were free of charge. In fact, Iran became a pioneer—the only country to require couples to take a class on modern contraception before receiving a marriage license.[25]

In addition to the direct health care interventions, a broad-based effort was launched to raise female literacy, boosting it from 25 percent in 1970 to more than 70 percent in 2000—an impressive achievement. Female school enrollment increased from 60 to 90 percent. Television was used to disseminate information on family planning throughout the country, taking advantage of the 70 percent of rural households with TV sets.

As a result of the impressive effort launched in 1989, family size in Iran dropped from seven children to fewer than three. From 1987 to 1994, Iran cut its population growth rate by half. Its overall population growth rate of 1.2 percent in 2004 is only slightly higher than that of the United States.[26]

If a country like Iran, with a strong tradition of Islamic fundamentalism, can move quickly toward population stability, other countries can too. Countries everywhere have little choice but to strive for an average of two children per couple. There is no feasible alternative. Any population that increases or decreases continually over the long term is not sustainable. The time has come for world leaders—including the Secretary-General of the United Nations, the President of the World Bank, and the President of the United States—to publicly recognize that the earth cannot easily support more than two children per family.

The costs of providing reproductive health and family planning services are not that high. At the International Conference on Population and Development held in 1994 in Cairo, it was estimated that a fully funded population and reproductive health program for the next 20 years would cost roughly $17 billion annually by 2000 and $22 billion by 2015. Developing countries agreed to cover two thirds of this, while industrial countries were to cover one third. Although we have passed the 10-year anniversary of the Cairo conference, developing countries have fallen short of their pledge by roughly 20 percent, while donor countries have fallen short by half, leaving a combined gap of roughly $6.6 billion per year.[27]

The United Nations estimated that meeting the needs of the 201 million women who do not have access to effective contraception could each year prevent 52 million unwanted pregnancies, 22 million induced abortions, and 1.4 million infant deaths. Some 142,000 pregnancy-related deaths could also be prevented. The costs to society of not filling the family planning gap are unacceptably high.[28]

Reinforcing these U.N. calculations are data from the grassroots showing how access to family planning services helps couples achieve their desired family size. Surveys in Honduras, for example, show poor women (often lacking family planning services) having twice as many children as they want, while women in high socioeconomic groups are quite successful at having the number of children they desire.[29]

Shifting to smaller families brings generous economic dividends. For Bangladesh, analysts concluded that $62 spent by the government to prevent an unwanted birth saved $615 in expenditures on other social services. Investing in reproductive health and family planning leaves more fiscal resources per child for education and health care, thus accelerating the escape from poverty. For donor countries, filling the entire $6.6 billion gap needed to ensure that couples everywhere have access to the services they want and need would yield strong social returns in improved education and health care.[30]

Better Health for All

While heart disease and cancer (largely the diseases of aging), obesity, and smoking dominate health concerns in industrial countries, in developing countries infectious diseases are the overriding health concern. Besides AIDS, the principal diseases of concern are diarrhea, respiratory illnesses, tuberculosis, malaria, and measles.

Many countries can no longer afford the vaccines for childhood diseases, such as measles, and are falling behind in their vaccination programs. Lacking the funds to invest today, they pay a far higher price tomorrow. There are not many situations where just a few pennies spent per youngster can make as much difference as vaccination programs can.[31]

Along with the eradication of hunger, ensuring access to a safe and reliable supply of water for the estimated 1 billion people who lack it is essential to better health for all. The realistic option in many cities now may be to bypass efforts to build costly water-based sewage removal and treatment systems and to opt instead for water-free waste disposal systems that do not disperse disease pathogens. (See the description of dry compost toilets in Chapter 11.) This switch would simultaneously help alleviate water scarcity, reduce the dissemination of disease agents in water systems, and help close the nutrient cycle—another win-win-win opportunity.

One of the most impressive health gains has come from a campaign led by UNICEF to treat the symptoms of diarrheal disease with oral rehydration therapy. This remarkably simple technique, which involves drinking a mild saline solution, has been extremely effective—reducing deaths from diarrhea among children from 4.6 million in 1980 to 1.5 million in 1999. Few investments have saved so many lives at such a low cost. In *Millions Saved,* Ruth Levine describes how Egypt used oral rehydration therapy to cut infant deaths from diarrhea by 82 percent from 1982 to 1989.[32]

Some leading sources of premature death are lifestyle-related. Cigarettes take a particularly heavy toll. The World Health Organization (WHO) estimates that 4.9 million people died in 2000 of tobacco-related illnesses, more than from any infectious disease. Today there are some 25 known diseases that are linked to tobacco use, including heart disease, stroke, respiratory illness, several forms of cancer, and male impotence. Cigarette smoke kills more people each year than all other air pollutants combined—nearly 5 million versus 3 million.[33]

Impressive progress is being made in reducing cigarette smoking. After a century-long buildup of the tobacco habit, the world is turning away from cigarettes, led by WHO's Tobacco Free Initiative. This gained further momentum from the Framework Convention on Tobacco Control, the first international accord to deal entirely with a health issue, which was adopted unanimously in Geneva in May 2003.[34]

Ironically, the country where tobacco originated is now leading the world away from it. In the United States, the number of cigarettes smoked per person has dropped from its peak of 2,872 in 1976 to 1,374 in 2003—a decline of 52 percent. Worldwide, where the downturn lags that of the United States by roughly a decade, usage has dropped from the historical high of 1,035 cigarettes smoked per person in 1986 to 856 in 2003, a fall of 17 percent. Media coverage of the health effects of smoking, mandatory health warnings on cigarette packs, and sharp increases in cigarette sales taxes have all contributed to the steady decline.[35]

Indeed, smoking is on the decline in nearly all the major cigarette-smoking countries, including such strongholds as France, China, and Japan. The number of cigarettes smoked per person has dropped 22 percent in France since peaking in 1984, 5 percent in China since 1989, and 20 percent in Japan since 1991.[36]

Following approval of the Framework Convention on Tobacco Control, a number of countries took strong steps in 2004 to reduce smoking. Ireland imposed a nationwide ban on smoking in workplaces, bars, and restaurants; India banned smoking in public places; Norway banned smoking in bars and restaurants; and Scotland banned smoking in public buildings. Bhutan, a Himalayan country of 1 million sandwiched between India and China, has prohibited tobacco sales entirely.[37]

In 2005, smoking was banned in public places in Bangladesh, in bars and restaurants in New Zealand, and in public places in Italy. In the United States, which already has stiff restrictions on smoking, the Union Pacific Corporation stopped hiring smokers in seven states as an economy measure to cut health care costs. General Mills imposes a $20-a-month surcharge on health insurance premiums for employees who smoke. Each of these measures helps the market to more accurately reflect the cost of smoking.[38]

The war against infectious diseases is being waged on a broad front. Perhaps the leading privately funded life-saving activity in the world today is the childhood immunization program. In an effort to fill the gap in this global program, the Bill and Melinda Gates Foundation has invested $1.5 billion through 2005 to protect children from infectious diseases.[39]

One of the international community's finest moments came with the eradication of smallpox, an effort led by WHO. This successful elimination of a feared disease, which required a worldwide immunization program, saves not only millions of lives but also hundreds of millions of dollars each year in smallpox vaccination programs and billions of dollars in health care expenditures. This achievement alone may justify the existence of the United Nations.[40]

Similarly, a WHO-led international coalition, including Rotary International, UNICEF, the U.S. Centers for Disease Control and Prevention, and Ted Turner's UN Foundation, has led a worldwide campaign to wipe out polio. Since 1988, Rotary International has contributed an extraordinary $500 million to this effort. Under this coalition-sponsored Global Polio Eradication Initiative, the number of polio cases worldwide dropped from some 350,000 per year in 1988 to just 800 in 2003.[41]

By mid-2003, pockets of polio remained only in Nigeria, Niger, Egypt, India, Pakistan, and Afghanistan, but then some of the Muslim-dominated states of northern Nigeria stopped vaccination because of a rumor that the vaccine would render people sterile or cause AIDS. By the end of 2004, when the misinformation was corrected, polio vaccinations were resumed in northern Nigeria. But during the interim, polio had become reestablished in several countries, apparently aided by the annual pilgrimage of Nigerian Muslims to Mecca. New infections appeared in Saudi Arabia, Yemen, Côte d'Ivoire, Burkina Faso, the Central African Republic, Chad, Mali, Sudan, Indonesia, and Somalia.[42]

These countries, once free of the disease, are scrambling now to contain and eradicate the new outbreak that as of September 2005 had grown to 1,260 cases. With two recently confirmed cases in Somalia, a failed state, the fear is that the virus may spread further not only in this country where there is no government to work with, but to other countries as well, making it extraordinarily difficult to eradicate.[43]

A 2001 WHO study analyzing the economics of health care in developing countries concluded that providing the most basic health care services, the sort that could be supplied by a village-level clinic, would yield enormous economic benefits for developing countries and for the world as a whole. The authors estimated that providing basic universal health care in developing countries will require donor grants totaling $27 billion in 2007, scaled up to $38 billion in 2015, or an average of $33 billion per year. In addition to basic services, this $33 billion includes funding for the Global Fund to Fight AIDS, Tuberculosis and Malaria and for universal childhood vaccinations.[44]

Curbing the HIV Epidemic

The key to curbing the AIDS epidemic, which has so disrupted economic and social progress in Africa, is education about prevention. We know how the disease is transmitted; it is not a medical mystery. In Africa, where once there was a stigma associated with even mentioning the disease, governments are beginning to design effective prevention education programs. The first goal is to reduce quickly the number of new infections, dropping it below the number of deaths from the disease, thus shrinking the number of those who are capable of infecting others.

Concentrating on the groups in a society who are most likely to spread the disease is particularly effective. In Africa, infected truck drivers who travel far from home for extended periods often engage in commercial sex, spreading HIV from one country to another. They are thus a target group for reducing infections. Sex workers are also centrally involved in the spread of the disease. In India, for example, the country's 2 million female sex workers have an average of two encounters per day, making them a key group to educate about HIV risks and the life-saving value of using a condom.[45]

Another target group is the military. After soldiers become infected, usually from engaging in commercial sex, they return to their home communities and spread the virus further. In Nigeria, where the adult HIV infection rate is 5 percent, President Olusegun Obasanjo requires free distribution of condoms to all military personnel. A fourth target group, intravenous drug users who share needles, figures prominently in the spread of the virus in the former Soviet republics.[46]

At the most fundamental level, dealing with the HIV threat requires roughly 10 billion condoms a year in the developing world and Eastern Europe. Including those needed for contraception adds another 2 billion. But of the 12 billion condoms needed, only 2.5 billion are being distributed, leaving a shortfall of 9.5 billion. At only 3¢ each, or $285 million, the cost of saved lives by supplying condoms is minuscule.[47]

The condom gap is huge, but the costs of filling it are small. In the excellent study *Condoms Count: Meeting the Need in the Era of HIV/AIDS*, Population Action International notes that "the costs of getting condoms into the hands of users—which involves improving access, logistics and distribution capacity, raising awareness, and promoting use—is many times that of the supplies themselves." If we assume that these costs are six times the price of the condoms themselves, filling this gap would still cost only $2 billion.[48]

Sadly, even though condoms are the only technology available to prevent the spread of HIV, the U.S. government is de-emphasizing their use, insisting that abstinence be given top priority. While encouraging abstinence is important, an effective campaign to curb the HIV epidemic cannot function without condoms.[49]

One of the few African countries to successfully lower the HIV infection rate after the epidemic became well established is

Uganda. Under the strong personal leadership of President Yoweri Museveni, the share of adults infected has dropped from a peak of 13 percent in the early 1990s to 4 percent in 2003. More recently, Zambia also appears to be making progress in reducing infection rates among young people as a result of a concerted national campaign led by church groups. Senegal, which acted early and decisively to check the spread of the virus, has an infection rate among adults of less than 1 percent today. It is a model for other African countries.[50]

The financial resources and medical personnel currently available to treat people who are already HIV-positive are severely limited compared with the need. For example, of the 4.7 million people who exhibited symptoms of AIDS in sub-Saharan Africa in June of 2005, only 500,000 were receiving the antiretroviral drug treatment that is widely available in industrial countries. However, this was up threefold from a year earlier. The increase is part of a worldwide effort by the World Health Organization to reach 3 million people in low- and middle-income countries by the end of 2005, known as the 3 by 5 Initiative.[51]

There is a growing body of evidence that the prospect of treatment encourages people to get tested for HIV. It also raises awareness and understanding of the disease and how it is transmitted. And if people know they are infected, they may try to avoid infecting others. To the extent that treatment extends life, and the average extension in the United States is about 15 years, it is not only the humanitarian thing to do, it also makes economic sense. Once society has invested in the rearing, education, and on-job training of an individual, the value of extending the working lifetime is high.[52]

Treating those with HIV infections is costly, but ignoring the need for treatment is a strategic mistake simply because treatment strengthens prevention efforts. Africa is paying a heavy cost for its delayed response to the epidemic. It is a window on the future of other countries, such as India and China, if they do not move quickly to contain the virus that is already well established within their borders.[53]

Reducing Farm Subsidies and Debt

Eradicating poverty involves much more than international aid programs. For many developing countries, farm subsidies in aid-giving countries and debt relief may be even more important. A successful export-oriented farm sector—taking advantage of low-cost labor and natural endowments of land, water, and climate to boost rural incomes and to earn foreign exchange—often offers a path out of poverty. Sadly, for many developing countries this path is blocked by the self-serving farm subsidies of affluent countries. Overall, the farm subsidies in the affluent countries at $279 billion are roughly four times the development assistance flows from these governments.[54]

The size of the agricultural budget of the European Union (EU) is staggering, accounting for over half of its total annual budget. It also looms large internationally. As the *Financial Times* points out, the cash subsidy to a dairy cow in the EU exceeds the EU development assistance per person in sub-Saharan Africa.[55]

Within affluent countries, the EU-25 in 2004 accounted for $133 billion of the $279 billion spent by affluent countries on farm subsidies. The United States spent $46 billion on farm subsidies. These encourage overproduction of farm commodities, which then are sent abroad with another boost from export subsidies. The result is depressed world market prices, particularly for sugar and cotton, the two commodities where developing countries have the most to lose.[56]

Although the European Union accounts for more than half of the $78 billion in development assistance from all countries, much of the economic gain from this assistance in the past was offset by the EU's annual dumping of some 6 million tons of sugar in the world market. This is one farm commodity where developing countries have a strong comparative advantage and should be permitted to capitalize on it. Fortunately, in 2005 the EU announced that it would reduce its sugar support price to farmers by 40 percent, thus discouraging the excess production that depressed the world market price when it was exported. The affluent world can no longer afford farm policies that permanently trap millions in poverty by cutting off a main avenue of escape.[57]

Help in raising world sugar prices may come from an unexpected quarter. Although it is too early to say for sure, rising oil prices may boost sugar prices as more and more sugarcane-based ethanol refineries are built. In effect, the price of sugar may track the price of oil upward, providing a strong economic boost for those developing-world economies where nearly all the world's sugarcane is produced.[58]

Recent developments may also lift world cotton prices. Although the U.S. government does not have explicit export subsidies, production subsidies provided to farmers enable them to export cotton at low prices. These subsidies to just 25,000 cotton farmers exceed U.S. financial aid to all of sub-Saharan Africa's 750 million people. And since the United States is the world's leading cotton exporter, its subsidies depress prices for all cotton exporters.[59]

U.S. cotton subsidies have faced a spirited challenge from four cotton-producing countries in Central Africa: Benin, Burkina Faso, Chad, and Mali. In addition, Brazil successfully challenged U.S. cotton subsidies within the framework of the World Trade Organization (WTO). To make its case, the Brazilian government hired a leading U.S. agricultural economist. Using U.S. Department of Agriculture data, Brazil convinced the WTO panel that U.S. cotton subsidies were depressing world prices and harming their cotton producers. In response, the panel ruled that the United States had to eliminate the subsidies.[60]

Along with eliminating harmful agricultural subsidies, debt forgiveness is another essential component of the broader effort to eradicate poverty. For example, with sub-Saharan Africa spending four times as much on debt servicing as it spends on health care, debt forgiveness can help boost living standards in this last major bastion of poverty.[61]

In July of 2005, heads of the G-8 group of industrial countries, meeting in Gleneagles, Scotland, agreed to the cancellation of the multilateral debt that a number of the poorest countries owed to the World Bank, the International Monetary Fund, and the African Development Bank. This initiative, immediately affecting 18 of the poorest debt-ridden countries (14 in Africa and 4 in Latin America), offers these countries a new lease on life. Up to another 20 of the poorest countries could benefit from this initiative if they can complete the qualification. A combination of public pressure by nongovernmental groups campaigning for debt relief in recent years and strong leadership from the U.K. government were the keys to this poverty reduction breakthrough.[62]

Although this was a giant step in the right direction, it eliminated only a minor share of the total debt of the poorest countries to international lending institutions. In addition to the 18 countries granted relief so far, there are at least 40 more countries with low incomes that desperately need help. The groups that are lobbying for debt relief, such as Oxfam International, believe it is inhumane to force those with incomes of scarcely a dollar per day to use part of that dollar to service debt. They pledge to keep the pressure on until all the debt of these poorest countries is cancelled.[63]

A Poverty-Eradication Budget

Many countries that have experienced rapid population growth for several decades are showing signs of demographic fatigue. Countries struggling with the simultaneous challenge of educating growing numbers of children, creating jobs for swelling ranks of young job seekers, and dealing with the environmental effects of population growth are stretched to the limit. When a major new threat arises—such as the HIV epidemic—governments often cannot cope.

Problems routinely managed in industrial societies are becoming full-scale humanitarian crises in developing ones. The rise in deaths in many African countries marks a tragic new development in world demography. In the absence of a concerted effort by national governments and the international community to accelerate the shift to smaller families, events in many countries could spiral out of control, leading to more death and to spreading political instability and economic decline.

There is an alternative to this bleak prospect, and that is to help countries that want to slow their population growth to do so quickly. This brings with it what economists call the demographic bonus. When countries move quickly to smaller families, growth in the number of young dependents—those who need nurturing and educating—declines relative to the number of working adults. In this situation, productivity rises, savings and investment climb, and economic growth accelerates.[64]

Japan, which cut its population growth in half between 1951 and 1958, was one of the first countries to benefit from the demographic bonus. South Korea and Taiwan followed, and more recently China, Thailand, Viet Nam, and Sri Lanka have benefited from earlier sharp reductions in birth rates. This effect lasts for only a few decades, but it is usually enough to launch a country into the modern era.[65]

The steps needed to eradicate poverty and accelerate the shift to smaller families are clear. They include filling several funding gaps, including those needed to reach universal primary education; to fight infectious diseases, such as AIDS, tuberculosis, and malaria; to provide reproductive health care; and to contain the HIV epidemic. Collectively, the initiatives discussed in this chapter are estimated to cost another $68 billion a year. (See Table 7–1.)[66]

The heaviest investments in this effort center on education and health, which are the cornerstones of both human capital development and population stabilization. Education includes both universal primary education and a global campaign to eradicate adult illiteracy. Health care includes the basic interventions involved in controlling infectious diseases, beginning with childhood vaccinations. Adopting the basic health care program outlined in the 2001 *Report of the Commission on Macroeconomics and Health* to the World Health Organization would save an estimated 8 million lives per year by 2010. These are the keys to breaking out of the poverty trap.[67]

Table 7–1. *Additional Annual Funding Needed to Reach Basic Social Goals*

Goal	Funding (billion dollars)
Universal primary education	12
Eradication of adult illiteracy	4
School lunch programs for 44 poorest countries	6
Assistance to preschool children and pregnant women in 44 poorest countries	4
Reproductive health and family planning	7
Universal basic health care	33
Closing the condom gap	2
Total	68

Source: See endnote 66.

As Jeffrey Sachs regularly reminds us, for the first time in history we have the technologies and financial resources to eradicate poverty. As noted earlier, the last 15 years have seen some impressive gains. For example, China has not only dramatically reduced the number living in poverty within its borders, but, with its trade and investment initiatives, it is helping poorer countries develop. China is investing substantial sums in Africa, investments often related to helping African countries develop their numerous mineral and energy resources, something that China needs.[68]

Helping low-income countries break out of the demographic trap is a highly profitable investment for the world's affluent nations. Industrial-country investments in education, health, and school lunches are in a sense a humanitarian response to the plight of the world's poorest countries. But more fundamentally, they are investments that will shape the world in which our children will live.

8

Restoring the Earth

The health of an economy cannot be separated from that of its natural support systems. More than half the world's people depend directly on croplands, rangelands, forests, and fisheries for their livelihoods. Many more depend on forest product industries, leather goods industries, cotton and woolen textile industries, and food processing industries for their jobs.[1]

A strategy for eradicating poverty will not succeed if an economy's environmental support systems are collapsing. If croplands are eroding and harvests are shrinking, if water tables are falling and wells are going dry, if rangelands are turning to desert and livestock are dying, if fisheries are collapsing, if forests are shrinking, and if rising temperatures are scorching crops, a poverty-eradication program—no matter how carefully crafted and well implemented—will not succeed.

In Chapter 5, we discussed the deforestation, soil erosion, and the utter devastation of Haiti's countryside. After looking at the desperate situation in Haiti, Craig Cox, Director of the U.S.-based Soil and Water Conservation Society, wrote, "I was reminded recently that the benefits of resource conservation—at the most basic level—are still out of reach for many. Ecological and social collapses have reinforced each other in a downward spiral into poverty, environmental degradation, social injustice, disease, and violence." Unfortunately, the situation Cox describes is no longer a rarity. It describes what lies ahead for more and more countries if we do not launch an earth restoration initiative.[2]

Restoring the earth will take an enormous international effort, one even larger and more demanding than the often-cited Marshall Plan that helped rebuild war-torn Europe and Japan. And such an initiative must be undertaken at wartime speed lest environmental deterioration translate into economic decline, just as it did for earlier civilizations that violated nature's thresholds and ignored its deadlines.

Protecting and Restoring Forests

Protecting the earth's 3.9 billion hectares of remaining forests and replanting those lost are both essential for restoring the earth's health, an important foundation for the new economy. Reducing rainfall runoff and the associated flooding and soil erosion, recycling rainfall inland, and restoring aquifer recharge depend on simultaneously reducing pressure on forests and on reforestation.[3]

There is a vast unrealized potential in all countries to lessen the demand pressure that is shrinking the earth's forest cover. In industrial nations the greatest opportunity lies in reducing the quantity of wood used to make paper, and in developing countries it depends on reducing fuelwood use.

The rates of paper recycling in the top 10 paper-producing countries range widely, from China and Italy on the low end, recycling 27 and 31 percent of the paper they use, to Germany and South Korea on the high end, at 72 and 66 percent. The recycling rate in Germany is high because the government has consistently emphasized paper recycling to reduce the flow to landfills. If every country recycled as much of its paper as Germany does, the amount of wood pulp used to produce paper worldwide would drop by one third.[4]

The United States, the world's largest paper consumer, is far behind Germany but is making some progress. Twenty years ago, roughly one fourth of the paper used in the United States was recycled. By 2003, the figure had reached 48 percent.[5]

The use of paper, perhaps more than any other single product, reflects the throwaway mentality that evolved during the last century. There is an enormous possibility for reducing paper use simply by replacing facial tissues, paper napkins, disposable diapers, and paper shopping bags with reusable cloth alternatives.

The largest single demand on trees—the need for fuel—accounts for just over half of all wood removed from forests. Some international aid agencies, including the U.S. Agency for International Development (AID), have begun to sponsor fuelwood efficiency projects. One of AID's more promising national projects is the distribution of 780,000 new, highly efficient wood cookstoves in Kenya. Investing public resources in replacing outmoded inefficient cookstoves can earn handsome dividends in forest protection and regeneration, including the restoration of forest services.[6]

Over the longer term, developing alternative cooking fuels is the key to reducing forest pressure in developing countries. As the world shifts from a fossil-fuel-reliant economy to one based on wind, solar, and geothermal energy, it will be much easier for developing countries without fossil fuels to develop indigenous sources of renewable energy. Replacing firewood with solar thermal cookers, with electric hotplates fed by wind-generated electricity, or with some other energy source will lighten the load on forests.

Kenya is also the site of a solar cooker project sponsored by Solar Cookers International. These inexpensive cookers, made from cardboard and aluminum foil and costing $10 each, cook slowly, much like a crockpot. Requiring three hours of sunshine to cook a complete meal, they can greatly reduce firewood use at little cost. They can also be used to pasteurize water, thus saving lives.[7]

Earlier definitions of sustainable forestry focused only on the sustained production of forest products, but they now include sustaining forest services such as flood control. Despite the high value to society of intact forests, only about 290 million hectares of global forest area are legally protected from logging. An additional 1.4 billion hectares are economically unavailable for harvesting because of geographic inaccessibility or low-value wood. Of the remaining area available for exploitation, 665 million hectares are undisturbed by humans and nearly 900 million hectares are semi-natural and not in plantations.[8]

Forests that are protected by national decree are often safeguarded not so much to preserve the long-term wood supply capacity as to ensure that the forest can continue to provide services. Countries that provide legal protection for forests often do so after they have suffered the consequences of extensive deforestation. The Philippines, for example, has banned logging in all remaining old-growth and virgin forests largely because the country has become so vulnerable to flooding, erosion, and landslides. The country was once covered by rich stands of tropical hardwood forests, but after years of massive clearcutting, it lost both the forest's products and its services and became a net importer of forest products.[9]

Reed Funk, professor of plant biology at Rutgers University, believes the vast areas of deforested land can be used to grow trillions of genetically improved trees for food, mostly nuts, and for fuel. Funk sees nuts used to supplement meat as a source of high-quality protein in developing-country diets. He also sees trees grown on this deforested land, much of it now wasteland, being used for conversion into ethanol for automotive fuel.[10]

Although nongovernmental organizations (NGOs) have worked for years to protect forests from clearcutting, the World Bank has only recently begun to systematically consider sustainable forestry. In 1998, the Bank joined forces with the World Wide Fund for Nature to form the Alliance for Forest Conservation and Sustainable Use; by 2005 they had helped designate 50 million hectares of new forest protected areas and certify 22 million hectares of forest. In mid-2005, the Alliance announced a goal of reducing global deforestation rates to zero by 2020.[11]

There are several forest product certification programs that link environmentally conscious consumers with sustainable management of the forest where the product originates. Some programs are national while others are international; some originate with importing countries and others with exporters.

The most rigorous international program, one that is certified by a group of NGOs, is the Forest Stewardship Council (FSC). Some 57 million hectares of forests in 65 countries are certified by FSC-accredited bodies as responsibly managed. Among the leaders in certified forest area are Sweden, with 10 million hectares; Poland, with 6 million hectares; the United States, with nearly 5 million hectares; and Brazil and South Africa, with 3 million and 2 million hectares respectively.[12]

Forest plantations can reduce pressures on the earth's remaining forests as long as they do not replace old-growth forest. As of 2000, the world had 187 million hectares in forest plantations, an area less than 5 percent of the total 3.9 billion hectares in forest and equal to nearly one fourth of the 700 million hectares planted in grain. Tree plantations produce mostly wood for paper mills or for wood reconstitution mills. Increasingly, reconstituted wood is substituting for natural wood in the world lumber market as industry adapts to a shrinking supply of large logs from natural forests.[13]

Production of roundwood on plantations is estimated at 414 million cubic meters per year, accounting for 12 percent of world wood production. This means that the lion's share, some 88 percent of the world timber harvest, comes from natural forest stands.[14]

Five countries account for two thirds of tree plantations. China, which has little original forest remaining, is the largest, with Russia and the United States following. India and Japan are fourth and fifth. Brazil is further back, but is expanding fast. As tree farming expands, it is shifting geographically to the moist tropics. In contrast to grain yields, which tend to rise with distance from the equator and the longer summer growing days, tree plantation yields rise with proximity to the equator and year-round growing conditions.[15]

In eastern Canada, the average hectare of forest plantation produces 4 cubic meters per year. In the southeastern United States, where U.S. plantations are concentrated, the yield is 10 cubic meters. But in Indonesia, it is 25 cubic meters, and in Brazil, newer plantations may be close to 30 cubic meters. While corn yields in the United States are nearly triple those in Brazil, timber yields are the reverse, favoring Brazil by nearly 3 to 1. To satisfy a given demand for wood, Brazil requires only one third as much land as the United States, which helps explain why growth in pulp capacity is now concentrated in equatorial regions.[16]

Projections of future growth show that plantations are constrained by land scarcity. They can sometimes be profitably established on already deforested, often degraded, land, but they are more likely to come at the expense of existing forests. There is also competition with agriculture, since land that is suitable for crops is also good for growing trees. Water scarcity is yet another constraint. Fast-growing plantations require abundant moisture.

Nonetheless, the U.N. Food and Agriculture Organization (FAO) projects that as plantation area expands and yields rise, the harvest could more than double during the next three decades. It is entirely conceivable that plantations could one day satisfy most of the world's demand for industrial wood, thus helping to protect the world's remaining forests.[17]

Historically, some highly erodible agricultural land in industrial countries has been reforested by natural regrowth. For example, New England, a geographically rugged region of the United States, was reforested beginning a century or so ago. Settled early by Europeans, this region was suffering from cropland productivity losses because soils were thin and the land was rocky, sloping, and vulnerable to erosion. As highly productive farmland opened up in the Midwest and the Great Plains during the nineteenth century, pressures on New England farmland lessened, permitting cropped land to return to forest. New England's forest cover has increased from a low of roughly one third two centuries ago to perhaps three fourths today, slowly regaining its original health and diversity.[18]

A somewhat similar situation exists now in parts of the former Soviet Union and in several East European countries. As central planning was replaced by market-based agriculture in the early 1990s, farmers on marginal land were forced to seek their livelihoods elsewhere. Precise figures are difficult to come by, but millions of hectares of farmland are now returning to forest.[19]

South Korea is in many ways a reforestation model for the rest of the world. When the Korean War ended, half a century ago, the mountainous country was largely deforested. Beginning around 1960, under the dedicated leadership of President Park Chung Hee, the South Korean government launched a national reforestation effort. Relying on the creation of village cooperatives, hundreds of thousands of people were mobilized to dig trenches and to create terraces for supporting trees on barren mountains. South Korea not only reclaimed denuded areas, it also supported the effort with the establishment of fuelwood forests. Se-Kyung Chong, researcher at the Korea Forest Research Institute, writes, "The result was a seemingly miraculous rebirth of forests from barren land."[20]

Today forests cover 65 percent of the country, an area of roughly 8 million hectares. While driving across South Korea in November 2000, it was gratifying for me to see the luxuriant stand of trees on mountains that a generation ago were bare. We can reforest the earth![21]

In Turkey, a mountainous country largely deforested over millennia, a leading environmental group, TEMA (Türkiye Erozyona Mücadele, Agaclandirma), has made reforestation its principal activity. Founded by two prominent Turkish businessmen, Hayrettin Karuca and Nihat Gokyigit, TEMA launched in 1998 a 10-billion-acorn campaign to restore tree cover and reduce runoff and soil erosion. During the years since, 850 million oak acorns have been planted. The program is also raising national awareness of the services that forests provide.[22]

China is engaging in its own reforestation effort. In addition to planting trees in the recently deforested upper reaches of the Yangtze River basin to control flooding, China is planting a belt of trees to protect land from the expanding Gobi Desert. This green wall, a modern version of the Great Wall, is projected to reach some 4,480 kilometers (2,800 miles) in length, stretching from outer Beijing through Inner Mongolia. An ambitious, long-term plan, it is expected to take 70 years to complete and to cost up to $8 billion.[23]

Shifting subsidies from building logging roads to planting trees would help protect forest cover worldwide. The World Bank has the administrative capacity to lead an international program that would emulate South Korea's success in blanketing mountains and hills with trees.

In addition, FAO and the bilateral aid agencies can work with individual farmers in national agroforestry programs to integrate trees wherever possible into agricultural operations. Well-chosen, well-placed trees provide shade, serve as windbreaks to check soil erosion, and can fix nitrogen, reducing the need for fertilizer.

Reducing wood use by developing more-efficient wood stoves and alternative means of cooking, systematically recycling paper, and banning the use of throwaway paper products all lighten pressure on the earth's forests. A global reforestation effort cannot succeed unless it is accompanied by the stabilization of population. With such an integrated plan, coordinated country by country, the earth's forests can be restored.

Conserving and Rebuilding Soils

In reviewing the literature on soil erosion, references to the "loss of protective vegetation" occur again and again. Over the last half-century, we have removed so much of that protective cover by clearcutting, overgrazing, and overplowing that we are fast losing soil accumulated over long stretches of geological time. Eliminating these excesses and the resultant decline in the earth's biological productivity depends on a worldwide effort to restore the earth's vegetative cover, an effort that is now under way in some countries.

The 1930s Dust Bowl that threatened to turn the U.S. Great Plains into a vast desert was a traumatic experience that led to revolutionary changes in American agricultural practices, including the planting of tree shelterbelts—rows of trees planted beside fields to slow wind and thus reduce wind erosion—and strip-cropping, the planting of wheat on alternate strips with fallowed land each year. Strip cropping permits soil moisture to accumulate on the fallowed strips, while the alternating planted strips reduce wind speed and hence erosion on the idled land.[24]

In 1985, the U.S. Congress, with strong support from the environmental community, created the Conservation Reserve Program (CRP) to reduce soil erosion and control overproduction of basic commodities. By 1990 there were some 14 million hectares (35 million acres) of highly erodible land in permanent vegetative cover under 10-year contracts. Under this program, farmers were paid to plant fragile cropland to grass or trees. The retirement of 14 million hectares under the CRP, together with the use of conservation practices on 37 percent of all cropland, reduced U.S. soil erosion from 3.1 billion tons to 1.9 billion tons during the 15 years from 1982 to 1997. The U.S. approach to controlling soil erosion by both converting highly erodible cropland back to grassland or trees and adopting soil conservation practices offers a model for the rest of the world.[25]

The conversion of cropland to nonfarm uses is often beyond the control of farmers, but the losses of soil and eroded land from severe erosion are not. Lowering soil losses caused by wind and water erosion below the gains in new soil formed by natural processes will take an enormous worldwide effort. Preserving the biological productivity of highly erodible cropland depends on planting it in grass or trees before it becomes wasteland. The first step in halting the decline in inherent land fertility is to pull back from this fast-deteriorating margin.[26]

Terracing, a time-tested method for dealing with water erosion, is common in rice paddies throughout the mountainous regions of Asia. On less steeply sloping land, contour strip farming, as found in the U.S. Midwest, works well.[27]

Another tool in the soil conservation toolkit—and a relatively new one—is conservation tillage, which includes both no-till and minimum tillage. In addition to reducing both wind and water erosion, this practice helps retain water, raises soil carbon content, and reduces the energy needed for crop cultivation.

Instead of the traditional cultural practices of plowing land, discing or harrowing it to prepare the seedbed, and then using a mechanical cultivator to control weeds in row crops, farmers simply drill seeds directly through crop residues into undisturbed soil, controlling weeds with herbicides. The only soil disturbance is the narrow slit in the soil surface where the seeds are inserted, leaving the remainder of the soil undisturbed, covered by crop residues and thus resistant to both water and wind erosion.[28]

In the United States, where farmers during the 1990s were required to implement a soil conservation plan on erodible cropland to be eligible for commodity price supports, the no-till area went from 7 million hectares in 1990 to 25 million hectares in 2004. Now widely used in the production of corn and soybeans in the United States, no-till has spread rapidly in the western hemisphere, covering 24 million hectares in 2004 in Brazil, 18 million hectares in Argentina, and 13 million in Canada. Australia, with 9 million hectares, rounds out the five leading no-till countries.[29]

Once farmers master the practice of no-till, its use can spread rapidly, particularly if governments provide economic incentives or require farm soil conservation plans for farmers to be eligible for crop subsidies. Recent FAO reports describe the early growth in no-till farming over the last few years in Europe, Africa, and Asia.[30]

Algeria, trying to halt the northward advance of the Sahara Desert, announced in December 2000 that it is concentrating its orchards and vineyards in the southern part of the country, hoping that these perennial plantings will halt the desertification of its cropland. In July 2005, the Moroccan government, responding to severe drought, announced that it was allocating $778 million to cancel farmers' debts and to convert cereal-planted areas into less vulnerable olive and fruit orchards."[31]

There are similar concerns about the expanding Sahara on the southern edge of the desert as well. President Olusegun Obasanjo of Nigeria has proposed planting a Great Green Wall of trees, a band five kilometers wide stretching 7,000 kilometers across Africa, in an effort to halt the desert's advance. Senegal, which is on the western end of this proposed wall and is losing 50,000 hectares of productive land each year, strongly supports the idea. No one knows how long this project would take, but Senegalese environment minister Modou Fada Diagne observes, "Poverty and desertification create a vicious cycle....Instead of waiting for the desert to come to us, we need to attack it."[32]

As noted earlier, China also is trying to halt the advance of deserts with its Great Green Wall. In addition, it is paying farmers in the threatened provinces to plant their cropland in trees. The goal is to plant trees on 10 million hectares of grainland, easily one tenth of China's current grainland area.[33]

In Inner Mongolia (Nei Monggol), efforts to halt the advancing desert and to reclaim the land for productive uses rely on planting desert shrubs to stabilize the sand dunes. And in many situations, sheep and goats have been banned entirely. In Helin County, south of the provincial capital of Hohhot, the planting of desert shrubs on abandoned cropland has now stabilized the soil on the county's first 7,000-hectare reclamation plot. Based on this success, the reclamation effort is being expanded.[34]

The Helin County strategy centers on replacing the large number of sheep and goats with dairy cattle, increasing the number of dairy animals from 30,000 in 2002 to 150,000 by 2007. The cattle are kept in enclosed areas, feeding on cornstalks, wheat straw, and the harvest from a drought-tolerant forage crop resembling alfalfa, which is grown on reclaimed land. Local officials estimate that this program will double incomes within the county during this decade.[35]

To relieve pressure on the country's rangelands, Beijing is asking herders to reduce their flocks of sheep and goats by 40 percent. But in communities where wealth is measured in livestock numbers and where most families are living in poverty, such cuts are not easy or, indeed, likely, unless alternative livelihoods are offered pastoralists along the lines proposed in Helin County.[36]

The only viable way to eliminate overgrazing on the two fifths of the earth's land surface classified as rangelands is to reduce the size of flocks and herds. Not only do the excessive numbers of cattle, and particularly sheep and goats, remove the vegetation, but their hoofs pulverize the protective crust of soil that is formed by rainfall and that checks wind erosion. In some situations, the only viable option is to keep the animals in enclosures, bringing the forage to them. India, which has successfully adopted this practice for its thriving dairy industry, is the model for other countries.[37]

Protecting the earth's remaining vegetation also warrants a ban on the clearcutting of forests in favor of selective harvesting, simply because with each clearcut there are heavy soil losses until the forest regenerates. Thus with each subsequent cutting, productivity declines further. Restoring the earth's tree and grass cover protects soil from erosion, reduces flooding, and sequesters carbon. It is one way we can restore the earth so that it can support our children and grandchildren.

Meeting Nature's Water Needs

There are many reasons for balancing water demand and supply. Failure to do so means that water tables will continue to fall, more rivers will run dry, and more lakes will disappear. If water tables are falling while energy prices are rising, irrigation water costs can rise to where farmers can no longer afford to irrigate. (Ways to raise irrigation efficiency are discussed in Chapter 9. Chapter 11 describes ways to reduce urban water waste.)

In *Rivers for Life: Managing Water for People and Nature*, Sandra Postel and Brian Richter cite South Africa's 1998 National Water Act as a model for other countries. The act focuses on two broad needs. The first is satisfying basic water needs of everyone for drinking, cooking, sanitation, and other essential purposes, which the legislation describes as a non-negotiable allocation. The second is the water needed to support river ecosystem functions "in order to conserve biodiversity and secure the valuable ecosystem services they provide to society."[38]

Establishing minimal river flows so as to satisfy the specific needs of downstream aquatic ecosystems such as floodplains, river deltas, and wetlands is not necessarily easy. For example, at times a strong flow is needed to meet the freshwater needs of an estuary. At other times, the needs of spawning fish may determine the ecological water needs.

A World Conservation Union–IUCN study in Australia notes that the Mowamba aqueduct has been permanently closed after 100 years of use in order to raise the flow of the Snowy River. This initial action, which raises the river flow from 3 percent of the natural level to 6 percent, is the first in a series of steps to bring the river flow back to 28 percent of the natural level and thus to restore its natural functions. In Australia's Murray-Darling basin, the enhanced flow of a river with releases from a storage facility in the basin helped to restore the natural wildlife population. The IUCN report noted, "the great egret bred for the first time since 1979, nine species of frog bred, as did native fish."[39]

Perhaps the best known and largest example of returning water to restore and support marine habitats occurred in California when the U.S. Congress passed legislation in 1992 that was designed to restore the overall health of the fish and wildlife habitat, including salmon runs, of the Sacramento-San Joaquin river system. Initially, as Sandra Postel reports in *Pillar of Sand*, Congress authorized the use of 800,000 acre-feet, nearly 1 billion cubic meters, or about 10 percent of the Central Valley Project's yearly water supply, for this purpose. Farmers who lost part of their irrigation water challenged the law.[40]

After several years of legal challenges and negotiations involving environmental groups, farmers, state government officials, and others, agreement was reached on an arrangement more or less consistent with the original congressional intent. The increased flow of the two rivers, which merge before emptying into San Francisco Bay, also helped protect the Bay's rich aquatic ecosystem, which is home to some 120 species of fish.[41]

Variations of these efforts to restore river flows to supply natural systems with the water they need are now commonplace. In the United States, literally hundreds of smaller dams are being demolished in an effort to restore river flows and natural systems, including spawning runs.[42]

In situations where growing water demand is exceeding the supply in more and more river basins, the challenge is to establish guidelines by which the various needs for water are met, recognizing that few will be fully met. Success hinges on having the institutions and a process by which water can be allocated among competing uses in a way that maximizes the contribution to society as a whole rather than to a small number of influential stakeholders at the expense of others.

Regenerating Fisheries

For decades governments tried to save specific fisheries by restricting the catch of individual species. Sometimes this worked; sometimes it failed and fisheries collapsed. In recent years, support for another approach—the creation of marine reserves or marine parks—has been gaining momentum. A network of marine reserves is defined as "a set of marine reserves within a biogeographic region, connected by larval dispersal and juvenile or adult migration." Reserves serve as natural hatcheries, helping to repopulate the surrounding area.[43]

In 2002, at the World Summit on Sustainable Development in Johannesburg, coastal nations pledged to create national networks of marine parks, which together could constitute a global network of such parks. At the World Parks Congress in Durban in 2003, delegates recommended protecting 20–30 percent of each marine habitat from fishing. This would be up from 0.5 percent of the oceans that are currently included in marine reserves of widely varying size. It compares with the 12 percent of the earth's land area that is in parks.[44]

A U.K. team of scientists led by Dr. Andrew Balmford of the Conservation Biology Group at Cambridge University analyzed the costs of operating marine reserves on a large scale based on data from 83 relatively small, well-managed reserves. They concluded that managing reserves that covered 30 percent of the world's oceans would cost $12–14 billion a year. This did not take into account the likely additional income from recovering fisheries, which would reduce the actual cost.[45]

At stake in the creation of a global network of marine reserves is the protection and possible increase of an annual oceanic fish catch worth $70–80 billion. Balmford said, "Our study suggests that we could afford to conserve the seas and their resources in perpetuity, and for less than we are now spending on subsidies to exploit them unsustainably."[46]

Coauthor of the U.K. study Callum Roberts, of the University of York, noted: "We have barely even begun the task of creating marine parks. Here in Britain a paltry one-fiftieth of one percent of our seas is encompassed by marine nature reserves and only one-fiftieth of their combined area is closed to fishing." Yet the seas are being devastated by unsustainable fishing, pollution, and mineral exploitation. The creation of the global network of marine reserves—"Serengetis of the seas," as some have dubbed them—would create more than 1 million jobs. Roberts went on to say, "If you put areas off limits to fishing, there is no more effective way of allowing things to live longer, grow larger, and produce more offspring."[47]

Jane Lubchenco, former President of the American Association for the Advancement of Science, strongly underlined Roberts' point when releasing a statement signed by 161 leading marine scientists calling for urgent action to create the global network of marine reserves. Drawing on the research of scores of marine parks, she said: "All around the world there are different experiences, but the basic message is the same: marine reserves work, and they work fast. It is no longer a question of *whether* to set aside fully protected areas in the ocean, but *where* to establish them."[48]

The signatories noted how quickly sea life improves once the reserves are established. A case study of a snapper fishery off the coast of New England showed that fishers, though they violently opposed the establishment of the reserve at first, now champion it because they have seen the local population of snapper increase 40-fold. In a study in the Gulf of Maine, all fishing methods that put ground fish at risk were banned within three marine reserves totaling 17,000 square kilometers. Unexpectedly, scallops flourished in this undisturbed environment, and their populations increased by 9–14 times within five years. This population buildup within the reserves also greatly increased the scallop population outside the reserves. The group of 161 scientists noted that within a year or two of establishing a marine reserve, population densities increased 91 percent, average fish size went up 31 percent, and species diversity rose 20 percent.[49]

While the creation of marine reserves is clearly the overriding priority in the long-standing effort to protect marine ecosystems, other measures are also required. One is to reduce the nutrient flows from fertilizer runoff and untreated sewage. These increased nutrient flows cause huge algal blooms that then die off and in the process of decomposition absorb all the free oxygen in the water, leading to the death of local sea life. Today there are some 146 dead zones, either seasonal or chronic, scattered in the world's oceans from the Gulf of Mexico to the Baltic Sea to the east coast of China.[50]

The Gulf of Mexico dead zone near the mouth of the Mississippi River is one of the best known. This New Jersey–size area substantially reduces the marine diversity and yield of this historically productive body of water. Better control of nutrient runoff can be achieved through the adoption of such farming practices as minimum tillage and no-till, through the precise application of fertilizer to meet crop needs, and through planting buffer and filter strips along the Mississippi River and its tributaries.[51]

In the end, there is a need for governments to eliminate fishery subsidies. There are now so many fishing trawlers that their catch potential is nearly double any yield the oceans can sustain. Managing a network of marine reserves governing 30 percent of the oceans would cost only $12–14 billion—substantially less than the $15–30 billion that governments dole out today as subsidies to fishers.[52]

Protecting Plant and Animal Diversity

The two steps essential to protecting the earth's extraordinary biological diversity are the stabilization of population and climate. If the world's population increases to 9 billion by midcentury, countless more plant and animal species may simply be crowded off the planet. If carbon dioxide levels and temperatures continue to rise, every ecosystem will change.

Aiming for the low U.N. population trajectory, which has world population peaking at 7.8 billion in 2041 and then gradually declining, is the most effective option for protecting earth's rich diversity of life. As it becomes more difficult to raise land productivity, continuing population growth will force farmers to clear ever more tropical forests in the Amazon and Congo basins and the outer islands of Indonesia.[53]

Water management at a time of growing water shortages is a key in protecting marine species. When rivers are drained dry to satisfy growing human needs for irrigation and for urban water, marine species cannot survive.

Perhaps the best known and most popular way of trying to protect plant and animal species is to create reserves. Millions of square kilometers have been set aside as parks. Indeed, some 12 percent of the earth's land area is now included in parks and nature preserves. With more resources, some of these parks in developing countries that now exist only on paper could become a reality.[54]

Some 15 years ago, Norman Myers and other scientists conceived the idea of biodiversity "hotspots"—areas that were especially rich biologically and thus deserving of special protection. This helped the World Wide Fund for Nature, Conservation International, The Nature Conservancy, and many other groups and governments to concentrate their preservation efforts. The 34 hotspots identified once covered nearly 16 percent of the earth's land surface, but largely because of habitat destruction they now cover less than 3 percent. Concentrating preservation efforts in these biologically rich regions was a step in the right direction.[55]

Some 30 years ago, the United States created the Endangered Species Act. This legislation prohibited any activities, such as clearing new land for agriculture and housing developments or draining wetlands, that would threaten an endangered species. There are numerous species in the United States, such as the bald eagle, that might now be extinct had it not been for this one piece of legislation.[56]

As a species humans have an enormous influence on the habitability of the planet for the millions of other species with which we share it. This influence brings with it an unprecedented responsibility.

The Earth Restoration Budget

Although we lack detailed data in some cases, we can roughly estimate how much it will cost to reforest the earth, protect the earth's topsoil, restore rangelands and fisheries, stabilize water tables, and protect biological diversity. Where data and information are lacking, we fill in with assumptions. The goal is not to have a set of precise numbers, but a set of reasonable estimates for an earth restoration budget. (See Table 8–1.)[57]

Calculating the cost of reforestation is complicated by the many approaches used. As noted, the big success story is South Korea, which over the last four decades has reforested its once denuded mountains and hills using locally mobilized labor. Other countries, including China, have tried extensive reforestation but mostly under more arid conditions and with much less success. Turkey has an ambitious NGO-led grassroots reforestation program, relying heavily on volunteer labor. So, too, does Kenya, where women's groups led by Nobel Peace Prize–winner Wangari Maathai have planted 30 million trees.[58]

In calculating reforestation costs, the focus is on developing countries since forested area is already expanding in the northern hemisphere's industrial countries. Meeting the growing fuelwood demand in these countries will require roughly an estimated 55 million additional hectares of forested area. Anchoring soils and restoring hydrological stability would require roughly another 100 million hectares located in thousands of watersheds in developing countries. Recognizing some overlap between these two, we will reduce the 155 million total to 150 million hectares. Beyond this, an additional 30 million hectares will be needed to produce lumber, paper, and other forest products.[59]

Only a small share of this tree planting will likely come from

Table 8–1. *Annual Earth Restoration Budget*

Activity	Funding (billion dollars)
Reforesting the earth	6
Protecting topsoil on cropland	24
Restoring rangelands	9
Restoring fisheries	13
Protecting biological diversity	31
Stabilizing water tables	10
Total	93

Source: See endnote 57.

plantations. Much of the planting will be on the outskirts of villages, along field boundaries, along roads, on small plots of marginal land, and on denuded hillsides. The labor for this will be local; some will be paid labor, some volunteer. Nearly all will be off-season labor. In China, farmers now planting trees where they once planted grain are compensated with grain from state-held stocks over a five-year period while the trees are becoming established.[60]

Reforestation is something of an uphill battle partly because the deforested land is often severely eroded and depleted of nutrients. Even the most dedicated nurturing does not guarantee high survival rates under marginal conditions.

If seedlings cost $40 per thousand, as the World Bank estimates, and if the typical planting rate is roughly 2,000 per hectare, then seedlings cost $80 per hectare. Labor costs for planting trees are high, but since much of the labor for planting these trees would consist of locally mobilized volunteers, we are assuming a total of $400 per hectare, including both seedlings and labor. With a total of 150 million hectares to be planted over the next decade, this will come to roughly 15 million hectares per year at $400 each for a total annual expenditure of $6 billion.[61]

Conserving the earth's topsoil by reducing erosion to the rate of new soil formation or below involves two principal steps. One is to retire the highly erodible land that cannot sustain cultivation—the estimated one tenth of the world's cropland that accounts for perhaps half of all erosion. For the United States, that has meant retiring 14 million hectares (nearly 35 million acres). The cost of keeping this land out of production is close to $50 per acre or $125 per hectare. In total, annual payments to farmers to plant this land in grass or trees under 10-year contracts approached $2 billion.[62]

The second initiative consists of adopting conservation practices on the remaining land that is subject to excessive erosion—that is, erosion that exceeds the natural rate of new soil formation. The initiative includes incentives to encourage farmers to adopt conservation practices such as contour farming, strip cropping, and, increasingly, minimum-till or no-till farming. These expenditures in the United States total roughly $1 billion per year.[63]

In expanding these estimates to cover the world, it is

assumed that roughly 10 percent of the world's cropland is highly erodible and should be planted to grass or trees before the topsoil is lost and it becomes barren land. In both the United States and China, the two leading food-producing countries, which account for a third of the world grain harvest, the official goal is to retire one tenth of all cropland. In Europe, it likely would be somewhat less than 10 percent, but in Africa and the Andean countries it could be substantially higher than that. For the world as a whole, converting 10 percent of cropland that is highly erodible to grass or trees seems a reasonable goal. Since this costs roughly $2 billion in the United States, which represents one eighth of the world cropland area, the total for the world would be roughly $16 billion annually.[64]

Assuming that the need for erosion control practices for the rest of the world is similar to that in the United States, we again multiply the U.S. expenditure by eight to get a total of $8 billion for the world as a whole. The two components together—$16 billion for retiring highly erodible land and $8 billion for adopting conservation practices—give an annual total for the world of $24 billion.[65]

For cost data on rangeland protection and restoration, we turn to the United Nations Plan of Action to Combat Desertification. This plan, which focuses on the world's dryland regions, containing nearly 90 percent of all rangeland, estimates that it would cost roughly $183 billion over a 20-year restoration period—or $9 billion per year. The key restoration measures include improved rangeland management, financial incentives to eliminate overstocking, and revegetation with appropriate rest periods, when grazing would be banned.[66]

This is a costly undertaking, but every dollar invested in rangeland restoration yields a return of $2.50 in income from the increased productivity of the rangeland ecosystem. From a societal point of view, countries with large pastoral populations, where the rangeland deterioration is concentrated, are invariably among the world's poorest. The alternative to action—ignoring the deterioration—brings not only a loss of land productivity, but ultimately millions of refugees, some migrating to nearby cities and others moving to other countries.[67]

The restoration of oceanic fisheries centers primarily on the establishment of a worldwide network of marine reserves, which would cover roughly 30 percent of the ocean's surface. For this exercise we use the detailed calculations by the U.K. team cited earlier in the chapter. Their estimated range of expenditures centers on $13 billion per year.[68]

For wildlife protection, the bill is somewhat higher. The World Parks Congress estimates that the annual shortfall in funding needed to manage and to protect existing areas designated as parks comes to roughly $25 billion a year. Additional areas needed, including those encompassing the biologically diverse hotspots not yet included in designated parks, would cost perhaps another $6 billion a year, yielding a total of $31 billion.[69]

There is one activity, stabilizing water tables, where we do not have an estimate, only a guess. The key to stabilizing water tables is raising water productivity, and for this we have the experience gained beginning a half-century ago when the world started to systematically raise land productivity. The elements needed in a comparable water model are research to develop more water-efficient irrigation practices and technologies, the dissemination of these research findings to farmers, and economic incentives that encourage farmers to adopt and use these improved irrigation practices and technologies.

The area for raising irrigation water productivity is much smaller than that for land productivity. Indeed, only about one fifth of the world's cropland is irrigated. In disseminating the results of irrigation research, there are actually two options today. One is to work through agricultural extension services, which were created to funnel new information to farmers on a broad range of issues, including irrigation. Another possibility is to work through the water users associations that have been formed in many countries. The advantage of the latter is that they are focused exclusively on water.[70]

Effectively managing underground water supplies requires knowledge of the amount of water being pumped and aquifer recharge rates. In most countries this information is simply not available. Finding out how much is pumped may mean installing meters on irrigation well pumps, much as has been done in Jordan and Mexico.[71]

In some countries, the capital needed to fund a program to

raise water productivity can come from cancelled subsidies that now often encourage the wasteful use of irrigation water. Sometimes these are power subsidies, as they are in India; other times they are subsidies that provide water at prices well below costs, as happens in the United States. In terms of additional resources needed worldwide, including the economic incentives for farmers to use more water-efficient practices and technologies, we assume it will take additional expenditures of $10 billion.[72]

Altogether, restoring the earth will require additional expenditures of $93 billion per year. Many will ask, Can the world afford this? But the only appropriate question is, Can the world afford to not make these investments?

9

Feeding Seven Billion Well

In April 2005, the World Food Programme and the Chinese government jointly announced that food aid shipments to China would stop after the end of the year. For a country where hundreds of millions of people were chronically hungry a generation ago, this was a landmark achievement. China's success in largely eradicating hunger can be traced to the wholesale reduction in poverty associated with the eightfold expansion in its economy since the economic reforms of 1978 and the 50-percent jump in its grain harvest between 1977 and 1986.[1]

While hunger has been disappearing in China, it has been spreading in sub-Saharan Africa and parts of the Indian subcontinent. As a result, the number of people who are hungry has increased from a recent historical low of 820 million in 2000 to 852 million in 2002.[2]

One key to the threefold expansion in the world grain harvest since 1950 was the rapid adoption in developing countries of high-yielding wheats and rices developed in Japan and hybrid corn from the United States. The spread of these highly productive seeds, combined with a tripling of irrigated area and a ninefold increase in world fertilizer use, tripled the world grain harvest. Growth in irrigation and fertilizer use essentially removed soil moisture and nutrient constraints on crop yields in much of the world.[3]

But now the world's farmers face enormous additional demand for farm products from the projected addition of some 70 million people a year, the desire by some 5 billion people to consume more livestock products, and the potential of millions of motorists turning to farm-produced fuel crops to supplement tightening supplies of gasoline and diesel fuel. On the supply side, farmers are faced with shrinking supplies of irrigation water, rising temperatures, the loss of cropland to nonfarm uses, rising fuel costs, and a dwindling backlog of yield-raising technologies. For those who like to be challenged, this is a good time to be a farmer or an agronomist.[4]

Rethinking Land Productivity

Efforts to raise cropland productivity are slowing as the backlog of unused agricultural technology shrinks. The loss of momentum in efforts to raise cropland productivity is worldwide. Between 1950 and 1990, world grain yield per hectare climbed by 2.1 percent a year. From 1990 to 2000, however, it rose only 1.2 percent annually. This is partly because the yield response to the additional application of fertilizer is diminishing and partly because irrigation water supplies are limited. During the current decade, the rise in land productivity may slow even more.[5]

This calls for fresh thinking on how to raise cropland productivity. One simple way of doing this, where soil moisture permits, is to increase the area multicropped—land that produces more than one crop per year. In North America and Western Europe, which in the past have restricted cropped area to control surpluses, there is some potential for double cropping that has not been fully exploited. Indeed, the tripling in the world grain harvest since 1950 is due in part to impressive increases in multiple cropping in Asia. Some of the more common combinations are wheat and corn in northern China, wheat and rice in northern India, and the double or triple cropping of rice in southern China, southern India, and nearly all the rice-growing countries in Southeast Asia.[6]

The double cropping of winter wheat and corn in the North China Plain boosted China's grain production to the U.S. level beginning two decades ago. Winter wheat grown there yields close to 4 tons per hectare. Corn averages 5 tons. Together these two crops, grown in rotation, can yield 9 tons of grain per hectare per year. China's double-cropped rice yields 8 tons per hectare.[7]

Forty years ago, North India produced only wheat, but with the advent of the earlier maturing high-yielding wheats and rices, the wheat could be harvested in time to plant rice. This wheat/rice combination is now widely used throughout the Punjab, Haryana, and parts of Uttar Pradesh. The wheat yield of 3 tons and rice yield of 2 tons combine for 5 tons of grain per hectare, helping to feed India's 1.1 billion people.[8]

The area that can be multicropped is limited by the supply of irrigation water and, in some areas, by a lack of enough labor to quickly harvest one crop and plant another. The loss of low-cost rural labor to industrialization can sharply reduce multiple cropping and therefore the harvested area. In Japan, for example, the grain harvested area peaked at nearly 5 million hectares in 1960 largely because the country's industrious farmers were harvesting two crops per year. As of 2005, Japan's harvested area had dropped to 2 million hectares in part because of cropland conversion to nonfarm uses, but mostly because of a steady decline in double cropping over the decades as rising wages in industry pulled workers away from agriculture. The cheap labor needed to cultivate small plots intensively has disappeared. Even a rice support price four times the world market level could not keep enough workers in agriculture to support extensive multicropping.[9]

Similarly, South Korea's harvested grain area has shrunk by half since peaking in 1965 primarily because of a decline in multiple cropping. Taiwan's has declined nearly two thirds since 1975. As industrialization progresses in China and India, their more prosperous regions may see similar declines in multiple cropping. In China, where incomes have quadrupled since 1980, this process already appears to be reducing production.[10]

In the United States, the lifting of planting area restrictions in 1996 opened new opportunities for multiple cropping. The most common U.S. double cropping combination is winter wheat with soybeans as a summer crop. Six percent of the soybean harvest comes from land that also produces winter wheat. Since soybeans fix nitrogen, this rotation reduces wheat farmers' need to apply fertilizer.[11]

A concerted U.S. effort to both breed earlier maturing varieties and develop cultural practices that would facilitate multiple cropping could substantially boost crop output. If China's farmers can extensively double crop wheat and corn, then U.S. farmers, at a similar latitude and with similar climate patterns, could do the same if agricultural research and farm policy were reoriented to support it.

Western Europe, with its mild winters and high-yielding winter wheat, might also be able to double crop more with a summer grain, such as corn, or with an oilseed crop. Elsewhere, Brazil and Argentina have an extended frost-free growing season that supports extensive multicropping, often wheat or corn with soybeans. The availability of chemical fertilizers also facilitates multiple cropping.[12]

In many countries, including the United States, most of those in Western Europe, and Japan, fertilizer use has reached a level where using more has little effect on crop yields. There are still some places, however, such as most of Africa, where additional fertilizer would help boost yields. Unfortunately, sub-Saharan Africa lacks the infrastructure to transport fertilizer economically to the villages where it is needed. As a result of nutrient depletion, grain yields in much of sub-Saharan Africa are falling.[13]

One encouraging response to this situation in Africa is the simultaneous planting of grain and leguminous trees. The trees start to grow slowly, permitting the grain crop to mature and be harvested. Then the trees grow quickly to several feet in height, dropping leaves that provide nitrogen and organic matter—both sorely needed in African soils. The wood is then cut and used for fuel. This simple, locally adapted technology, developed by scientists at the International Centre for Research in Agroforestry in Nairobi, has enabled farmers to double their grain yields within a matter of years as soil fertility builds.[14]

Despite local advances, the overall loss of momentum in expanding food production is unmistakable. It will force us to think about both limiting the growth in demand and using the existing harvest more productively. On the demand side, achieving an acceptable worldwide balance between food and people may now depend on stabilizing population as soon as possible, reducing the unhealthily high consumption of livestock products in industrial countries, and restricting the conversion of food crops to automotive fuels. But we must also think more broadly about land productivity, considering not only the individual crop but how to increase multiple cropping and how to get more out of existing harvests.

Raising Water Productivity

Since it takes 1,000 tons of water to produce 1 ton of grain, it is not surprising that 70 percent of world water use is devoted to irrigation. Thus, raising irrigation water efficiency is central to raising water productivity overall. Using more water-efficient irrigation technologies and shifting to crops that use less water can help expand the irrigated area, even with a limited water supply. Eliminating water subsidies and energy subsidies that encourage wasteful water use allows water prices to rise to market levels. Higher water prices encourage all water users to use water more efficiently. Institutionally, local rural water users associations that directly involve those using the water in its management have raised water productivity in many countries.[15]

The world now needs to launch an effort to raise water productivity similar to the one that nearly tripled grainland productivity during the last half of the twentieth century. Land productivity is typically measured in tons of grain per hectare or bushels per acre. A comparable indicator for irrigation water is kilograms of grain produced per ton of water. Worldwide, that average is now roughly 1 kilogram of grain per ton of water used.[16]

Some data have been compiled on water irrigation efficiency at the international level for surface water projects—that is, dams that deliver water to farmers through a network of canals. Crop usage of irrigation water never reaches 100 percent simply because some irrigation water evaporates from the land surface, some percolates downward, and some runs off.[17]

Water policy analysts Sandra Postel and Amy Vickers found that "surface water irrigation efficiency ranges between 25 and 40 percent in India, Mexico, Pakistan, the Philippines, and Thailand; between 40 and 45 percent in Malaysia and Morocco; and between 50 and 60 percent in Israel, Japan, and Taiwan." Irrigation water efficiency is affected not only by the type and condition of irrigation systems but also by soil type, temperature, and humidity. In arid regions with high temperatures, the evaporation of irrigation water is far higher than in humid regions with lower temperatures.[18]

In a May 2004 meeting, China's Minister of Water Resources Wang Shucheng outlined for me in some detail the plans to raise China's irrigation efficiency from 43 percent in 2000 to 51 percent in 2010 and then to 55 percent in 2030. The steps he described included raising the price of water, providing incentives for adopting more irrigation-efficient technologies, and developing the local institutions to manage this process. Reaching these goals, he felt, would assure China's future food security.[19]

Raising irrigation water efficiency typically means shifting from the less efficient flood or furrow system to overhead sprinklers or to drip irrigation, the gold standard of irrigation efficiency. Switching from flood or furrow to low-pressure sprinkler systems reduces water use by an estimated 30 percent, while switching to drip irrigation typically cuts water use in half.[20]

As an alternative to furrow irrigation, a drip system also raises yields because it provides a steady supply of water with minimal losses to evaporation. Since drip systems are both labor-intensive and water-efficient, they are well suited to countries with underemployment and water shortages, allowing farmers to raise their water productivity by using labor, which is often in surplus in rural communities.[21]

A few small countries—Cyprus, Israel, and Jordan—rely heavily on drip irrigation. Among the big three agricultural producers, this more-efficient technology is used on less than 1 percent of irrigated land in India and China and roughly 4 percent of such land in the United States.[22]

In recent years, the tiniest small-scale drip-irrigation systems—virtually a bucket with flexible plastic tubing to distribute the water—have been developed to irrigate a small vegetable garden with roughly 100 plants (covering 25 square meters). Somewhat larger drum systems irrigate 125 square meters. In both cases, the containers are elevated slightly, so that gravity distributes the water. Somewhat larger drip systems using plastic lines that can be moved easily are also becoming popular. These simple systems can pay for themselves in one year. By simultaneously reducing water costs and increasing yields, they can dramatically raise incomes of smallholders.[23]

Sandra Postel believes that the combination of these drip technologies at various scales has the potential to profitably irrigate 10 million hectares of India's cropland, or nearly one tenth of the total. She sees a similar potential for China, which is now also expanding its drip irrigation area to save scarce water.[24]

Institutional shifts—specifically, moving the responsibility for managing irrigation systems from government agencies to local water users associations—can facilitate a more efficient use of water. Farmers in many countries are organizing locally so they can assume this responsibility. Since local people have an economic stake in good water management, they tend to do a better job than a distant government agency.[25]

Mexico is a leader in this movement. As of 2002, farmers associations managed more than 80 percent of Mexico's publicly irrigated land. One advantage of this shift for the government is that the cost of maintaining the irrigation system is assumed locally, reducing the drain on the treasury. This also means that associations often need to charge more for irrigation water. Even so, for farmers the production gains from managing their water supply more than outweigh this additional expenditure.[26]

In Tunisia, where water users associations manage both irrigation and residential water, the number of associations increased from 340 in 1987 to 2,575 in 1999. Many other countries now have such bodies managing their water resources. Although the early groups were organized to deal with large publicly developed irrigation systems, some recent ones have been formed to manage local groundwater irrigation as well. They assume responsibility for stabilizing the water table with the goal of avoiding aquifer depletion and the economic disruption that it brings to the community.[27]

Low water productivity is often the result of low water prices. In most countries, prices are irrationally low, belonging to an era when water was an abundant resource. As water becomes scarce, it needs to be priced accordingly. Provincial governments in northern China are raising water prices in small increments to discourage waste. A higher water price affects all water users, encouraging investment in more water-efficient irrigation technologies, industrial processes, and household appliances.[28]

What is needed now is a new mindset, a new way of thinking about water use. For example, shifting to more water-efficient crops wherever possible also boosts water productivity. Rice production is being phased out around Beijing because rice is such a thirsty crop. Similarly, Egypt restricts rice production in favor of wheat.[29]

Any measures that raise crop yields on irrigated land also raise the productivity of irrigation water. Similarly, anything that increases the efficiency with which grain is converted into animal protein in effect increases water productivity.

For people consuming unhealthy amounts of livestock products, moving down the food chain means not only a healthier diet and reduced health care costs but also a reduction in water use. In the United States, where annual consumption of grain as food and feed averages some 800 kilograms (four fifths of a ton) per person, a modest reduction in the consumption of meat, milk, and eggs could easily cut grain use per person by 100 kilograms. Given that there are now nearly 300 million Americans, such a reduction would cut grain use by 30 million tons and irrigation water use by 30 billion tons.[30]

Reducing water use to a level that can be sustained by aquifers and rivers worldwide involves a wide range of measures not only in agriculture but throughout the economy. The more obvious steps, in addition to more water-efficient irrigation practices and more water-efficient crops, include adopting more water-efficient industrial processes and using more water-efficient household appliances. One of the less conventional steps is to shift from outdated coal-fired power plants, which require vast amounts of water for thermal cooling, to wind power—something long overdue in any case because of climate disruption. Recycling urban water supplies is another obvious step to consider in countries facing acute water shortages.

Producing Protein More Efficiently

The second way to raise both land and water productivity is to produce animal protein more efficiently. With some 38 percent (about 730 million tons) of the world grain harvest used to produce animal protein, the potential for more-efficient grain use is large.[31]

World meat consumption increased from 47 million tons in

1950 to 260 million tons in 2005, more than doubling consumption per person from 17 kilograms to 40 kilograms. Consumption of milk and eggs has also risen. In every society where incomes have risen, meat consumption has too, perhaps reflecting a taste that evolved over 4 million years of hunting and gathering.[32]

As both the oceanic fish catch and the production of beef on rangelands have leveled off, the world has shifted to grain-based production of animal protein to expand output. And as the demand for animal protein climbs, the mix of protein products consumed is shifting toward those that convert grain into protein most efficiently, the lower-cost products. Health concerns have also prompted a shift from beef and pork to poultry and fish.

The efficiency with which various animals convert grain into protein varies widely. With cattle in feedlots, it takes roughly 7 kilograms of grain to produce a 1-kilogram gain in live weight. For pork, the figure is close to 4 kilograms of grain per kilogram of weight gain, for poultry it is just over 2, and for herbivorous species of farmed fish (such as carp, tilapia, and catfish), it is less than 2. As the market shifts production to the more grain-efficient products, it raises the productivity of both land and water.[33]

Global beef production, most of which comes from rangelands, grew less than 1 percent a year from 1990 to 2005. Growth in the number of cattle feedlots was minimal. Pork production grew by 2.5 percent annually, and poultry by nearly 5 percent. The rapid growth in poultry production, going from 41 million tons in 1990 to 80 million tons in 2005, enabled poultry to eclipse beef in 1995, moving it into second place behind pork. (See Figure 9–1.) World pork production, half of it in China, overtook beef production in 1979 and has continued to widen the lead since then. World beef production, constrained by both grazing limits and the inefficient feedlot conversion by cattle, is continuing to expand, but slowly. Indeed, within the next decade or so, fast-growing, highly grain-efficient aquacultural output may overtake beef production.[34]

The big winner in the animal protein sweepstakes has been aquaculture, largely because herbivorous fish convert feed into protein so efficiently. Aquacultural output expanded from 13 million tons in 1990 to 42 million tons in 2003, growing by more

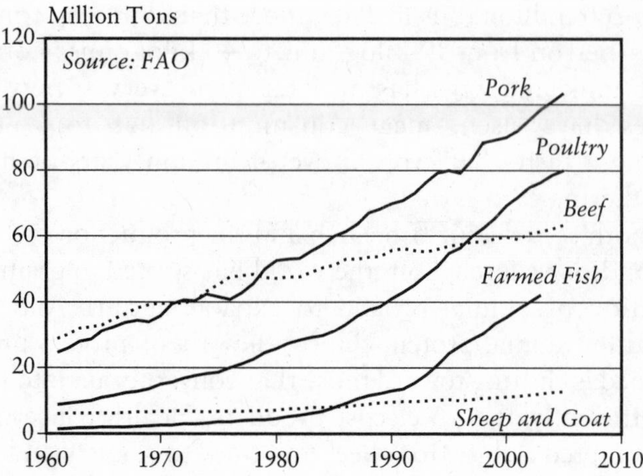

Figure 9–1. *W M P b T , 1961–2005*

than 10 percent a year. China, the leading producer, accounts for an astounding two thirds of global output. Aquacultural production in China is dominated by finfish (mostly carp), which are produced inland in freshwater ponds, lakes, reservoirs, and rice paddies, and by shellfish (mostly oysters, clams, and mussels), which are produced mostly in coastal regions.[35]

China's aquaculture is often integrated with agriculture, enabling farmers to use agricultural wastes, such as pig or duck manure, to fertilize ponds, thus stimulating the growth of plankton on which the fish feed. Fish polyculture, which commonly boosts pond productivity over that of monocultures by at least half, is widely practiced in both China and India.[36]

As land and water for fish ponds become even scarcer, China's fish farmers are feeding fish more grain concentrates, including soybean meal, to raise pond productivity. Using this technique, China's farmers raised the annual pond yield per hectare from 2.4 tons of fish in 1990 to 4.1 tons in 1996.[37]

In the United States, catfish, which require less than 2 kilograms of feed per kilogram of live weight, are the leading aquacultural product. U.S. annual catfish production of 600 million pounds (about two pounds per person) is concentrated in four states: Mississippi, Louisiana, Alabama, and Arkansas. Mississippi, with easily 60 percent of U.S. output, is the catfish capital of the world.[38]

Public attention has focused on aquacultural operations that are environmentally inefficient or disruptive, such as the farming of salmon, a carnivorous species, and shrimp. These operations account for 3.6 million tons of output, less than 9 percent of the global farmed fish total, but they are growing fast. Salmon are inefficient in that they are fed other fish, usually as fishmeal, which comes either from fish processing plant wastes or from low-value fish caught specifically for this purpose. Shrimp farming often involves the destruction of coastal mangrove forests to create areas for the shrimp.[39]

World aquaculture is dominated by herbivorous species—mainly carp in China and India, but also catfish in the United States and tilapia in several countries—and shellfish. This is where the great growth potential for efficient animal protein production lies.[40]

When we think of soybeans in our daily diet, it is typically as tofu, veggie burgers, or other meat substitutes. But most of the world's fast-growing soybean harvest is consumed indirectly in the beef, pork, poultry, milk, eggs, and farmed fish that we eat. Although not a visible part of our diets, the incorporation of soybean meal into feed rations has revolutionized the world feed industry, greatly increasing the efficiency with which grain is converted into animal protein.[41]

In 2005, the world's farmers produced 220 million tons of soybeans—1 ton for every 9 tons of grain produced. Of this, some 15 million tons were consumed directly as tofu or meat substitutes. The bulk of the remaining 205 million tons, after some was saved for seed, was crushed in order to extract 33 million tons of soybean oil, separating it from the highly valued, high-protein meal. By 2006, perhaps 2 million tons (7 percent) of these 33 million tons will be heading to service stations as biodiesel.[42]

The 144 million tons of soybean meal that remain after the oil is extracted is fed to cattle, pigs, chicken, and fish, enriching their diets with high-quality protein. Combining soybean meal with grain in roughly one part meal to four parts grain dramatically boosts the efficiency with which grain is converted into animal protein, sometimes nearly doubling it.[43]

The world's three largest meat producers—China, the United States, and Brazil—now all rely heavily on soybean meal as a protein supplement in feed rations. In the United States, which has long used soybean meal to upgrade livestock and poultry feed, the soybean meal share of feed rations climbed from 8 percent in 1964 to roughly 18 percent in recent years.[44]

For Brazil, where the shift began in the late 1980s, soybean meal now makes up roughly 21 percent of the feed mix. In China, the realization that feed efficiency could be dramatically boosted with soymeal came several years later. Between 1991 and 2002, the soymeal component of feed there jumped from 2 percent to 20 percent. For fish, whose protein demands are particularly high, China incorporated some 5 million tons of soymeal into the 16 million tons of grain-based fish feed used in 2003.[45]

With this phenomenal growth, soybean meal both replaced some grain in feed and increased the efficiency with which the remaining grain was converted into livestock products. This also helps explain why the share of the world grain harvest used for feed has not increased over the last 20 years even though production of meat, milk, eggs, and farmed fish has climbed. And it explains why world soybean production jumped from 16 million tons in 1950 to 220 million tons in 2005, a 13-fold increase. While the potential for raising feed efficiency with soybean meal has now been largely realized in key food-producing countries, there are still many developing countries that have not yet fully exploited its potential.[46]

New Protein Production Systems

Mounting pressures on the earth's land and water resources to produce livestock, poultry, and fish feed have led to the evolution of some promising new animal protein production models, one of which is milk production in India. Since 1970, India's milk production has increased more than fourfold, jumping from 21 million to 95 million tons. In 1998, India overtook the United States to become the world's leading producer of milk and other dairy products. (See Figure 9–2.)[47]

The spark for this explosive growth came in 1965 when an enterprising young Indian, Dr. Verghese Kurien, organized the National Dairy Development Board, an umbrella organization of dairy cooperatives. The dairy coop's principal purpose was to market the milk from tiny herds that typically averaged two

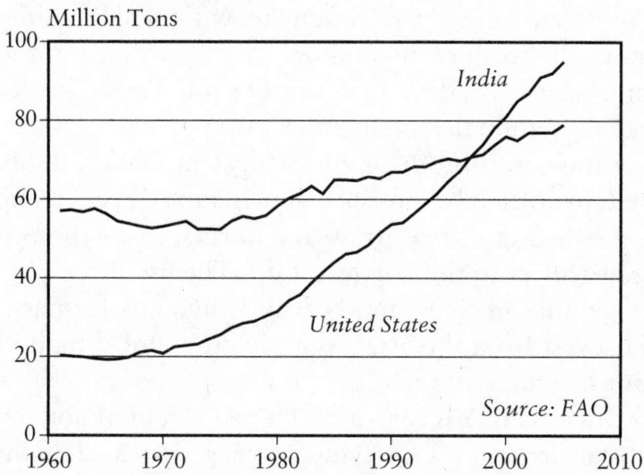

Figure 9– . *Milk P od ion in ndi nd he ni ed S e , 1961–2005*

to three cows each, providing the link between the growing appetite for dairy products and the millions of village families who had only a small marketable surplus.[48]

Creating the market for milk spurred the fourfold growth in output. In a country where protein shortages stunt the growth of so many children, expanding the milk supply from less than half a cup per person a day 30 years ago to more than a cup represents a major advance.[49]

What is so remarkable is that India has built the world's largest dairy industry almost entirely on roughage—wheat straw, rice straw, corn stalks, and grass collected from the roadside. Even so, the value of the milk produced each year now exceeds that of the rice harvest.[50]

A second new protein production model, one that also relies on ruminants, has evolved in four provinces of central Eastern China—Hebei, Shangdong, Henan, and Anhui—where double cropping of winter wheat and corn is common. Once the winter wheat ripens in early summer, it is harvested quickly so the seedbed can be prepared for the corn. Similarly, once the corn is harvested in the fall, the seedbed is quickly prepared to sow the wheat. Although the wheat straw and cornstalks are often used as fuel for cooking, villagers are shifting to other sources of energy for this, which lets them feed the straw and cornstalks to cattle. Supplementing this roughage with small amounts of nitrogen in the form of urea allows the microflora in the complex four-stomach digestive system of cattle to convert roughage into animal protein efficiently.[51]

These four crop-producing provinces in China, dubbed the Beef Belt by officials, produce much more beef, using crop residues—wheat and rice straw and corn stalks—than the vast grazing provinces in the northwest do. The use of crop residues to produce milk in India and beef in China lets farmers reap a second harvest from the original grain crop, boosting both land and labor productivity.[52]

Over time, China has also developed a remarkably efficient fish polyculture using four types of carp that feed at different levels of the food chain, in effect emulating natural aquatic ecosystems. Silver carp and bighead carp are filter feeders, eating phytoplankton and zooplankton respectively. The grass carp, as its name implies, feeds largely on vegetation, while the common carp is a bottom feeder, living on detritus. These four species thus form a small ecosystem, with each filling a particular niche. This multi-species system, which converts feed into high-quality protein with remarkable efficiency, yielded some 13 million tons of carp in 2002.[53]

While poultry production has grown rapidly in China, as in other developing countries, it has been dwarfed by the phenomenal growth of aquaculture. Today aquacultural output in China—at 29 million tons—is double that of poultry, making it the first major country where aquaculture has eclipsed poultry farming. The great economic and environmental attraction of this system is the efficiency with which it produces animal protein.[54]

Although these three new protein models have evolved in India and China (both densely populated), similar systems can be adopted in other countries as population pressures intensify, as demand for meat and milk increase, and as farmers seek new ways to convert plant products into animal protein.

The world desperately needs more new protein production techniques such as these. A half-century ago, when there were only 2.5 billion people in the world, almost everyone wanted to move further up the food chain. Now that number may have doubled. Meat consumption is growing twice as fast as population, egg consumption is growing nearly three times as fast, and growth in the demand for fish—both from the oceans and from fish farms—is also outpacing that of population. Against this backdrop of growing world demand, human ingenuity in producing animal protein in ever-larger quantities and ever more efficiently will be tested.[55]

While the world has had many years of experience in feeding an additional 70 million or more people each year, it has no experience with some 5 billion people wanting to move up the food chain at the same time. For a sense of what this translates into, consider what has happened in China since the economic reforms in 1978. As the fastest-growing economy in the world since then, China has in effect telescoped history, showing how diets change when incomes rise rapidly.[56]

As recently as 1978, meat consumption was low in China, consisting mostly of modest amounts of pork. Since then, consumption of pork, beef, poultry, and mutton has climbed severalfold, pushing China's total meat consumption far above that of the United States. As incomes rise in other countries, consumers there will also want more animal protein. Considering the effect of this expanding demand on global land and water resources, along with the more traditional demand from population growth, provides a better sense of future pressures on the earth. If world grain supplies tighten in the years ahead, the competition for grain between people wanting more biofuels, those living high on the food chain, and those on the bottom rungs of the economic ladder will become both more visible and a possible source of tension within and among societies.[57]

Moving Down the Food Chain

One of the questions I am most often asked on a speaking tour is, "How many people can the earth support?" I answer with another question: "At what level of food consumption?" At the U.S. level of 800 kilograms per person per year for food and feed, the 2-billion-ton annual world harvest of grain would support 2.5 billion people. At the Italian level of consumption of close to 400 kilograms per year, the current harvest would support 5 billion people. At the nearly 200 kilograms of grain consumed per year by the average Indian, it would support a population of 10 billion.[58]

In every society where incomes rise, people move up the food chain, eating more animal protein as beef, pork, poultry, milk, eggs, and seafood. The mix of animal protein products varies with geography and culture, but the shift to more animal protein as purchasing power increases appears to be universal.

As consumption of livestock products, poultry, and farmed fish rises, grain use per person also rises. Of the roughly 800 kilograms of grain consumed per person each year in the United States, about 100 kilograms is eaten directly as bread, pasta, and breakfast cereals. But the bulk of the grain is consumed indirectly in the form of livestock, poultry, and farmed fish. By contrast, in India, where people consume just under 200 kilograms of grain per year, or roughly a pound per day, nearly all grain is eaten directly to satisfy basic food energy needs. Little is available for conversion into livestock products.[59]

Of the three countries just cited, life expectancy is highest in Italy even though U.S. expenditures on medical care per person are much higher. Those who live very low on the food chain or very high on the food chain do not live as long as those in an intermediate position. The Mediterranean diet includes meats, cheeses, and seafood, but in moderation. Nutritionally, this is the healthiest way to eat.[60]

What this means is that those living high on the food chain, such as the average American or Canadian, can consume less grain and improve health at the same time. For those who live in low-income countries like India, where diets are dominated by a starchy staple such as rice, sometimes supplying 60 percent or more of total caloric intake, eating more animal products can improve health and raise life expectancy.[61]

In addition to having the affluent sector move down the food chain by consuming fewer livestock products, the world is turning to the more grain-efficient forms of animal protein. Together these two steps have helped hold the share of the world grain harvest used for feed constant at roughly 38 percent for the last two decades.[62]

It is widely assumed that moving from animal protein to high-quality proteins from plant sources, such as beans or tofu made from soybeans, is more land-efficient. But this is not always the case. For example, as noted earlier, with poultry it takes just over 2 kilograms of grain to produce 1 kilogram of additional live weight. For catfish, it is less than 2 kilograms of grain per kilogram of weight gain. An acre of land in Iowa can thus produce 140 bushels of corn or 35 bushels of the much lower-yielding soybean. Feeding the corn to chickens or catfish can yield more high-quality protein than growing soybeans and consuming them directly, say as tofu.[63]

It takes a good deal of land to produce soy protein, largely because plants require more metabolic energy to produce high-quality plant protein than to produce starch. But because poultry and catfish are so efficient at converting grain, eating them is more land- and water-efficient than eating soybeans is.[64]

Some countries are moving down the food chain by turning to the more grain-efficient protein sources such as aquaculture. China, with its huge aquacultural output, may be the first country where the farmed fish harvest exceeds the wild fish catch.[65]

With incomes now rising in densely populated Asia, other countries are following China's lead. Among them are India, Thailand, and Viet Nam. Viet Nam, for example, devised a plan in 2001 of developing 700,000 hectares of land in the Mekong Delta for aquaculture, with the goal of producing 1.7 million tons of fish and shrimp by 2005. It now appears likely to exceed this goal.[66]

Action on Many Fronts

Historically, the responsibility for food security rested largely with the ministry of agriculture. During the last half of the last century, ensuring adequate supplies of grain in the world market was a relatively simple matter. Whenever the world grain harvest fell short and prices started to rise, the U.S. Department of Agriculture would simply return to production part of the cropland idled under supply control programs, thus boosting output and stabilizing prices. This era ended in 1996 when the United States dismantled its annual cropland set-aside program.[67]

Ministries of agriculture bear the primary responsibility for expanding food production to satisfy the world's seemingly insatiable appetite. The fast-growing demand from the addition of 70 million mouths to feed each year, from 5 billion people wanting to move up the food chain, and now, for the first time, from the insatiable demand for farm commodities to produce automotive fuel will pose an unprecedented challenge to ministries of agriculture. At the same time they are faced with a dwindling backlog of unused agricultural technology, shrinking supplies of irrigation water, and the prospect of crop-withering heat waves. Demand growth and supply constraints together will challenge agricultural leaders as never before.

In this chapter, we have discussed some of the newer measures that can be used to raise land and water productivity. Adoption of these and other actions are obviously important, but in the new world we have entered, the policies of other ministries also bear heavily on the food security prospect.

Now with our finite planet being pushed to its limits and beyond, the capacity of health and family planning ministries to educate the public about the consequences of population growth and to meet family planning needs has become a food security issue. Whether individual couples decide to have one, two, or three children directly affects world food security.

In today's world, decisions made in ministries of energy on whether to stay with fossil fuels and continue to drive the earth's temperature upward or to shift to renewable energy sources and stabilize the earth's temperature could have a greater effect on long-term food security than any actions taken by ministries of agriculture.

And in much of the world, water is a more serious constraint on food production than land. The success, or lack thereof, of water ministries in raising water productivity will directly affect future food security and food prices.

Similarly, in a world where cropland is scarce and becoming more so, decisions made in ministries of transportation on whether to develop auto-centered systems or more diversified transport systems that rely heavily on less land-intensive transport forms, including light rail, buses, and bicycles will also affect world food security. Policies adopted by the ministers of transportation in land-scarce countries like China and India directly affect world food security.

More broadly, how far governments go in encouraging the use of scarce agricultural resources to produce commodities to be converted into automotive fuel will directly affect efforts to eradicate hunger. The question is how effective governments will be in managing this emerging competition between cars and people for food commodities.

In our increasingly crowded world, the responsibility for

food security extends far beyond the ministry of agriculture, involving all ministries in the effort to fully realize the earth's sustainable food production potential. At the same time, there are many agricultural successes that can be imported by countries struggling to improve their food security. Encouragingly, the two big breakthroughs in expanding animal protein supplies—the dramatic gains in milk production in India and fish farming in China—can be replicated in many other developing countries.

10

Stabilizing Climate

Some time ago, I had a call from my son Brian, who had come across a huge new wind farm as he was driving on one of the interstate highways in west Texas. He described the rows of wind turbines receding toward the horizon. Interspersed among them were oil wells. The wind turbines were turning and the oil wells were pumping. My son was fascinated by the juxtaposition of the old and the new, the past and the future. I said, "If you return 30 years from now, the wind turbines will still be turning, but it is unlikely that the oil wells will be pumping." What he was looking at in a nutshell was the energy transition, the shift from the age of fossil fuels to renewables.

The energy transition is gaining momentum. When the Kyoto Protocol was negotiated in 1997, the proposed 5-percent reduction in carbon emissions from 1990 levels in industrial countries by 2012 seemed like an ambitious goal. Now it is widely seen as an outmoded, grossly inadequate goal. National governments, local governments, corporations, and environmental groups are coming up with plans to cut carbon emissions much further than was agreed to in Kyoto by turning to renewables and raising energy efficiency. Some individuals and groups are even beginning to think about how to cut carbon emissions by 70 percent, the amount that scientists say will be needed to stabilize climate.[1]

In July 2005, the European Commission proposed a new plan to cut energy use 20 percent by 2020 and to increase the renewable share of Europe's energy supply to 12 percent by 2010. Together, these two initiatives will reduce Europe's carbon emissions by nearly one third. Among the long list of measures to boost energy efficiency in these countries are replacing old, inefficient refrigerators, switching to high-efficiency light bulbs, and insulating roofs. Reaching the renewables goal requires a rather conservative addition of 15,000 megawatts of wind power, a fivefold expansion of ethanol production, and a threefold increase in biodiesel production. The Europeans' proposed 20-percent cut in energy use by 2020 contrasts sharply with the projected growth of 10 percent under a business-as-usual scenario.[2]

The proposed plan, which is scheduled for final approval in 2006, is designed to save 60 billion euros by 2020. It is also designed to stimulate economic growth, create new jobs, and, by reducing energy outlays, enhance European competitiveness in world markets. The 25-member European Union is second only to the United States in energy consumption.[3]

In 2005 the Japanese government also announced a national campaign to dramatically boost energy efficiency in its economy, already one of the world's most efficient. It urged its people to replace older, inefficient appliances and to buy hybrid cars. The *New York Times* described this as "all part of a patriotic effort to save energy and fight global warming." It noted that the large manufacturing firms were jumping on the energy efficiency bandwagon as a way of increasing sales of their latest high-efficiency models.[4]

Beyond this initial effort, Japan has set goals for boosting appliance efficiency even further, cutting energy use of television sets by 17 percent, of personal computers by 30 percent, of air conditioners by 36 percent, and of refrigerators by a staggering 72 percent. Scientists are working on a vacuum-insulated refrigerator that will use only one eighth as much electricity as those marketed a decade ago.[5]

At the nongovernmental level, a plan developed for Canada by the David Suzuki Foundation and the Climate Action Network would halve carbon emissions by 2030 and would do it only with investments in energy efficiency that are profitable. And in early April 2003, the World Wildlife Fund released a peer-reviewed analysis by a team of scientists that proposed reducing carbon emissions from U.S. electric power generation 60 percent by 2020. This proposal centers on a shift to more energy-efficient power generation equipment, the use of more-efficient household appliances and industrial motors and other equipment, and in some situations a shift from coal to natural gas as an energy source. If implemented, it would result in national savings averaging $20 billion a year from now until 2020.[6]

In Ontario, Canada's most populous province, the ministry of energy plans to phase out the province's five large coal-fired power plants by 2009. The first, Lakeview Generating Station, was closed in April 2005; three more will close by the end of 2007, and the last will be shut down in early 2009. All three major political parties support the plan to replace coal with wind, natural gas, and efficiency gains. Jack Gibbons, director of the Ontario Clean Air Alliance, which endorses the ministry's plan, says of coal burning, "It's a nineteenth century fuel that has no place in twenty-first century Ontario."[7]

Corporations are also getting involved. U.S.-based Interface, the world's largest manufacturer of industrial carpeting, cut carbon emissions by two thirds in its Canadian affiliate during the 1990s. It did so by examining every facet of its business—from electricity consumption to trucking procedures. Founder and chairman Ray Anderson says, "Interface Canada has reduced greenhouse gas emissions by 64 percent from the peak, and made money in the process, in no small measure because our customers support environmental responsibility." The Suzuki plan to cut Canadian carbon emissions in half by 2030 was inspired by the profitability of the Interface initiative.[8]

Although stabilizing atmospheric carbon dioxide levels is a staggering challenge, it is entirely doable. With advances in wind turbine design, the evolution of gas-electric hybrid cars, advances in solar cell manufacturing, and gains in the efficiency of household appliances, we now have the basic technologies needed to shift quickly from a fossil-fuel-based to a renewable-energy-based economy. Cutting world carbon emissions in half by 2015 is entirely within range. Ambitious though this goal might seem, it is commensurate with the threat that climate change poses.

Raising Energy Productivity

The enormous potential for raising energy productivity becomes clear in comparisons of energy use among countries. Some nations in Europe have essentially the same living standard as the United States yet use scarcely half as much energy per person. But even the countries that use energy most efficiently are not close to realizing the full potential for doing so.[9]

When the Bush administration released a new energy plan in April 2001 that called for construction of 1,300 new power plants by 2020, Bill Prindle of the Washington-based Alliance to Save Energy responded by pointing out how the country could eliminate the need for those plants and save money in the process. He ticked off several steps that would reduce the demand for electricity: Improving efficiency standards for household appliances would eliminate the need for 127 power plants. More stringent residential air conditioner efficiency standards would eliminate 43 power plants. Raising commercial air conditioner standards would eliminate the need for 50 plants. Using tax credits and energy codes to improve the efficiency of new buildings would save another 170 plants. Similar steps to raise the energy efficiency of existing buildings would save 210 plants. These five measures from the longer list suggested by Prindle would not only eliminate the need for 600 power plants, they would also save money. Although these calculations were made in 2001, they are still valid simply because there has been so little progress in raising U.S. energy efficiency since then.[10]

Of course, each country will have to fashion its own plan for raising energy productivity. Nevertheless, there are a number of common components. Some are quite simple but highly effective, such as using more energy-efficient household appliances, eliminating the use of incandescent light bulbs, shifting to gas-electric hybrid cars, and redesigning urban transport systems to raise efficiency and increase mobility.

Although there was an impressive round of efficiency gains in household appliances after the oil price jumps during the 1970s, the world generally lost interest as oil prices declined after 1980. Rising oil and natural gas prices are rekindling interest in this issue. Fortuitously, engineering advances since then have brought another wave of efficiency gains, such as those mentioned for Japan, that promise to substantially reduce electricity use. If national governments raise appliance efficiency standards to fully exploit the latest technologies, it would sharply cut carbon emissions worldwide.

One simple energy-saving step is to replace all remaining incandescent light bulbs with compact fluorescent lamps (CFLs), which use only one third as much electricity and last 10 times as long. In the United States, where 20 percent of all electricity is used for lighting, if each household replaced the still widely used incandescents with compact fluorescents, electricity for lighting would be easily cut in half. The combination of greater longevity and lower electricity use greatly outweighs the higher costs of the CFLs, yielding a risk-free investment return of some 25–40 percent a year. Worldwide, replacing incandescent light bulbs with CFLs in, say, the next three years would facilitate the closing of hundreds of climate-disrupting coal-fired power plants.[11]

A second obvious area for raising energy efficiency is automobiles. If over the next decade the United States, for example, were to shift from the current fleet of cars powered with gasoline engines to gas-electric hybrids with the fuel efficiency of the Toyota Prius, gasoline use could easily be cut in half. Sales of hybrid cars, introduced into the U.S. market in 1999, reached an estimated 88,000 in 2004. Higher gasoline prices and mounting climate change worries are driving sales upward. With U.S. auto manufacturers coming onto the market with several new models, hybrid vehicle sales are projected to exceed 1 million by 2008.[12]

Another attractive way to raise energy efficiency is to redesign urban transport systems, moving from the existing system centered on single-occupant automobiles to a more diverse bicycle- and pedestrian-friendly system that would include well-developed light-rail subway systems complemented with buses. Such a system would increase mobility, reduce energy use and air pollution, and provide more opportunities for exercise, a win-win-win situation. Taking automobiles off the street would facilitate the conversion of parking lots into parks, creating more friendly cities.

Harnessing the Wind

World wind-generating capacity, growing at 29 percent a year, has jumped from less than 5,000 megawatts in 1995 to more than 47,000 megawatts in 2004, a ninefold increase. (See Figure 10–1.) Wind's annual growth rate of 29 percent compares with 1.7 percent for oil, 2.5 percent for natural gas, 2.3 percent for coal, and 1.9 percent for nuclear power. There are six reasons why wind is growing so fast. It is abundant, cheap, inexhaustible, widely distributed, clean, and climate-benign. No other energy source has all these attributes.[13]

Europe is leading the world into the age of wind energy. Germany, which overtook the United States in 1997, leads the world with 16,600 megawatts of generating capacity. Spain, a rising wind power in southern Europe, overtook the United States in 2004. Denmark, which now gets an impressive 20 percent of its electricity from wind, is also the world's leading manufacturer and exporter of wind turbines.[14]

In its 2005 projections, the Global Wind Energy Council showed Europe's wind-generating capacity expanding from 34,500 megawatts in 2004 to 75,000 megawatts in 2010 and

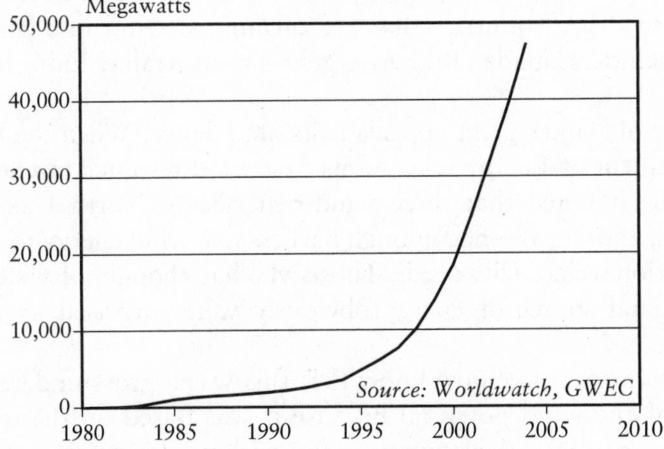

Figure 10–1. *World Wind Energy Generating Capacity, 1980–2004*

230,000 megawatts in 2020. By 2020, just 15 years from now, wind-generated electricity is projected to satisfy the residential needs of 195 million consumers, half of Europe's population.[15]

After developing most of its existing 34,500 megawatts of capacity on land, Europe is now tapping offshore wind as well. A 2004 assessment of the region's offshore potential by the Garrad Hassan wind energy consulting group concluded that if governments move aggressively to develop their vast offshore resources, wind could be supplying all of Europe's residential electricity by 2020.[16]

The United Kingdom, moving fast to develop its offshore wind capacity, accepted bids in April 2001 for sites designed to produce 1,500 megawatts of wind-generating capacity. In December 2003, the government took bids for 15 additional offshore sites with a generating capacity that could exceed 7,000 megawatts. Requiring an investment of over $12 billion, these offshore wind farms could satisfy the residential electricity needs of 10 million of the country's 60 million people. At the end of 2004, the United Kingdom had an offshore generating capacity of 124 megawatts, with an additional 180 megawatts under construction.[17]

The push to develop wind in Europe is spurred by concerns about climate change. The record heat wave in Europe in August 2003 that scorched crops and claimed 49,000 lives has accelerated the replacement of climate-disrupting coal with clean energy sources. Other countries that are turning to wind in a major way include Canada, Brazil, Argentina, Australia, India, and China.[18]

One of wind's great appeals is its abundance. When the U.S. Department of Energy released its first wind resource inventory in 1991, it noted that three wind-rich states—North Dakota, Kansas, and Texas—had enough harnessable wind energy to satisfy national electricity needs. Those who had thought of wind as a marginal source of energy obviously were surprised by this finding.[19]

In retrospect, we now know that this was a gross underestimate of the wind potential because it was based on the technologies of 1991. Advances in wind turbine design since then enable turbines to operate at lower wind speeds, to convert wind into electricity more efficiently, and to harness a much larger wind regime. In 1991, wind turbines may have averaged scarcely 40 meters in height. Today, new turbines are 100 meters tall, perhaps tripling the harvestable wind. We now know that the United States has enough harnessable wind energy to meet not only national *electricity* needs, but national *energy* needs.[20]

When the wind industry began in California in the early 1980s, wind-generated electricity cost 38¢ per kilowatt-hour. Since then it has dropped to 4¢ or below at prime wind sites. And some U.S. long-term supply contracts have been signed for 3¢ per kilowatt-hour. Wind farms at prime sites may be generating electricity at 2¢ per kilowatt-hour by 2010, making it one of the world's cheapest sources of electricity.[21]

Low-cost electricity from wind can be used to electrolyze water to produce hydrogen, which provides a way of both storing and transporting wind energy. At night, when the demand for electricity drops, the hydrogen generators can be turned on to build up reserves. Once in storage, hydrogen can be used to fuel power plants. Wind-generated hydrogen can thus become a backup for wind power, with hydrogen-powered electricity generation kicking in when wind power ebbs. Wind-generated hydrogen can also serve as an alternative to natural gas, especially if rising prices make gas prohibitively costly for electricity generation.

The principal cost for wind-generated electricity is the upfront capital outlay for initial construction. Since wind is a free fuel, the only ongoing cost is for turbine maintenance. Given the recent volatility of natural gas prices, the stability of wind power prices is particularly appealing. With the near certainty of even higher costs of natural gas in the future, natural-gas-fired plants may one day be used only as a backup for wind-generated electricity.

The United States is lagging in developing wind energy simply because the wind production tax credit (PTC) of 1.5¢ per kilowatt-hour, which was adopted in 1992 to establish parity with subsidies to fossil fuel, has lapsed three times in five years. Uncertainty about the tax credit has disrupted planning throughout the wind power industry. With the two-year extension of the PTC in mid-2005, however, through the end of 2007, growth in wind power investments is escalating rapidly.[22]

Given wind's enormous potential and the associated benefits

of climate stabilization, it is time to consider an all-out effort to develop wind resources. Instead of doubling every 30 months or so, perhaps we should be doubling wind electric generation each year for the next several years, much as the number of computers linked to the Internet doubled each year from 1985 to 1995. Costs would then drop precipitously, giving electricity generated from wind an even greater advantage over fossil fuels.[23]

Energy consultant Harry Braun points out that since wind turbines are similar to automobiles in the sense that each has an electrical generator, a gearbox, an electronic control system, and a brake, they can be mass-produced on assembly lines. Indeed, the slack in the U.S. automobile industry is sufficient to produce a million wind turbines per year. The lower cost associated with mass production could drop the cost of wind-generated electricity below 2¢ per kilowatt-hour. Assembly-line production of wind turbines at "wartime" speed would quickly lower urban air pollution, carbon emissions, and the prospect of oil wars.[24]

The economic incentives to spur such growth could come in part from simply restructuring global energy subsidies—shifting the $210 billion in annual fossil fuel subsidies to the development of wind and other renewable sources of energy. The investment capital could come from private capital markets but also from companies already in the energy business. Shell, for example, has become a major player in the world wind energy economy. In 2002, General Electric, one of the world's largest corporations, entered the wind business, becoming overnight a major wind turbine manufacturer.[25]

These goals may seem farfetched, but here and there around the world ambitious efforts are beginning to take shape. In the United States, a 3,000-megawatt wind farm is in the early planning stages. Located in South Dakota near the Iowa border, it is being initiated by Clipper Wind, led by James Dehlsen, a wind energy pioneer in California. Designed to feed power into the industrial Midwest around Chicago, this project is not only large by wind power standards, it is one of the largest energy projects of any kind in the world today. In the eastern United States, Cape Wind is planning a 420-megawatt wind farm off the coast of Cape Cod, Massachusetts.[26]

Some 24 states now have commercial-scale wind farms feeding electricity into the U.S. grid. Although there is occasionally a NIMBY problem ("not in my backyard"), the PIMBY response ("put it in my backyard") is much more pervasive. This is not surprising, since a single large turbine can easily generate $100,000 worth of electricity in a year.[27]

The competition among farmers in places like Iowa or ranchers in Colorado for wind farms is intense. Farmers, with no investment on their part, typically receive $3,000–5,000 a year in royalties from the local utility for siting a single, large, advanced-design wind turbine, which occupies a quarter-acre of land. This land would produce 40 bushels of corn worth $120 or, in ranch country, beef worth perhaps $15.[28]

In addition to the additional income, tax revenue, and jobs that wind farms bring, money spent on electricity generated from wind farms stays in the community, creating a ripple effect throughout the local economy. Within a matter of years, thousands of ranchers could be earning far more from electricity sales than from cattle sales.

The question is not whether wind is a potentially vast source of climate-benign energy that can be used to stabilize climate. It is. But will we develop it fast enough to head off economically disruptive climate change?

Hybrid Cars and Wind Power

With the price of oil over $60 a barrel at this writing in September 2005, with political instability in the Middle East on the rise, with little slack in the world oil economy, and with temperatures rising, the world needs a new energy economy. Fortunately, the foundation for a new transportation energy economy has been laid with two new technologies—the gas-electric hybrid engines pioneered by Toyota and advanced-design wind turbines.[29]

These technologies deployed together can dramatically reduce world oil use. As noted earlier, the United States could easily cut its gasoline use in half by converting the U.S. automobile fleet to hybrid cars as efficient as the Toyota Prius. No change in the number of vehicles, no change in miles driven—just doing it with the most efficient propulsion technology on the market.[30]

In fact, there are now several gas-electric hybrid car models on the market in addition to the Prius, including the Honda Insight and a hybrid version of the Honda Civic. According to the Environmental Protection Agency, the Prius—a midsize car on the cutting-edge of automotive technology—gets an astounding 55 miles per gallon in combined city/highway driving compared with 22 miles per gallon for the average new passenger vehicle. No wonder there are lists of eager buyers willing to wait several months for delivery.[31]

Recently, Ford released a hybrid model of its Escape SUV, and Honda released a hybrid version of its popular Accord sedan. General Motors will offer hybrid versions of several of its cars beginning with the Saturn VUE in 2006, followed by the Chevy Tahoe and Chevy Malibu.[32]

Earlier in this chapter we outlined how to cut U.S. gasoline use in half by shifting to gas-electric hybrid vehicles over the next decade. As we shift to these cars, the stage is set for the second step to reduce gasoline use, namely the use of wind-generated electricity to power automobiles. If we add to the gas-electric hybrid a second battery to increase its electricity storage and a plug-in capacity so the batteries can also be recharged from the grid, motorists could then do their commuting, grocery shopping, and other short-distance travel largely with electricity, saving gasoline for the occasional long trip. Even more exciting, recharging batteries with off-peak wind-generated electricity would cost the equivalent of gasoline at 50¢ per gallon. This modification of hybrids could reduce remaining gasoline use by perhaps another 40 percent (or 20 percent of the original level of use), for a total reduction of gasoline use of 70 percent.[33]

These are not the only technologies that can dramatically cut gasoline use. Amory Lovins, a highly regarded pioneer in devising ways of reducing energy use, observes that most efforts to reduce automotive fuel efficiency focus on designing more-efficient engines, largely overlooking the potential of fuel savings from reducing vehicle weight. He notes that substituting advanced polymer composites for steel in constructing the body of automobiles can "roughly double the efficiency of a normal-weight hybrid without materially raising its total manufacturing cost." If we build gas-electric hybrids using the new advanced polymer composites, then we can cut the remaining 30 percent of fuel use by another half, for a total reduction of 85 percent.[34]

Unlike the widely discussed fuel cell/hydrogen transportation model, the gas-electric hybrid/wind model does not require a costly new infrastructure, since the network of gasoline service stations and the electricity grid are already in place. To fully exploit this technology, the United States would need to integrate its weak regional grids into a strong national one, which it needs to do anyway to reduce the risk of blackouts. This, combined with the building of thousands of wind farms across the country, would allow the nation's fleet of automobiles to run largely on wind energy.[35]

One of the few weaknesses of wind energy—its irregularity—is largely offset with the use of plug-in gas-electric hybrids, since the vehicle batteries become a storage system for wind energy. Beyond this, there is always the tank of gasoline as a backup.

The combination of gas-electric hybrids with a second storage battery and a plug-in capacity, the development of wind resources, and the use of advanced polymer composites to reduce vehicle weight has been discussed in a U.S. context but it is a model that can be used throughout the world. It is particularly appropriate for countries that are richly endowed with wind energy, such as China, Russia, Australia, Argentina, and many of those in Europe.[36]

Moving to the highly efficient plug-in gas-electric hybrids, combined with the construction of thousands of wind farms across the country to feed electricity into a strong, integrated national grid, could cut U.S. gasoline use by 85 percent. It would also rejuvenate farm and ranch communities and shrink the U.S. balance-of-trade deficit. Even more important, it could cut automobile carbon emissions by some 85 percent, making the United States a model for other countries.

Converting Sunlight to Electricity

Wind is not the only vast untapped source of energy. When a team of three scientists at Bell Labs in Princeton, New Jersey, discovered in 1952 that sunlight striking a silicon surface could generate electricity, they opened the door to another near limitless source of energy—photovoltaic (or solar) cells. "No country uses as much energy as is contained in the sunlight that strikes its buildings each day," writes Denis Hayes, former Director of the U.S. government's Solar Energy Research Institute.[37]

Sales of solar cells worldwide jumped by a phenomenal 57 percent in 2004, pushing the generating capacity installed during the year to 1,200 megawatts. With this addition, world solar-cell generating capacity, which has doubled in the last two years, now exceeds 4,300 megawatts, roughly the equivalent of 13 coal-fired power plants. (See Figure 10–2.) A decade ago the United States had roughly half of the world market, but this has now dropped to 12 percent as Japan and Germany have surged ahead with ambitious solar programs.[38]

Solar cells are used either in stand-alone systems or in systems that can feed into the grid. In its early years, the solar cell industry was dominated by non-grid uses to supply electricity to communication satellites and in remote sites such as national forests or parks, offshore lighthouses, summer homes in isolated mountain regions, or islands.

Over the last decade, solar cell installations that feed into the grid have grown rapidly in response to incentives offered by governments, and they now account for more than three fourths of

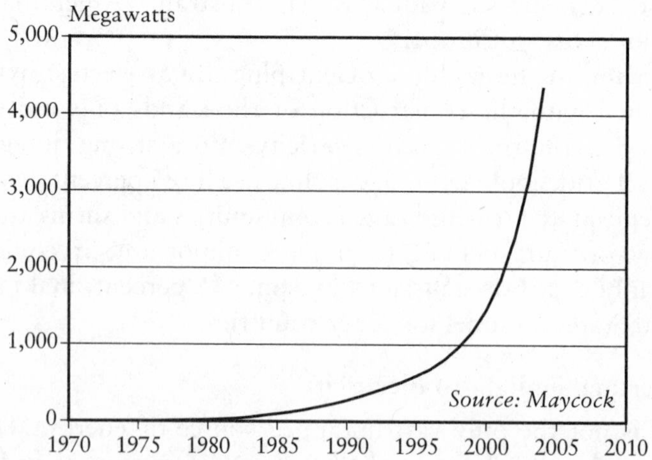

Figure 10–2. *World Photovoltaic Generating Capacity, 1971–2004*

all new installations. Two-way meters that enable utility customers to feed surpluses into the grid for a fixed rate have spurred rapid growth in solar cell use. The U.S. Energy Policy Act of 2005 established two-way metering for any customer requesting it. Some countries have established a fixed price for utilities to pay for electricity fed into the grid. In Germany, this has been set well above the market price to reflect the value of clean electricity and to get the fledgling solar cell industry off the ground.[39]

The residential use of solar cells is expanding at a breakneck pace in some countries. In Japan, where companies have commercialized a solar roofing material, the idea of making the roof the power plant for the home is increasingly popular. This, combined with Japan's 70,000 Roofs Program launched in 1994 to subsidize the installations, got the country off to a fast start, making it a world leader in solar-generated electricity.[40]

In 1998, Germany initiated a 100,000 Roofs Program, which gave consumers 10-year loans for buying photovoltaic systems at reduced interest rates. This ended in 2003 when the goal of 100,000 solar roofs was reached. With this fast-growing market, solar cell costs now have fallen to where German manufacturers are quite competitive internationally.[41]

Within the United States, California is also providing attractive incentives for the residential installation of solar cells. In a climate where peak capacity on hot summer days presses against the limits of the grid, solar cells are seen as an alternative to fossil fuel plants, mostly gas-fired, that operate only during the peak daytime demand. Happily, solar cells generate the most electricity during the hottest times of the day, making them ideal for satisfying peak power demands.[42]

Solar cell installations may be even more economical in large buildings. In Manchester, England, a 40-story office building in need of renovation will be covered with photovoltaic material. With three sides of this 400-foot building covered with this material, the building has a huge generating surface. An official of the building owner and occupant, the Co-operative Insurance Society, noted with a smile that it would produce enough electricity each year to make 9 million cups of tea.[43]

In recent years, a vast new off-grid solar cell market has opened up in developing-country villages, where the cost of building a centralized power plant and a grid to deliver relatively small amounts of electricity to individual consumers is prohibitive. With solar cells costs falling, however, it is now often cheaper to provide electricity from solar cell installations than from a centralized source.

In Andean villages, solar installations are replacing candles as a source of lighting. For villagers who are paying for the installation over 30 months, the monthly payment is roughly equal to the cost of a month's supply of candles. Once the solar cells are paid for, the villagers then have an essentially free source of power—one that can supply electricity for decades. Similarly, in villages in India, where light now comes from kerosene lamps, soaring oil prices mean that kerosene from imported oil may now cost far more than solar cells.[44]

Today more than 1 million homes in villages in the developing world are getting their electricity from solar cells, but this represents less than 1 percent of the 1.7 billion people who do not yet have electricity. The principal obstacle to the spread of solar cell installations in villages is not the cost per se, but the lack of small-scale credit programs to finance them. If this credit shortfall is quickly overcome, village purchases of solar cells will soar.[45]

The future of solar cells is promising. Japan, for example, where residential installations exceeded 1,000 megawatts at the end of 2004, plans to get 10 percent of its electricity from solar cells by 2030. Germany now has 700 megawatts of installed capacity and is growing fast. The United States, a distant third, introduced a solar tax credit in the Energy Policy Act of 2005. The first such credit in 20 years, it promises to rejuvenate the U.S. solar industry.[46]

The cost of solar cells has been dropping for several decades and is expected to continue falling for the indefinite future. With each doubling of cumulative production, the manufacturing economics of scale drop the price an additional 20 percent. In addition, technologies for producing solar cells that convert more sunlight into electricity and do so at a lower cost are being worked on at numerous research facilities in several countries.[47]

In addition to generating electricity from solar cells, solar energy can also be concentrated to boil water and produce steam, driving a turbine to generate electricity. There are various designs used in solar-thermal power plants, including power towers, which consist of an elevated facility containing water that is heated by an array of mirrors, and solar troughs. Usually computer-controlled, the mirrors shift their position as the sun moves across the sky to maximize the sunlight focused on the boiler. Some 350 megawatts of generating capacity in California has been operating very successfully since nine solar trough plants were built in the mid-1980s and early 1990s. New initiatives to develop solar thermal power plants are now under way in Spain.[48]

One of the most popular ways of harnessing solar energy is the use of rooftop solar thermal collectors for both water and space heating. Janet Sawin of Worldwatch Institute reports that the global installations of 150 million square meters, excluding the one fourth that is used for swimming pools, supplies water or space heating for 32 million households.[49]

For years both Israel and Cyprus, countries rich in sunlight, have been encouraging solar water heaters as a way of reducing the need for imported fossil fuels. Germany, which has 5.4 million square meters of solar water heating panels, ranks second in installed capacity. This panel area totals 540 hectares or roughly 1,300 acres.[50]

China, far and away the world leader in this technology, is planning to quadruple its current 52 million square meters of collectors by 2015, reports Sawin. Spain, a leading manufacturer of solar thermal panels, is making a bid for industry leadership by requiring the inclusion of rooftop solar water heaters on all new buildings, residential and commercial, beginning in 2005. A two-meter panel on a single-family residence can reduce annual water heating bills by 70 percent. In effect, Spain is substituting its abundant sunlight for imported oil.[51]

The technologies for converting sunlight into electricity or using it to heat water and building space are now well developed. And the economics are falling into place. What is needed to accelerate this is a set of incentives in all countries that reflects the value to society of reducing dependence on oil and of reducing carbon emissions.

Energy from the Earth

When we think of renewable energy, we typically think of those sources that derive directly or indirectly from the sun. But the earth itself is a source of heat energy (mostly from radioactivity deep within the earth), which gradually escapes either through conduction or through hot springs and geysers that bring internal heat to the earth's surface. Geothermal energy is inexhaustible and will last as long as the earth itself.

Aside from being an ideal source for base load (continuous) power, geothermal energy is environmentally attractive for several reasons. Its carbon dioxide, sulfur dioxide, and nitrogen oxide emissions are negligible to non-existent. Water use for geothermal electric generation is 1 percent that of natural-gas-fired power plants.[52]

The potential of geothermal energy is extraordinary. Japan alone has an estimated geothermal electric-generating capacity of 69,000 megawatts, enough to satisfy one third of its electricity needs. Among the countries rich in geothermal energy are those bordering the Pacific in the so-called Ring of Fire. These include (on the east) Chile, Peru, Ecuador, Colombia, all the Central American countries, Mexico, the western United States, and Canada and (on the west) Russia, China, South Korea, Japan, the Philippines, Indonesia, Australia, and New Zealand. Other geothermally rich countries include those along the Great Rift of Africa and the Eastern Mediterranean. Fortunately, many countries now have enough experience and engineering capacity to tap this vast resource.[53]

Like solar energy, geothermal energy is used both to generate electricity and to directly heat buildings, greenhouses, and aquacultural ponds. It is also used as a source of heat for industrial processes. After Italy pioneered the use of geothermal energy to generate electricity in 1904, the practice spread to some 25 countries. The global capacity of 8,400 megawatts in 2003 represents a 44-percent growth over the 5,800 megawatts available in 1990.[54]

Two countries—the United States with 2,000 megawatts and the Philippines with 1,900 megawatts—account for almost half of world generating capacity. In the Philippines, geothermal provides a world-leading 27 percent of the country's electricity supply. California, the most populous state, gets 5 percent of its electricity from geothermal power plants. Most of the remaining geothermal power generation is concentrated in five countries: Italy, Mexico, Indonesia, Japan, and New Zealand.[55]

The direct use of geothermal heat for various heating purposes worldwide is even larger, equivalent to 12,000 megawatts of electricity generation. Its use in heat pumps, which extract and concentrate heat from warm water for various uses, is the largest single use. More than 30 countries tap geothermal energy for heating.[56]

Iceland and France are the leaders. In Iceland, 93 percent of the country's homes are heated with geothermal energy, saving over $100 million per year in avoided oil imports. Geothermal energy accounts for more than one third of Iceland's total energy use. Following the two oil price hikes in the 1970s, some 70 geothermal heating facilities were constructed in France, providing both heat and hot water for an estimated 200,000 residences. In the United States, individual homes are supplied directly with geothermal heat in Reno, Nevada, and in Klamath Falls, Oregon. Other countries that have extensive geothermally based district-heating systems include China, Japan, and Turkey.[57]

Geothermal energy is an ideal source of heat for greenhouses, particularly in northern climes. Russia, Hungary, Iceland, and the United States all use geothermally heated greenhouses to produce fresh vegetables in winter. With rising oil prices boosting fresh produce transport costs, this option will likely become more popular in the years ahead.[58]

Some 16 countries use geothermal energy for aquaculture. Among these are China, Israel, and the United States. In California, for example, 15 fish farms produce tilapia, striped bass, and catfish with warm water from underground. This warmer water enables fish to grow without interruption during the winter and to mature more quickly. Collectively these California farms produce 4.5 million kilograms of fish per year.[59]

The number of countries turning to geothermal energy both for electricity and for direct use is increasing rapidly. So, too, is the range of uses. Once the value of geothermal energy is discovered, its use is often quickly diversified. Romania, for example, uses its geothermal energy for district heating, for greenhouses, and to supply hot water for homes and factories. With heat pumps, the earth can serve as both a heat source and a sink to provide heating in winter and cooling in summer.[60]

Geothermal energy is widely used for bathing and swimming. Japan, for example, has 2,800 spas, 5,500 public bathhouses, and 15,600 hotels and inns that use hot geothermal water. Iceland uses geothermal energy to heat some 100 public swimming pools, most of them year-round open-air pools. Hungary heats 1,200 swimming pools with geothermal energy.[61]

Indonesia, with more than 222 million people, could easily get all of its electricity from geothermal energy. Situated on the western edge of the Pacific, with 500 volcanoes, 128 of them active, Indonesia has a master plan for 11 geothermal power plants with a generation capacity of just over 300 megawatts each—a total of 3,400 megawatts. This plan was derailed by the Asian financial crisis of 1997, but supporters are now attempting to revive it. With its oil production falling, Indonesia needs to quickly develop alternative sources of energy. Unlike investments in oil, those in geothermal energy are tapping an energy source that can last forever.[62]

Cutting Carbon Emissions Fast

By far the cheapest and fastest way to cut carbon emissions is to raise the efficiency of energy use. Not only is it cheap, it is often profitable. The other option is to develop renewable sources of energy. Within this framework, perhaps the most complex question is which alternative automotive fuels to develop. Until recently, the only widely considered alternative to oil since the initial oil price hikes in the 1970s was biofuels. Now with the advent of gas-electric hybrid plug-ins, wind-generated electricity becomes an appealing option because of its abundance and low cost.

The frugal use of land by wind is impressive. Within the United States, a quarter-acre of land in the corn belt can be used to site an advanced-design wind turbine that will produce $100,000 worth of electricity per year or it can be used to produce 40 bushels of corn that will yield 100 gallons of ethanol worth perhaps $200. If the goal is to minimize competition from the automotive fuel economy for food resources, wind-generated electricity is the obvious choice.[63]

Among the various ethanol sources, sugarcane is by far the most efficient in both land and energy use. The ethanol yield of sugarcane per acre is roughly 650 gallons, whereas for corn in the United States it is 350 gallons, scarcely half as much. The net energy yield of 8 for sugarcane offers an overwhelming advantage over that of the 1.5 for corn.[64]

The palm oil yield of over 500 gallons of biodiesel per acre compares very favorably with 56 gallons per acre for soybeans. The downside of sugarcane and palm oil as feedstocks is that both are grown in tropical and subtropical regions, which means their production will likely be expanded by clearing tropical forests.[65]

The most efficient automotive fuel option appears to the gas-electric hybrid with a plug-in capacity and wind energy as the principal fuel. Since nearly all the basic food commodities can be converted into automotive fuel, either ethanol or biodiesel, there is a risk that rising oil prices will stimulate massive investments in biofuels production, using food staples as the feedstock. This could set up direct competition between affluent motorists and low-income food consumers for the same foodstuffs, including wheat, rice, corn, soybeans, and sugarcane. Avoiding this potential competition between supermarkets and service stations for the same food commodities will depend on governments establishing policies to protect food consumers.

In a world facing disruptive climate change, each country will need to fashion its own carbon reduction strategy in light of its unique complement of renewable sources of energy and its most promising potentials for raising energy efficiency. Yet, many technologies for cutting carbon emissions, such as energy-efficient household appliances and gas-electric hybrid vehicles, are common to all societies.

Among countries, Iceland may be the only one that currently has a strategy to phase out the use of fossil fuels, including oil, entirely. Currently it heats 85 percent of all its buildings, residential and commercial, with geothermal energy. In addition, 82 percent of its electricity comes from hydropower, with most of the remaining electricity geothermally generated. It is now using its cheap hydroelectricity to electrolyze water and produce hydrogen. With its first hydrogen station in operation in Reykjavik, the country is turning to fuel-cell-powered buses. Next it plans to convert its automobiles to fuel cells and then eventually to do the same with its fishing fleet, which lies at the heart of its economy.[66]

The biggest single gain in carbon emission reductions could come in the U.S. automotive sector where, as described earlier, the potential exists for cutting gasoline use by a staggering 85 percent. This model applied worldwide could help the world adjust to the coming decline in oil production.[67]

For the United States, its rich endowment of low-cost wind energy suggests that wind will likely emerge as the centerpiece of the new energy economy. It can supply electricity for heating, cooling, cooking, powering automobiles, and even producing steel, using energy-efficient electric arc furnaces for steel smelting. The United States, which gets 7 percent of its electricity from existing hydroelectric facilities, also has a substantial geothermal potential in the western states and an enormous solar cell potential throughout the country.[68]

Germany plans to cut its carbon emissions dramatically by continuously raising energy efficiency and harnessing renewable energy resources, with an emphasis on wind. By 2050, Germany plans to reduce overall energy use by 37 percent as it uses the latest technologies to raise energy efficiency. Of the remaining 63 percent, 45 percent will come from renewables. This means a cut of 65 percent in overall carbon emissions. Germany will rely heavily on wind and solar cells for electricity generation and solar thermal panels for water and space heating.[69]

Indonesia's energy future lies in its vast resources of geothermal energy. With more than enough geothermal energy to satisfy all its electricity needs, Indonesia can also develop its abundance of solar and wind resources and use electricity to fuel hybrid vehicles. With 11 percent of its electricity coming from hydro, Indonesia has a wide range of renewable energy sources.[70]

For Spain, bathed in sunlight year-round, solar cells and solar panels will figure prominently in supplying electricity, heating, and cooling. Spain is also moving fast to develop its rich endowment of wind energy.[71]

Brazil is unique in that self-sufficiency in automotive fuel in the form of sugarcane-based ethanol could be only a few years away. Along with a generous supply of hydropower, wind and solar cells will also supply electricity. Solar panels will heat water. Brazil could be one of the first large countries to substantially phase out fossil fuels.[72]

For China, hydropower already supplies 15 percent of its electricity, but the big potential lies with wind. China could easily double its current electricity generation from wind alone. Like the United States, a combination of gas-electric hybrids with a second storage battery and a plug-in capacity and a heavy investment in harnessing its abundant wind resources can minimize the use of gasoline and reduce dependence on coal.[73]

In the United Kingdom, wind-generated electricity, primarily from offshore wind farms, holds enormous potential. This, combined with wave power (of which it has an abundance) and solar panels for water heating, can meet much of the country's energy needs.[74]

For Argentina, where hydropower already supplies 42 percent of electricity, wind could easily supply the remainder. Its large Patagonian region has some of the richest wind resources found anywhere. Argentina also has the potential for solar electricity and solar panels.[75]

During the last century, the world became increasingly dependent on a small handful of countries in the Middle East for its energy. During this century, the world is turning to local energy resources. The last century saw the globalization of the energy economy, while today we are seeing its localization. Whereas "one size fit all" in the last century, in the twenty-first century each country will fashion an energy strategy that fits its own renewable energy resources and its potential for raising energy efficiency.

For countries everywhere, particularly developing countries, the economic good news in the energy transition is that it is much more labor-intensive than the use of fossil fuels. Even though Germany is still early in the energy transition, renewable energy industries already employ more workers than the long-standing fossil fuel and nuclear industries. In a world where unemployment is widespread, this is welcome news indeed.[76]

Furthermore, in contrast to investments in oil and gas fields and coal mines, where depletion and abandonment were inevitable, the new energy sources are inexhaustible. While wind turbines, solar cells, and solar thermal panels will all need repair and occasional replacement, the initial investment can last indefinitely. This well will not go dry.

11

Designing Sustainable Cities

As I was being driven through Tel Aviv from my hotel to a conference center a few years ago, I could not help but note the overwhelming presence of cars and parking lots. Tel Aviv, expanding from a small settlement a half-century ago to a city of some 3 million today, evolved during the automobile era. It occurred to me that the ratio of parks to parking lots may be the best single indicator of the livability of a city—an indication of whether the city is designed for people or for cars.[1]

The world's cities are in trouble. In Mexico City, Tehran, Bangkok, Shanghai, and hundreds of other cities, the quality of daily life is deteriorating. Breathing the air in some cities is equivalent to smoking two packs of cigarettes per day. In the United States, the number of hours commuters spend going nowhere sitting in traffic-congested streets and highways climbs higher each year, raising frustration levels.[2]

In response to these conditions, we are seeing the emergence of a new urbanism. One of the most remarkable modern urban transformations has occurred in Bogotá, Colombia, where Enrique Peñalosa served as Mayor for three years, beginning in 1998. When he took office he did not ask how life could be improved for the 30 percent who owned cars; he wanted to know what could be done for the 70 percent—the majority—who did not own cars.[3]

Peñalosa realized that a city that is a pleasant environment for children and the elderly would work for everyone. In just a few years, he transformed the quality of urban life with his vision of a city designed for people. Under his leadership, the city banned the parking of cars on sidewalks, created or renovated 1,200 parks, introduced a highly successful bus-based rapid transit system, built hundreds of kilometers of bicycle paths and pedestrian streets, reduced rush hour traffic by 40 percent, planted 100,000 trees, and involved local citizens directly in the improvement of their neighborhoods. In doing this, he created a sense of civic pride among the city's 8 million residents, making the streets of Bogotá in this strife-torn country safer than those in Washington, D.C.[4]

Enrique Peñalosa observes that "high quality public pedestrian space in general and parks in particular are evidence of a true democracy at work." He further observes: "Parks and public space are also important to a democratic society because they are the only places where people meet as equals.…In a city, parks are as essential to the physical and emotional health of a city as the water supply." He notes this is not obvious from most city budgets, where parks are deemed a luxury. By contrast, "roads, the public space for cars, receive infinitely more resources and less budget cuts than parks, the public space for children. Why," he asks, "are the public spaces for cars deemed more important than the public spaces for children?"[5]

In espousing this new urban philosophy, Peñalosa is not alone. The reform he initiated in Bogotá is being carried on by his successor, Antanas Mockus. Now government planners everywhere are experimenting, seeking ways to design cities for people not cars. Cars promise mobility, and they provide it in a largely rural setting. But in an urbanizing world there is an inherent conflict between the automobile and the city. After a point, as their numbers multiply, automobiles provide not mobility but immobility.[6]

Some cities in industrial and developing countries alike are dramatically increasing urban mobility by moving away from the car. Jaime Lerner, the former mayor of Curitiba, Brazil, was one of the first to design and adopt an alternative transportation system, one that does not mimic those in the West but that is inexpensive and commuter-friendly. Since 1974 Curitiba's transportation system has been totally restructured. Although one third of the people own cars, these play a minor role in urban transport. Busing, biking, and walking totally dominate, with two thirds of all trips in the city by bus. The city's population has doubled since 1974, but its car traffic has declined by a remarkable 30 percent.[7]

Aside from the growth of population itself, urbanization is the dominant demographic trend of our time. In 1900, 150 million people lived in cities. By 2000, it was 2.9 billion people, a 19-fold increase. By 2007 more than half of us will live in cities—making us, for the first time, an urban species.[8]

In 1900 there were only a handful of cities with a million people. Today 408 cities have at least that many inhabitants. And there are 20 megacities with 10 million or more residents. Tokyo's population of 35 million exceeds that of Canada. Mexico City's population of 19 million is nearly equal to that of Australia. New York, São Paulo, Mumbai (formerly Bombay), Delhi, Calcutta, Buenos Aires, and Shanghai follow close behind.[9]

The Ecology of Cities

Cities require a concentration of food, water, energy, and materials that nature cannot provide. Concentrating these masses of materials and then dispersing them in the form of garbage, sewage, and as pollutants in air and water is challenging city managers everywhere.

Most of today's cities are not healthy places to live. Urban air everywhere is polluted. Typically centered on the automobile and no longer bicycle- or pedestrian-friendly, cities deprive people of needed exercise, creating an imbalance between caloric intake and caloric expenditures. As a result, obesity is reaching epidemic proportions in cities in developing as well as industrial countries. With more than 1 billion people overweight worldwide, epidemiologists now see this as a public health threat of historic proportions—a growing source of heart disease, high blood pressure, diabetes, and a higher incidence of several forms of cancer.[10]

The evolution of modern cities is tied to advances in transport, initially for ships and trains, but it was the internal combustion engine combined with cheap oil that provided the mobility of people and freight that fueled the phenomenal urban growth of the twentieth century. As the world urbanized, energy use climbed.

Early cities relied on food and water from the surrounding countryside, but today cities often depend on distant sources even for such basic amenities. Los Angeles, for example, draws much of its water supply from the Colorado River, some 970 kilometers (600 miles) away. Mexico City's burgeoning population, living at 3,000 meters, must now depend on the costly pumping of water from 150 kilometers away and must lift it a kilometer or more to augment its inadequate water supplies. Beijing is planning to draw water from the Yangtze River basin nearly 1,500 kilometers away.[11]

Food comes from even greater distances, as is illustrated by Tokyo. While Tokyo still depends for its rice on the highly productive farmers in Japan, with their land carefully protected by government policy, its wheat comes largely from the Great Plains of North America and from Australia. Its corn supply comes largely from the U.S. Midwest. Soybeans come from the U.S. Midwest and the Brazilian *cerrado*.[12]

The oil that provides much of the energy to move resources into and out of cities itself often comes from distant oil fields. Rising oil prices will affect cities, but they will affect even more the suburbs that many cities have spawned.

It is widely assumed that urbanization will continue. But this is not necessarily so. The growing scarcity of water and the high cost of the energy invested in transporting water over long distances may itself begin to constrain urban growth. For example, some 400 cities in China are already facing a chronic shortage of water.[13]

Against this backdrop, Richard Register, author of *Ecocities: Building Cities in Balance with Nature*, says it is time to fundamentally rethink the design of cities. He agrees with Peñalosa that cities should be designed for people, not for cars. He goes even further, talking about pedestrian cities—communities designed so that people do not need cars because they can walk to most of the places they need to go or take public transportation.[14]

Register also says that a city should be seen as a functioning system not in terms of its parts but in terms of its whole. He makes a convincing case that cities should be integrated into local ecosystems rather than imposed on them.[15]

He describes with pride an after-the-fact integration into the local ecosystem of San Luis Obispo, a California town of 50,000 north of Los Angeles: "[It] has a beautiful creek restoration project with several streets and through-building passageways lined with shops that connect to the town's main commercial street, and people love it. Before closing a street, turning a small parking lot into a park, restoring the creek and making the main street easily accessible to the 'nature' corridor, that is, the creek, the downtown had a 40% vacancy rate in the storefronts; and now it has zero. Of course it's popular. You sit at your restaurant by the creek...where fresh breezes rustle the trees in a world undisturbed by car noise and blazing exhaust." San Luis Obispo is surrounded by both natural and agricultural landscapes.[16]

For Register, the design of the city and its buildings become a part of the local landscape, capitalizing on the local ecology. For example, buildings are designed to be heated and cooled by nature as much as possible. Later in this chapter we discuss how cities can largely live on recycled water that is cleaned and used again and again. The "flush and forget" water system will become too costly for many water-short cities in a world after oil. Urban food production, particularly fresh fruits and vegetables, will expand in vacant lots and on rooftops as oil prices rise.[17]

In the years ahead, urbanization could slow or even be reversed. In a world of land, water, and energy scarcity, the value of each resource may increase substantially, shifting the terms of trade between the countryside and cities. Ever since the beginning of the Industrial Revolution, the terms of trade have favored cities because they control capital and technology, the scarce resources. But if land and water become the scarcest resources, then those in rural areas who control them may sometimes have the upper hand. With a new economy based on renewable energy, a disproportionate share of that energy, particularly wind energy and biofuels, will come from nearby rural areas.[18]

Beyond resource shortages, the evolution of the Internet, which is changing how we think about distance and mobility, could also affect urbanization. Among other things, the potential for telecommuting may reduce the advantages of living in a city. Internet commerce, offering more options than any shopping mall, may also lessen the role of urban shopping centers as sources of a wide variety of goods and services.

Redesigning Urban Transport

Urban transport systems based on a combination of rail lines, bus lines, bicycle pathways, and pedestrian walkways offer the best of all possible worlds in providing mobility, low-cost transportation, and a healthy urban environment. Megacities regularly turn to underground rail systems to provide mobility. Whether it is these rail systems, light-rail surface systems, or both depends in part on city size and geography. For cities of intermediate size, light rail is often an attractive option.

A rail system provides the foundation for a city's transportation system. Rails are geographically fixed, providing a permanent means of transportation that people can count on. Once in place, the nodes on such a system become the obvious places to concentrate office buildings, high-rise apartment buildings, and shops.

As noted earlier, some of the most innovative public transportation systems, those that move huge numbers of people from cars into buses, have been developed in Curitiba and Bogotá. The success of Bogotá's bus rapid transit (BRT) system TransMilenio, which relies on special express lanes to move people quickly through the city, is being replicated not only in six other Colombian cities, but in cities elsewhere too: Beijing, Mexico City, São Paulo, Seoul, Taipei, and Quito. Several more cities in Africa and China are also planning BRT systems. Even industrial-country cities, such as Ottawa and—much to everyone's delight—Los Angeles, are now considering BRT systems.[19]

Many cities are reducing traffic congestion and air pollution by charging cars to enter the city. Singapore, long a leader in urban transport innovation, has imposed a tax on all roads leading into the city center. Electronic sensors identify each car, and then debit the owner's credit card. This system has reduced the number of automobiles in Singapore, providing its residents with both more mobility and cleaner air than in most other cities.[20]

Singapore has been joined by London and by several Norwegian cities, including Oslo, Bergen, and Trondheim. In London—where the average speed of an automobile a few years ago was about the same as that of a horse-drawn carriage a century ago—a congestion tax was adopted in early 2003. The £5 charge on all motorists driving into the center city between 7 a.m. and 6:30 p.m. immediately reduced the number of vehicles, permitting traffic to flow more freely while cutting pollution and noise.[21]

During the first year after the new tax was introduced, the number of people using buses to travel into the central city climbed by 38 percent. Since the congestion charge, the daily flow of cars into central London has been reduced by 65,000–70,000, a drop of 18 percent, while delays have dropped by 30 percent. The number of bicycles and mopeds has increased by 17 percent, and vehicle speeds on key thoroughfares have increased by 21 percent, from 8.7 to 10.6 miles per hour.[22]

Contrary to the fear about falling profits, 65 percent of businesses in London's inner city have not noticed any effect on their bottom line. A substantial majority of business owners think the reduced vehicle flow has had a positive effect on the city's image. A similar tax is now being considered in Cardiff for adoption within the near future. Other cites considering the measure include Stockholm, São Paulo, San Francisco, Milan, and Barcelona. French officials are looking at a congestion charge to deal with the suffocating air pollution in Paris. This highly successful use of taxes to restructure urban transport systems is discussed in terms of restructuring the overall economy in Chapter 12.[23]

The bicycle, a form of personal transportation, has many attractions. It alleviates congestion, lowers air pollution, reduces obesity, increases physical fitness, does not emit climate-disrupting carbon dioxide, and has a price within reach for the billions of people who cannot afford an automobile. Bicycles increase mobility while reducing congestion and the area of land paved over. Six bicycles can typically fit into the road space used by one car. For parking, the advantage is even greater, with 20 bicycles occupying the space required to park a car.[24]

The bicycle is not only a flexible means of transportation, it is an ideal way of restoring a balance between caloric intake and expenditure. The opportunity to exercise is valuable in its own right. Regular exercise of the sort provided by cycling to work reduces cardiovascular disease, osteoporosis, and arthritis and strengthens the immune system. Millions of people pay a monthly fee to use a fitness center, which they often drive to, where they ride stationary bikes, trying to achieve the same benefits.

Few methods of reducing carbon emissions are as effective as substituting the bicycle for the automobile on short trips. A bicycle is a marvel of engineering efficiency, one where an investment in 13 kilograms (28 pounds) of metal and rubber boosts the efficiency of individual mobility by a factor of three. On my bike I estimate that I get easily 7 miles (11 kilometers) per potato. An automobile, which requires 1–2 tons of material to transport even one person, is extraordinarily inefficient in comparison.

The capacity of the bicycle to provide mobility for low-income populations was dramatically demonstrated in China. In 1976, China produced 6 million bicycles. After the reforms in 1978 that led to rapid economic growth, rising incomes, and a market economy in which people could exercise their preferences, annual bicycle production started climbing, eventually soaring over 40 million in 1988. Once the market was largely saturated, production dropped somewhat and remained between 30 million and 40 million a year through the 1990s. Since 1999, production has taken off once again, rising to 79 million bicycles in 2004. The vast surge to 545 million bicycle owners in China since 1978 provided the greatest increase in human mobility in history. Bicycles took over city streets and rural roads. Although China's 7 million passenger cars are getting a lot of attention, especially in the large cities, it is the bicycle that provides personal mobility.[25]

Many cities are turning to bicycles for numerous uses. In the United States, more than 80 percent of police departments serving populations of 50,000–249,999 and 96 percent of those serving more than 250,000 residents now have routine patrols by bicycle. Officers on bikes are more productive in cities partly because they are more mobile and can reach the scene of an accident or crime more quickly and more quietly than officers in cars. They typically make 50 percent more arrests per day than officers in squad cars. For fiscally sensitive officials, the cost of operating a bicycle is trivial compared with that of a police car.[26]

Urban bicycle messenger services are common in the world's larger cities. Bicycles deliver small parcels in cities more quickly than motor vehicles can and at a much lower cost. As the information economy unfolds and as e-commerce expands, the need for quick, reliable, urban delivery services is escalating. For many competitive Internet marketing firms, quick delivery wins customers. In a city like New York, this means bicycle delivery. An estimated 300 bicycle messenger firms are operating in New York City, competing for $700 million worth of business annually. In large cities, the bicycle is becoming an integral part of the support system for e-commerce.[27]

The key to realizing the potential of the bicycle is to create a bicycle-friendly transport system. This means providing both bicycle trails and designated street lanes for bicycles. These should be designed to serve both commuters and those biking for recreation. In addition, bicycle use is enhanced by the provision of parking facilities and showers at workplaces. Among the industrial-country leaders in designing bicycle-friendly transport systems are the Dutch, the Danes, and the Germans.[28]

The Netherlands, the unquestioned leader among industrial countries in encouraging bicycle use, has incorporated a vision of the role of bicycles into a Bicycle Master Plan. In addition to creating bike lanes and trails in all its cities, the system also often gives cyclists the advantage over motorists in right-of-ways and at traffic lights. Some traffic signals permit cyclists to move out before cars. Roughly 30 percent of all urban trips in the Netherlands are on bicycle. This compares with 1 percent in the United States.[29]

Spain, one of the latest countries to climb on the bicycle bandwagon, began converting abandoned railway lines into recreational paths in 1993. The 52 new "greenways" include 1,300 kilometers of bicycle trails throughout the country.[30]

Within the Netherlands a nongovernmental group called Interface for Cycling Expertise (I-ce) has been formed to share the Dutch experience in designing a modern transport system that prominently features bicycles. It is working with other groups in Brazil, Colombia, Ghana, India, Kenya, South Africa, Sri Lanka, Tanzania, and Uganda to facilitate bicycle use. Roelof Wittink, head of I-ce, observes, "If you plan only for cars then drivers will feel like the King of the Road. This reinforces the attitude that the bicycle is backward and used only by the poor. But if you plan for bicycles it changes the public attitude."[31]

Both the Netherlands and Japan have made a concerted effort to integrate bicycles and rail commuter services by providing bicycle parking at rail stations, making it easier for cyclists to commute by train. In Japan, the use of bicycles for commuting to rail transportation has reached the point where some stations have invested in vertical, multi-level parking garages for bicycles, much as is often done for automobiles.[32]

The combination of rail and bicycle, and particularly their integration into a single, overall transport system, makes a city eminently more livable than one that relies almost exclusively on private automobiles. Noise, pollution, congestion, and frustration are all lessened. We and the earth are both healthier.

Farming in the City

While attending a conference on the outskirts of Stockholm in the fall of 1974, I walked past a community garden near a high-rise apartment building. It was an idyllic Indian summer afternoon, with many people tending gardens a short walk from their residences. More than 30 years later I can still recall the setting because of the aura of contentment surrounding those working in their gardens. Nearly all were elderly; they were absorbed in producing not only vegetables, but in some cases flowers as well. I remember thinking, "This is the mark of a civilized society."

In June 2005, the U.N. Food and Agriculture Organization (FAO) reported that urban and peri-urban farms—those within or immediately adjacent to a city—supply food to some 700 million urban residents worldwide. These are mostly small plots—vacant lots, yards, even rooftops.[33]

Within and near the city of Dar es Salaam, capital of Tanzania, there are some 650 hectares of land producing vegetables. This land supplies not only the city's fresh produce but a livelihood for 4,000 farmers who intensively farm their small plots year-round. On the far side of the continent, an FAO project has urban residents in Dakar, Senegal, producing up to 30 kilograms of tomatoes per square meter each year with continuous cropping in rooftop gardens.[34]

In Hanoi, 80 percent of the fresh vegetables come from farms in and immediately adjacent to the city. These urban farms also produce 50 percent of the pork and the poultry consumed in the city. Half of the city's freshwater fish are produced by enterprising urban fish farmers. Some 40 percent of the egg supply is produced within the city or in its shadow. Urban farmers ingeniously recycle human and livestock waste to nourish plants and to fertilize fish ponds.[35]

People living in wetlands in the region of East Calcutta in India manage wastewater fish ponds that cover nearly 3,500 hectares. The city's sewage water is kept in ponds and moved through various stages so the bacteria can break down the organic waste. This, in turn, supports the rapid growth of algae that supply food for the various local strains of herbivorous fish. This system provides a steady supply of fish for the city, fish that are consistently of better quality than any entering the Calcutta market.[36]

The magazine *Urban Agriculture* describes how Shanghai has in effect created a nutrient recycling zone around the city. In order to have enough land to recycle the city's night soil, the boundary was extended to include 300,000 hectares of surrounding farmland. This land supplies 60 percent of this megacity's vegetables. Half of Shanghai's pork and poultry and 90 percent of its milk and eggs come from the extended city and the immediately surrounding region.[37]

In Caracas, Venezuela, a government-sponsored FAO-assisted project has created 4,000 microgardens of one square meter each in the city's barrios, many of them located within a few steps of family kitchens. As soon as one crop is mature, it is harvested and immediately replaced with new seedlings. Each square meter, continuously cropped, can produce 330 heads of lettuce, 18 kilograms of tomatoes, or 16 kilograms of cabbage per year.[38]

Venezuela's goal is to have 100,000 microgardens in the country's urban areas and 1,000 hectares of urban compost-based gardens nationwide. Leonardo Gil Mora, vice minister of integrated rural development, points out that "in the barrios as in Venezuela in general, people are the most important thing we have. Through urban agriculture, we hope to increase the poor's self-confidence, and so increase their participation in society."[39]

There is a long tradition of community gardens in European cities. As a visitor flies into Paris, numerous community gardens can be seen on the outskirts of the city. These small plots produce not only high-quality food but a sense of well-being and community.

As a result of a national campaign in Cuba to expand urban farming after the loss of Soviet support more than a decade ago, Havana now produces half of the vegetables its residents consume. The city-state of Singapore has 10,000 urban farmers who produce four fifths of the poultry and a fourth of all the vegetables eaten there. A 2003 *Urban Agriculture* study reports that 14 percent of London's 7.6 million residents produce some of their own food. For Vancouver, Canada's largest west coast city, the comparable figure is an impressive 44 percent.[40]

In the U.S. city of Philadelphia, community gardeners were asked why they gardened. Some 20 percent did it for recreational reasons, 19 percent said it improved their mental health, and 17 percent their physical health. Another 14 percent did it because they wanted the higher-quality fresh produce that a garden could provide, 10 percent did it for spiritual reasons, and 7 percent said it was mostly economic—cost and convenience. Urban gardens are social gathering places that engender a sense of community. In addition, those who garden three to four times a week get the same physical benefits as people who engage in moderate walking or bicycling.[41]

In some countries, such as the United States, there is a huge unrealized potential for urban gardening. A survey indicated that Chicago has 70,000 vacant lots, and Philadelphia, 31,000. Nationwide, vacant lots in cities would total in the hundreds of thousands. The Urban Agriculture report summarizes why urban agriculture is so desirable. It has "a regenerative effect…when vacant lots are transformed from eyesores—weedy, trash-ridden dangerous gathering places—into bountiful, beautiful, and safe gardens that feed people's bodies and souls."[42]

Given the near inevitable rise in future oil prices, the economic benefits of expanding urban agriculture, even in affluent societies, will become much more obvious. Aside from supplying more fresh produce, it will help millions discover the social benefits and the psychological well-being that urban gardening can bring.

Reducing Urban Water Use

The one-time use of water to disperse human and industrial wastes is an outmoded practice, made obsolete by new technologies and water shortages. Water enters the city, is contaminated with human and industrial wastes, and leaves the city dangerously polluted. Toxic industrial wastes discharged into rivers and lakes or into wells also permeate aquifers, making water—both surface and underground—unsafe for drinking. And their toxic wastes are destroying marine ecosystems, including local fisheries. The time has come to manage waste without discharging it into the local environment, allowing water to be recycled indefinitely and reducing both urban and industrial demand dramatically.

The current engineering concept for dealing with human waste is to use vast quantities of water to wash it away, preferably into a sewer system where it will be treated before being discharged into the local river. The "flush and forget" system is water-intensive, it disrupts the nutrient cycle, most of humanity cannot afford it, and it is a major source of disease in developing countries.

As water scarcity spreads, the viability of water-based sewage systems will diminish. Water-based sewage systems take nutrients originating in the soil and typically dump them into rivers, lakes, or the sea. Not only are the nutrients lost from agriculture, but the nutrient overload has led to the death of many rivers and to the formation of some 146 dead zones in ocean coastal regions. Sewer systems that dump untreated sewage into rivers and streams are a major source of disease and death.[43]

Sunita Narain of the Centre for Science and Environment in India argues convincingly that a water-based disposal system with sewage treatment facilities is neither environmentally nor economically viable for India. She notes that an Indian family of five, producing 250 liters of excrement in a year and using a water flush toilet, requires 150,000 liters of water to wash away its wastes.[44]

As currently designed, India's sewer system is actually a pathogen-dispersal system. It takes a small quantity of contaminated material and uses it to make vast quantities of water unfit for human use, often simply discharging it into nearby rivers or streams. Narain says both "our rivers and our children are dying." India's government, like that of many other developing countries, is hopelessly chasing the goal of universal water-based sewage systems and sewage treatment facilities—unable to close the huge gap between services needed and provided, but unwilling to admit that it is not an economically viable option. Narain concludes that the "flush and forget" approach is not working.[45]

This dispersal of pathogens is a huge public health challenge. Worldwide, poor sanitation and personal hygiene claim 2.7 million lives per year, second only to the 5.9 million claimed by hunger and malnutrition.[46]

Fortunately, there is a low-cost alternative: the composting toilet. This is a simple, waterless, odorless toilet linked to a small compost facility. Table waste can also be incorporated into the composter. The dry composting converts human fecal material into a soil-like humus, which is essentially odorless and is scarcely 10 percent of the original volume. These compost facilities need to be emptied every year or so, depending on design and size. Vendors periodically collect the humus and can market it as a soil supplement, thus ensuring that the nutrients and organic matter return to the soil, reducing the need for fertilizer.[47]

This technology reduces residential water use, thus cutting water bills and lowering the energy needed to pump and purify water. As a bonus, it also reduces garbage flow if table waste is incorporated, eliminates the sewage water disposal problem, and restores the nutrient cycle. The U.S. Environmental Protection Agency now lists several brands of dry toilets approved for use. Pioneered in Sweden, these toilets work well under the widely varying conditions where they are now used, including Swedish apartment buildings, U.S. private residences, and Chinese villages.[48]

At the household level, water can be saved by using appliances that are more water-efficient, including showerheads, flush toilets, dishwashers, and clothes washers. Some countries are adopting water efficiency standards and labeling for appliances, much as has been done for energy efficiency. When water costs rise, as they inevitably will, investments in composting toilets and more water-efficient household appliances will become increasingly attractive to individual homeowners.

For cities, the most effective single step to raise water productivity is to adopt a comprehensive water treatment/recycling system, reusing the same water continuously. With this system, only a small percentage of water is lost to evaporation each time it cycles through. Given the technologies that are available today, it is quite possible to recycle urban water supplies comprehensively, largely removing cities as a claimant on scarce water resources.

Some cities faced with shrinking water supplies and rising water costs are beginning to recycle their water supplies. Singapore, for example, which buys its water from Malaysia at an ever higher price, is beginning to recycle water, reducing the amount it imports. For some cities, the continuous recycling of water may become a condition of their survival.[49]

Individual industries facing the same water-related issues as cities are beginning to move away from the use of water to disperse industrial waste. Some companies segregate effluent streams, treating each individually with the appropriate chemicals and membrane filtration, preparing the water for reuse. Peter Gleick, senior author and editor of the biannual report *The World's Water,* writes: "Indeed, some industries, such as paper and pulp, industrial laundries, and metal finishing, are beginning to develop 'closed-loop' systems where all the waste water is reused internally, with only small amounts of fresh water needed to make up for water incorporated into the product or lost in evaporation." Industries are moving faster than cities, but the technologies they are developing can also be used in urban water recycling.[50]

Saving water in cities depends primarily on two household appliances: toilets and showers, which together account for over half of indoor use. Whereas traditional flush toilets used 6 gallons (or 22.7 liters) per flush, the legal U.S. maximum for new toilets is 1.6 gallons (6 liters). An Australian-produced toilet with a dual-flush two-button technology uses only 1 gallon for a liquid waste flush and 1.6 gallons for a solid waste flush. Shifting from a showerhead flowing at 5 gallons per minute to a 2.5 gallons-per-minute model cuts water use nearly in half. With washing machines, a horizontal axis design developed in Europe uses 40 percent less than the traditional top-loading U.S. models. In addition, this European model now being marketed internationally also uses less energy.[51]

The existing water-based waste disposal economy is not viable. There are too many households, factories, and feedlots to simply try and wash waste away on our crowded planet. To do so is ecologically mindless and outdated—an approach that belongs to an age when there were many fewer people and far less economic activity.

The Challenge of Urban Slums

Between 2000 and 2050, there is little population growth projected for the industrial countries or for the rural developing world. Thus nearly all of the projected world population growth of some 3 billion people by 2050 will be added to the cities of developing countries, much of it in squatter settlements.[52]

Squatter settlements—whether they are *favelas* in Brazil, *barriadas* in Peru, or *gecekondu* in Turkey—typically consist of an urban residential area inhabited by very poor people who do not have any land. They simply "squat" on vacant land, either private or public.[53]

Life in these settlements is characterized by grossly inadequate housing and a lack of access to urban services. As Hari Srinivas, coordinator of the Global Development Research Center, writes, these rural-urban migrants undertake the "drastic option of illegally occupying a vacant piece of land to build a rudimentary shelter" simply because it is their only option. They are often treated if not by apathy then by antipathy by government agencies, who view them as invaders and trouble. Some see squatter settlements as a social "evil," something that needs to be eradicated.[54]

Urban slums include not only squatter settlements but also severely rundown older parts of cities, which are also overcrowded and often lacking in rudimentary urban services, such as sewage disposal.

One of the best ways to make rural/urban migration manageable is to improve conditions in the countryside. This means not only providing basic social services, such as basic health care and education for children, as outlined in Chapter 7, but also encouraging industrial investment in small towns throughout the country rather than just in prime cities, such as Mexico City or Bangkok. Such policies will slow the flow into cities to a more orderly pace.

The evolution of cities in developing countries is often shaped by the unplanned nature of squatter settlements. Letting squatters settle wherever they can—on steep slopes, on river floodplains, or in other high-risk areas—makes it difficult to provide basic services such as transport, water, and sewerage. Curitiba, on the cutting edge of the new urbanism, has designated tracts of land for squatter settlements. By setting aside these planned tracts, the process can at least be structured in a way that is consistent with the official development plan of the city.[55]

Among the simplest services that can be provided in a squatter settlement are community composting toilets. Beyond these, taps that provide safe running water at intervals throughout the squatter settlement can go a long way to control the spread of disease in overcrowded settlements. Regular bus service can enable workers living in the settlements to travel to their place of work. If the Curitiba approach is widely followed, parks and other commons areas can be incorporated into the community from the beginning.

Some political elites simply want to bulldoze squatter settlements away, but this treats the symptoms of urban poverty, not the cause. People who lose what little they have been able to invest in housing are not richer as a result of the demolition, but poorer, as is the city itself. The preferred option by far is in situ upgrading of housing. The key to this is providing security of tenure to the squatters and small loans, enabling them to make incremental improvements over time.[56]

Upgrading slums depends on local governments that respond to them rather than ignoring them. Progress in eradicating poverty and creating stable, progressive communities depends on establishing constructive links with governments. In some cases, government-supported micro-credit lending facilities can help not only establish a link between the city government and the squatter communities but offer hope to the residents.[57]

Although political leaders might hope that these settlements will be driven away or demolished, the reality is that they will likely expand over the next several decades. The challenge is to integrate them into urban life in a humane way that provides hope through the potential for upgrading. The inevitable alternative is mounting resentment, spreading social friction, and violence.

Cities for People

As the new century begins, it is becoming evident to urban dwellers, whether in industrial or developing countries, that there is an inherent conflict between the automobile and the city. Urban air pollution, often from automobiles, claims millions of lives. Congestion also takes a direct economic toll in rising costs in time and gasoline.

Another cost of cities that are devoted to cars is a psychological one, a deprivation of contact with the natural world—an "asphalt complex." There is a growing body of evidence that there is an innate human need for contact with nature. Both ecologists and psychologists have been aware of this for some time. Ecologists, led by Harvard University biologist E.O. Wilson, have formulated the "biophilia hypothesis," which argues that those who are deprived of contact with nature suffer psychologically and that this deprivation leads to a measurable decline in well-being.[58]

Meanwhile psychologists have coined their own term—ecopsychology—in which they make the same argument. Theodore Roszak, a leader in this field, cites a study that documents humans' dependence on nature by looking at the rate of recovery of patients in a hospital in Pennsylvania. Those whose rooms overlooked gardens with grass, trees, flowers, and birds recovered from illnesses more quickly than those who were in rooms overlooking the parking lot.[59]

One of the arguments for community gardens is that in addition to providing food, they also provide greenery and a sense of community. Working with soil and watching things grow has a therapeutic effect.

Throughout the modern era, budget allocations for transportation in most countries—and in the United States, in particular—have been heavily biased toward the construction and maintenance of highways and streets. Creating more livable cities and the mobility that people desire depends on reallocating budgets to emphasize the development of rail- or bus-based public transport and bicycle support facilities.

The exciting news is that there are signs of change, daily indications of an interest in redesigning cities for people, not for cars. One encouraging trend comes from the United States. Rising public transit ridership nationwide of 2.1 percent a year since 1996 indicates that people are gradually abandoning their cars for buses, subways, and light rail. Sharp rises in gasoline prices in 2005 are encouraging still more commuters to abandon their cars and take the bus or subway or get on a bicycle.[60]

Mayors and city planners the world over are beginning to rethink the role of the car in urban transportation systems. A group of eminent scientists in China challenged Beijing's decision to promote an automobile-centered transportation system. They noted a simple fact: China does not have enough land to accommodate the automobile and to feed its people. What is true for China is also true for India and dozens of other densely populated developing countries.[61]

Some cities are far better at planning their growth than others. They plan transport systems that provide mobility, clean air, and exercise—a sharp contrast to cities that offer congestion, unhealthy air, and little opportunity for exercise. When 95 percent of a city's workers depend on the automobile for commuting, as in Atlanta, Georgia, the city is in trouble. By contrast, in Amsterdam only 40 percent of workers commute by car; 35 percent commute by bike or walk, while 25 percent use public transit. Copenhagen's commuting patterns are almost identical to Amsterdam's. In Paris, just under half of commuters rely on cars. Even though these European cities are older, often with narrow streets, they have far less congestion than Atlanta.[62]

Not surprisingly, car-dependent cities have more congestion and less mobility than those that offer a wider range of commuting options. The very vehicle whose great promise was personal mobility is in fact virtually immobilizing entire urban populations, making it difficult for rich and poor alike to move about.

Existing long-term transportation strategies in many developing countries assume that everyone will one day be able to own a car. Unfortunately, given the constraints of land available for cars, not to mention those imposed by oil reserves, this is simply not realistic. These countries will provide more mobility if they support public transportation and the bicycle.

If developing-country governments continue to invest most of the public resources available for transportation in support of the automobile, they will end up with a system built for the small fraction of their people who own cars. Recognition now that most people will never own automobiles can lead to a fundamental reorientation of transport planning and investment.

There are many ways to restructure the transportation system so that it satisfies the needs of all people, not just the affluent, so that it provides mobility, not immobility, and so that it improves health rather than damaging it. One is to eliminate the subsidies, often indirect, that many employers provide for parking. For example, parking subsidies in the United States that are worth an estimated $85 billion a year obviously encourage people to drive to work.[63]

In 1992, California mandated that employers match parking subsidies with cash that can be used by the recipient either to pay public transport fares or to buy a bicycle. In firms where data were collected, this shift in policy reduced automobile use by some 17 percent. At the national level, a provision was incorporated into the 1998 Transportation Equity Act of the 21st Century to change the tax code so that those who used public transit or vanpools would enjoy the same tax-exempt subsidies as those who received free parking. What societies should be striving for is not parking subsidies, but parking taxes—taxes that begin to reflect the cost to the community of traffic congestion and the deteriorating quality of life as cities are taken over by cars and parking lots.[64]

Scores of cities are declaring car-free areas, among them Stockholm, Vienna, Prague, and Rome. Paris enjoys a total ban on cars along stretches of the Seine River on Sundays and holidays and is looking to make much of the central city traffic-free starting in 2012.[65]

In addition to ensuring that subways are functional and affordable, the idea of making them attractive, even cultural centers, is gaining support. In Moscow, with works of art in the stations, the subway system is justifiably referred to as Russia's crown jewel. In Washington, D.C., Union Station, which links the city's subway system with intercity train lines, is an architectural delight. Since its restoration was completed in 1988, it has become a social gathering place, with shops, conference rooms, and a rich array of restaurants.

One of the more innovative steps to encourage the use of public transportation comes from State College, a small town in central Pennsylvania that is home to Pennsylvania State University. To reduce traffic congestion on campus and to address the lack of sufficient parking, Penn State decided in 1999 that it would provide $1 million to the bus-based local transit system in exchange for unlimited free rides for its students, faculty, and staff. As a result, bus ridership in State College jumped by 240 percent in one year, requiring the transit company to invest heavily in new buses to accommodate the additional passengers. This initiative by the university has created a far more pleasant, attractive campus—an asset in recruiting both students and faculty.[66]

As the new century begins, the world is reconsidering the urban role of automobiles in one of the most fundamental shifts in transportation thinking in a century. The challenge is to redesign communities, making public transportation the centerpiece of urban transport and augmenting it with sidewalks, jogging trails, and bikeways. This also means replacing parking lots with parks, playgrounds, and playing fields. We can design an urban lifestyle that systematically restores health by incorporating exercise into daily routines while reducing air pollution and obesity.

III

An Exciting New Option

12

Building a New Economy

In Chapter 1 we concluded that the western economic model—the fossil-fuel-based, automobile-centered, throwaway economy—was not viable for the world. Instead, the new economy will be powered by renewable sources of energy, will have a more diverse transport system—relying more on rail, buses, and bicycles and less on cars—and will recycle materials comprehensively.

We can describe this new economy in some detail. The question is how to get from here to there quickly enough to avoid economic decline and collapse. In our favor, we do have some assets that earlier civilizations did not, including archeological records, more advanced scientific knowledge, and, most important, a sense of how to use economic policy to achieve social goals.

The key to building a global economy that can sustain economic progress is the creation of an honest market, one that tells the ecological truth. The market is an incredible institution, allocating resources with an efficiency that no central planning body can match. It easily balances supply and demand, and it sets prices that readily reflect both scarcity and abundance.

The market does, however, have some fundamental weaknesses. It does not incorporate into prices the indirect costs of providing goods or services into prices, it does not value nature's services properly, and it does not respect the sustainable-yield thresholds of natural systems. It also favors the near term over the long term, showing little concern for future generations.

Throughout most of recorded history, the indirect costs of economic activity were so small that they were rarely an issue and, even then, only at the local level. But with the sevenfold global economic expansion since 1950, the failure to address these market shortcomings and the irrational economic distortions they create could be fatal.[1]

As noted in Chapter 1, accounting systems that do not tell the truth can be costly. Faulty corporate accounting systems that leave costs off the books have driven some of the world's largest corporations into bankruptcy. Unfortunately, our faulty global economic accounting system has potentially far more serious consequences. Our modern economic prosperity is achieved in part by running up ecological deficits, costs that do not show up on the books, but costs that someone will eventually pay.

Once we calculate the indirect costs of a product or service, we can incorporate them into market prices in the form of a tax, offsetting them with income tax reductions. If we can get the market to tell the truth, then we can avoid being blindsided by faulty accounting systems that lead to bankruptcy. As Øystein Dahle, former Vice President of Exxon for Norway and the North Sea, has pointed out: "Socialism collapsed because it did not allow the market to tell the economic truth. Capitalism may collapse because it does not allow the market to tell the ecological truth."[2]

Shifting Taxes

The need for tax shifting—lowering income taxes while raising levies on environmentally destructive activities—in order to get the market to tell the truth has been widely endorsed by economists. For example, a tax on coal that incorporated the increased health care costs associated with breathing polluted air, the costs of damage from acid rain, and the costs of climate disruption would encourage investment in renewable sources of energy such as wind or geothermal. With this concept in hand, it is a short step to tax shifting. A number of countries in Western Europe are already shifting taxes in a process known there as environmental tax reform, to achieve the environmental goals outlined in preceding chapters.[3]

Among the various environmentally damaging activities taxed in Europe are carbon emissions, the generation of garbage (so-called landfill taxes), and the excessive number of cars in cities. A four-year plan adopted in Germany in 1999 systematically shifted taxes from labor to energy. By 2001, this plan had lowered fuel use by 5 percent. It had also accelerated growth in the renewable energy sector, creating some 45,400 jobs by 2003 in the wind industry alone, a number that is projected to rise to 103,000 by 2010.[4]

In 2001, Sweden launched a bold new 10-year environmental tax shift designed to convert 30 billion kroner ($3.9 billion) of taxes on income to taxes on environmentally destructive activities. Much of this shift of $1,100 per household is levied on road transport, including substantial hikes in vehicle and fuel taxes. Electricity is also picking up part of the shift. As of 2005, Sweden is running slightly ahead of its 10-year tax-shifting plan, making it the world leader in environmental tax reform.[5]

Among the other European countries with strong tax reform efforts are Spain, Italy, Norway, the United Kingdom, and France. There are isolated cases elsewhere. A number of countries, including Malaysia, Thailand, and Turkey, have used a tax on lead emissions to eliminate lead as an additive in gasoline. The United States imposed a stiff tax on chlorofluorocarbons to phase them out in accordance with the Montreal Protocol of 1987 and its subsequent updates. At the municipal level, when Victoria, the capital of British Columbia, adopted a trash tax of $1.20 per bag of garbage in 1992, it reduced its daily trash flow 18 percent within one year.[6]

Cities that are being suffocated by cars are using stiff entrance taxes to reduce congestion. First adopted by Singapore some two decades ago, this tax was later introduced by Oslo, Melbourne, and, most recently, London. The London tax of £5, or nearly $9, first enacted in February 2002 by Mayor Ken Livingstone, was raised to £8, more than $14, in July 2005. The resulting revenue will be invested in improving the bus network, which carries 2 million passengers a day. The goal of this congestion tax is a total restructuring of the London transport system to reduce congestion, air pollution, and carbon emissions and to increase mobility.[7]

While London and other cities are taxing cars that enter the central city, others are simply imposing a tax on automobile ownership. In Denmark, the tax on the purchase of a new car is larger than the price of the car itself. A new $25,000 car costs the buyer more than $50,000! In 2000, partial rebates were introduced for energy-efficient vehicles. Other governments are moving in this direction. *New York Times* reporter Howard French writes that Shanghai, which is being suffocated by automobiles, "has raised the fees for car registrations every year since 2000, doubling over that time to about $4,600 per vehicle—more than twice the city's per capita income."[8]

For some products where the costs to society are large and obvious, pressure is mounting to impose taxes. By far the most dramatic example of this was the agreement negotiated between the tobacco industry and all the state governments in the United States. After numerous state governments launched litigation to force tobacco companies to reimburse them for the Medicare costs of treating smoking-related illnesses, the industry decided to negotiate a package reimbursement, agreeing in November 1998 to pay the 50 state governments some $251 billion—nearly $1,000 for every person in the United States. This landmark agreement was, in effect, a retroactive tax on cigarettes smoked in the past, one designed to cover indirect costs. To pay this enormous bill, companies boosted cigarette prices, further discouraging smoking.[9]

A study by the Centers for Disease Control and Prevention (CDC) in the United States calculated the social costs of smoking cigarettes at $7.18 per pack. This not only justifies raising taxes on cigarettes, which claim 4.9 million lives per year worldwide, but it also provides guidelines for how much to raise them. In 2002, a year in which state governments faced fiscal deficits, 21 states in the United States raised cigarette taxes. Perhaps the biggest jump came in New York City, where smokers paid an additional 39¢ in state tax and $1.42 in city tax—a total increase of $1.81 per pack. Since a 10-percent price increase typically reduces smoking by 4 percent, the health benefits of this tax increase should be substantial.[10]

If the cost to society of smoking a pack of cigarettes is $7.18, how much is the cost to society of burning a gallon of gasoline? Fortunately, as noted in Chapter 1, the International Center for Technology Assessment has done a detailed analysis, entitled "The Real Price of Gasoline." The group calculates several indirect costs, including oil industry tax breaks, oil supply protection costs, oil industry subsidies, and health care costs of treating auto exhaust-related respiratory illnesses. The total of these indirect costs centers around $9 per gallon, somewhat higher than the social cost of smoking a pack of cigarettes. Add this external or social cost to the roughly $2 per gallon average price of gasoline in the United States in early 2005, and gas would cost $11 a gallon. These costs are real. Someone bears them. Now that these costs have been calculated, they can be used to set tax rates on gasoline, just as the CDC analysis is being used to raise taxes on cigarettes.[11]

Asia's two leading economies—Japan and China—are now considering the adoption of carbon taxes. For the last few years, many members of the Japanese Diet have wanted to launch an environmental tax shift, but industry has opposed a carbon tax. China, which is experiencing near-record explosive growth in energy use and carbon emissions, is working on an environmental tax restructuring that will discourage fossil fuel use. Wang Fengchun, an official with the National People's Congress, says, "Taxation is the most powerful tool available in a market economy in directing a consumer's buying habits. It is superior to government regulations." If Chinese policymakers can engineer an environmental tax reform, it will be a landmark development not only for China but for the world.[12]

Environmental tax shifting usually brings a double dividend. In reducing taxes on income—in effect, taxes on labor—labor becomes less costly, creating additional jobs while protecting the environment. This was the principal motivation in the German four-year shift of taxes from income to energy. By reducing the air pollution from smokestacks and tailpipes, the incidence of respiratory illnesses, such as asthma and emphysema, is reduced—and thus overall health care costs are as well.[13]

With forests, ecologists can calculate the values of services that trees provide. Once these are determined, they can be incorporated into the price of trees as a stumpage tax of the sort that Bulgaria and Lithuania have adopted. Anyone wishing to cut a tree would have to pay a tax equal to the value of the services provided by that tree, such as flood control. The market for lumber would then be telling the ecological truth. The effect of this is to reduce tree cutting and to encourage wood reuse and paper recycling.[14]

Tax shifting also helps countries gain the lead in producing new equipment, such as new energy technologies or those used for pollution control. For example, the Danish government's tax incentives for wind-generated electricity have enabled Denmark, a country of only 5 million people, to become the world's leading manufacturer of wind turbines.[15]

Some 2,500 economists, including eight Nobel Prize winners in economics, have endorsed the concept of tax shifts. Harvard economics professor N. Gregory Mankiw wrote in *Fortune* magazine: "Cutting income taxes while increasing gasoline taxes would lead to more rapid economic growth, less traffic congestion, safer roads, and reduced risk of global warming—all without jeopardizing long-term fiscal solvency. This may be the closest thing to a free lunch that economics has to offer."[16]

The Economist strongly endorses environmental tax shifting: "On environmental grounds, never mind energy security, America taxes gasoline too lightly. Better than a one-off increase, a politically more feasible idea, and desirable in its own terms, would be a long-term plan to shift taxes from incomes to emissions of carbon." In Europe and the United States, polls indicate that at least 70 percent of voters support environmental tax reform once it is explained to them.[17]

Tradable permits are sometimes a sensible alternative to environmental taxes. Both are economic instruments that can be used to reach environmental goals. The principal difference between them is that with permits, governments set the amount of a given activity that is allowed, such as the harvest from a fishery, and let the market set the price of the permits as they are auctioned off. With environmental taxes, in contrast, the price of the environmentally destructive activity is set by government in the tax rate, and the market determines the amount of the activity that will occur at that price. Both economic instruments can be used to discourage environmentally irresponsible behavior.[18]

The decision of when to use which instrument is not always clearcut. Governments have much more experience with environmental taxes than with tradable permits. It is also clear that such taxes work under a wide range of conditions. Still, permits have been used successfully in widely differing situations, ranging from restricting the catch in an Australian fishery to reducing sulfur emissions in the United States.

For example, concerned about the overfishing of its lobster fishery, the government of Australia estimated the sustainable yield of lobsters and then issued permits totaling that amount. Fishers could then bid for these permits. In effect, the government decided how many lobsters could be taken each year and let the market decide how much the permits were worth. Since the permit trading system was adopted in 1986, the fishery has stabilized and appears to be operating on a sustainable basis.[19]

Perhaps the most ambitious effort to date to use tradable permits was the U.S. effort to reduce sulfur emissions from power plants by half from 1990 to 2000. The goal was reached in 1995, well ahead of schedule and at a minimal cost. One of the weaknesses of tradable permits is that in some communities emissions might not be reduced at all.[20]

Although tradable permits are popular in the business community, permits are administratively more complicated and not as well understood as taxes. Edwin Clark, former senior economist with the White House Council on Environmental Quality, observes that tradable permits "require establishing complex regulatory frameworks, defining the permits, establishing the rules for trades, and preventing people from acting without permits." In contrast to restructuring taxes, something with which there is wide familiarity, tradable permits are a concept not widely understood by the public, making it more difficult to generate broad public support.[21]

Shifting Subsidies

Each year the world's taxpayers provide an estimated $700 billion of subsidies for environmentally destructive activities, such as fossil fuel burning, overpumping aquifers, clearcutting forests, and overfishing. An Earth Council study, *Subsidizing Unsustainable Development*, observes that "there is something unbelievable about the world spending hundreds of billions of dollars annually to subsidize its own destruction."[22]

Iran provides a classic example of extreme subsidies when it prices oil for internal use at one tenth the world price, strongly encouraging car ownership and gas consumption. The World Bank reports that if this $3.6-billion annual subsidy were phased out, it would reduce Iran's carbon emissions by a staggering 49 percent. It would also strengthen the economy by freeing up public revenues for investment in the country's economic development. Iran is not alone. The Bank reports that removing energy subsidies would reduce carbon emissions in Venezuela by 26 percent, in Russia by 17 percent, in India by 14 percent, and in Indonesia by 11 percent.[23]

Some countries are eliminating or reducing these climate-disrupting subsidies. Belgium, France, and Japan have phased out all subsidies for coal. Germany reduced its coal subsidy from $5.4 billion in 1989 to $2.8 billion in 2002, meanwhile lowering its coal use by 46 percent. It plans to phase out this support entirely by 2010. China cut its coal subsidy from $750 million in 1993 to $240 million in 1995. More recently, it has imposed a tax on high-sulfur coals.[24]

A study by the U.K. Green Party, "Aviation's Economic Downside," describes the extent of subsidies currently given to the U.K. airline industry. The giveaway begins with $17 billion in tax breaks, including a total exemption from the federal tax. External or indirect costs that are not paid, such as treating illness from breathing the air polluted by planes, the costs of climate change, and so forth, adds nearly $7 billion to the tab. The subsidy in the United Kingdom totals $391 per resident. This is also an inherently regressive tax policy simply because a substantial share of the U.K. population cannot afford to fly very often if at all, yet they help subsidize this high-cost mode of transportation for their more affluent compatriots.[25]

While some leading industrial countries have been reducing subsidies to fossil fuels—notably coal, the most climate disrupting of all fuels—the United States has been increasing its support for the fossil fuel and nuclear industries. A Green Scissors report from 2002, a study supported by a coalition of environmental groups, calculated that over the past 10 years subsidies for the energy industry totaled $33 billion. Of that, the oil and gas industry got $26 billion, coal $3 billion, and nuclear $4 billion. At a time when there is a need to conserve oil resources, U.S. taxpayers are subsidizing their depletion.[26]

The environmental tax shifting just described reduces taxes on wages and encourages investment in such activities as wind electric generation and recycling, thus simultaneously boosting employment and lessening environmental destruction. Eliminating environmentally destructive subsidies reduces both the burden on taxpayers and the destructive activities themselves.

Subsidies are not inherently bad. Many technologies and industries were born of government subsidies. Jet aircraft developed with military R&D expenditures led to modern commercial airliners. The Internet was the result of publicly funded links among computers in government laboratories and research institutes. And the combination of the federal tax deduction and a robust state tax deduction in California gave birth to the modern wind power industry.[27]

But just as there is a need for tax shifting, there is also a need for subsidy shifting. A world facing the prospect of economically disruptive climate change, for example, can no longer justify subsidies to expand the burning of coal and oil. Shifting these subsidies to the development of climate-benign energy sources such as wind, solar, biomass, and geothermal power is the key to stabilizing the earth's climate. Shifting subsidies from road construction to rail construction could increase mobility in many situations while reducing carbon emissions.

In a troubled world economy facing fiscal deficits at all levels of government, exploiting these tax and subsidy shifts with their double and triple dividends can help balance the books and save the economy's environmental support systems. Tax and subsidy shifting promise both gains in economic efficiency and reductions in environmental destruction, a win-win situation.

Ecolabeling: Voting with Our Wallets

Yet another instrument for environmental restructuring of the economy is ecolabeling. Labeling products that are produced with environmentally sound practices lets consumers vote with their wallets. Ecolabeling is now used to enable consumers to identify energy-efficient household appliances, forest products from sustainably managed forests, fishery products from sustainably managed fisheries, and "green" electricity from renewable sources.

Among these ecolabels are those awarded by the Marine

Stewardship Council (MSC) for seafood. In March 2000, the MSC launched its fisheries certification program when it approved the Western Australia Rock Lobster fishery. Also earning approval that day was the West Thames Herring fishery. In September 2000, the Alaska salmon fishery became the first American fishery to be certified. Among the key players in the seafood processing and retail sectors supporting the MSC initiative were European-based Unilever, Youngs-Bluecrest, and Sainsbury's.[28]

To be certified, a fishery must demonstrate that it is being managed sustainably. Specifically, according to the MSC: "First, the fishery must be conducted in a way that does not take more fish than can be replenished naturally or [that] kills other species through harmful fishing practices. Secondly, the fishery must operate in a manner that ensures the health and diversity of the marine ecosystem on which it depends. Finally, the fishery must respect local, national, and international laws and regulations for responsible and sustainable fishing." By mid-2005 there were over 46 certified fisheries worldwide supplying some 2 million tons of seafood.[29]

The MSC's counterpart for forest products is the Forest Stewardship Council (FSC), which was founded in 1993 by the World Wide Fund for Nature (WWF) and other groups. It provides information on forest management practices within the forest products industry. Some of the world's forests are managed to sustain a steady harvest in perpetuity; others are clearcut, decimated overnight in the quest for quick profits. The FSC issues labels only for products from the former, whether it be lumber sold at a hardware store, furniture in a furniture store, or paper in a stationery store.[30]

Headquartered in Oaxaca, Mexico, the FSC accredits national organizations that verify that forests are being sustainably managed. In addition to this on-the-ground monitoring, the accredited organizations must also be able to trace the raw product through the various stages of processing to the consumer. The FSC sets the standards and provides the FSC label, the stamp of approval, but the actual work is done by national organizations.[31]

The FSC has established nine principles that must be satisfied if forests are to qualify for its label. The central requirement is that the forest be managed in a way that ensures that its yield can be sustained indefinitely. This means careful selective cutting, in effect mimicking nature's management of a forest by removing the more mature, older trees over time.[32]

The FSC label provides consumers with the information they need to support responsible forestry through their purchases of forest products. By identifying timber companies and retailers that are participating in the certification program, socially minded investors also have the information they need for responsible investing.

In March 1996, the first certified wood products were introduced into the United Kingdom. Since then, the certification process has grown worldwide. As of August 2005, some 57 million hectares of forests in 65 countries had been certified under the auspices of the FSC.[33]

To support this certification program, forest and trade networks have been set up in some 35 countries, including Austria, Brazil, Canada, France, Germany, the Nordic countries, Russia, Spain, Switzerland, the United Kingdom, and the United States. These networks are part of the vast support group of companies that adhere to the FSC standards in their marketing. The world's three largest wood buyers—Home Depot, Lowe's, and Ikea—all preferentially buy FSC-certified wood.[34]

In June 2001, the Natural Resources Ministry in Moscow announced that it was introducing national mandatory certification of wood. Although a small portion of its timber harvest was already certified, buyers' discrimination against the rest of the harvest costs Russia $1 billion in export revenues. The ministry estimates that its uncertified wood sells for 20–30 percent less than certified wood from competing countries.[35]

Another commodity that is getting an environmental label is electricity. In the United States, many state utility commissions are requiring utilities to offer consumers a green power option. This is defined as power from renewable sources other than hydroelectric, and it includes wind power, solar cells, solar thermal energy, geothermal energy, and biomass. Utilities simply enclose a return card with the monthly bill, giving consumers the option of checking a box if they would prefer to get green power. The offer specifies the additional cost of the green power, which typically is from 3 to 15 percent.[36]

Utility officials are often surprised by how many consumers sign up for green power. Many are apparently prepared to pay more for their electricity in order to help stabilize the climate for future generations. Local governments, including, for example, those in Santa Monica, Oakland, and Santa Barbara in California, have signed up to use green power exclusively. This includes the power they use for municipal buildings as well as that required to operate various municipal services, such as street lights and traffic signals. Other city and state governments committed to purchasing a portion of their electricity from green sources include Chicago, Portland, New Jersey, and New York.[37]

Many corporations are signing up as well. Johnson & Johnson, Whole Foods Market, and Staples all rank among the top 25 green power purchasers, according to the Environmental Protection Agency's Green Power Partnership. Literally scores of companies in California and Texas are subscribing.[38]

The net effect of these growing numbers of green power proponents is a tidal wave of demand that is forcing many utilities to scramble for an adequate supply of green electricity. One reason wind farms are springing up in so many states is that this is one of the fastest ways to bring new green power online. While green power marketing is now quite advanced in the United States, it is now also well established in Japan, where the rapidly growing purchases of green power threatened to outrun the supply in 2004, forcing utilities to quickly invest in more wind turbines.[39]

Other types of ecolabeling include the efficiency labels put on household appliances that achieve a certain electricity efficiency standard. These have been in effect in many countries since the energy crisis of the late 1970s. There are also green labels provided by environmental or governmental groups at the national level. Among the better-known environmental seal of approval programs are Germany's Blue Angel, Canada's Environmental Choice, and the U.S. Environmental Protection Agency's Energy Star.[40]

A New Materials Economy

In nature, one-way linear flows do not long survive. Nor, by extension, can they long survive in the expanding economy that is a part of the earth's ecosystem. The challenge is to redesign the materials economy so that it is compatible with nature.

The throwaway economy that has been evolving over the last half-century is an aberration, now itself headed for the junk heap of history.

The potential for reducing materials use has been examined over the last decade in three specific studies. The first—*Factor Four*, by Ernst von Weizsäcker, an environmentalist and leader in the German Bundestag—argued that modern industrial economies could function very effectively with a level of virgin raw material use only one fourth that of today. This was followed a few years later by the Factor Ten Institute organized in France under the leadership of Friedrich Schmidt-Bleek. Its research concludes that resource productivity can be raised by a factor of 10, which is well within the reach of existing technology and management given the appropriate policy incentives.[41]

In 2002, American architect William McDonough and German chemist Michael Braungart teamed up to coauthor a book entitled *Cradle to Cradle: Remaking the Way We Make Things*. Waste and pollution are to be avoided at any cost. "Pollution," says McDonough, "is a symbol of design failure."[42]

One of the keys to reducing materials use is recycling steel, the use of which dwarfs that of all other metals combined. Steel use is dominated by the automobile, household appliance, and construction industries. Among steel-based products in the United States, automobiles are the most highly recycled. Cars today are simply too valuable to be left to rust in out-of-the-way junkyards.[43]

The recycling rate for household appliances is estimated at 90 percent. For steel cans, the U.S. recycling rate in 2003 of 60 percent can be traced in part to municipal recycling campaigns launched in the late 1980s.[44]

In the United States, roughly 71 percent of all steel produced in 2003 was from scrap, leaving 29 percent to be produced from virgin ore. Steel recycling started climbing more than a generation ago with the advent of the electric arc furnace, a method of producing steel from scrap that uses only one third the energy of that produced from virgin ore. And since it does not require any mining, it completely eliminates one source of environmental disruption. In the United States, Italy, and Spain, electric arc furnaces used for recycling now account for half or more of all steel production.[45]

It is easier for mature industrial economies with stable populations to get most of their steel from recycled scrap, simply because the amount of steel embedded in the economy is essentially fixed. The number of household appliances, the fleet of automobiles, and the stock of buildings is increasing little or not at all. For countries in the early stages of industrialization, however, the creation of infrastructure—whether factories, bridges, high-rise buildings, or transportation, including automobiles, buses, and rail cars—leaves little steel for recycling.

In the new economy, electric arc steel minimills that efficiently convert scrap steel into finished steel will largely replace iron mines. Advanced industrial economies will come to rely primarily on the stock of materials already in the economy rather than on virgin raw materials. For metals such as steel and aluminum, the losses through use will be minimal. With the appropriate policies, metal can be used and reused indefinitely.

In recent years, the construction industry has begun deconstructing old buildings, breaking them down into their component parts so they can be recycled and reused. For example, when PNC Financial Services in Pittsburgh took down a seven-story downtown building, the principal products were 2,500 tons of concrete, 350 tons of steel, 9 tons of aluminum, and foam ceiling tiles. The concrete was pulverized and used to fill in the site, which is to become a park. The steel and aluminum were recycled. And the ceiling tiles went back to the manufacturer to be recycled. This recycling saved some $200,000 in dump fees. By deconstructing a building instead of simply demolishing it, most of the material in it can be recycled.[46]

Germany and, more recently, Japan are requiring that products such as automobiles, household appliances, and office equipment be designed so that they can be easily disassembled and their component parts recycled. In May 2001, the Japanese Diet enacted a tough appliance recycling law, one that prohibits discarding household appliances, such as washing machines, televisions, or air conditioners. With consumers bearing the cost of disassembling appliances in the form of a disposal fee to recycling firms, which can come to $60 for a refrigerator or $35 for a washing machine, the pressure to design appliances so they can be more easily and cheaply disassembled is strong.[47]

With computers becoming obsolete every few years as technology advances, the need to be able to quickly disassemble and recycle them is a paramount challenge in building an eco-economy.

In addition to measures that encourage the recycling of materials are those that encourage the reuse of products such as beverage containers. Finland, for example, has banned the use of one-way soft drink containers. Canada's Prince Edward Island has adopted a similar ban on all nonrefillable beverage containers. The result in both cases is a sharply reduced flow of garbage to landfills.[48]

A refillable glass bottle used over and over requires about 10 percent as much energy per use as an aluminum can that is recycled. Cleaning, sterilizing, and relabeling a used bottle requires little energy, but recycling cans made from aluminum, which has a melting point of 660 degrees Celsius (1,220 degrees Fahrenheit), is an energy-intensive process. Banning nonrefillables is a win-win-win option—cutting material and energy use, garbage flow, and air and water pollution.[49]

There are also transport fuel savings, since the containers are simply back-hauled to the original bottling plants or breweries. If nonrefillable containers are used, whether glass or aluminum, and they are recycled, then they must be transported to a manufacturing facility where they can be melted down, refashioned into containers, and transported back to the bottling plant or brewery.

Even more fundamental than the design of products is the redesign of manufacturing processes to eliminate the discharge of pollutants entirely. Many of today's manufacturing processes evolved at a time when the economy was much smaller and when the volume of pollutants was not overwhelming the ecosystem. More and more companies are now realizing that this cannot continue and some, such as Dupont, have adopted zero emissions as a goal.[50]

Another way to reduce waste is to systematically cluster factories so that the waste from one process can be used as the raw material for another. NEC, the large Japanese electronics firm, is one of the first multinationals to adopt this approach for its various production facilities. In effect, industrial parks are being designed, both by corporations and governments, specifically to combine factories that have usable waste products. Now in industry, as in nature, one firm's waste becomes another's sustenance.[51]

Government procurement policies can be used to dramatically boost recycling. For example, when the Clinton administration issued an Executive Order in 1993 requiring that all government-purchased paper contain 20 percent or more post-consumer waste by 1995 (increasing to 25 percent by 2000), it created a strong incentive for paper manufacturers to incorporate wastepaper in their manufacturing process. Since the U.S. government is the world's largest paper buyer, this provided a burgeoning market for recycled paper.[52]

New technologies that are less material-dependent also reduce materials use. Cellular phones, which rely on widely dispersed towers or on satellites for signal transmission, now totally dominate telephone use in developing countries, thus sparing them investment in the millions of miles of copper wires that the industrial countries made.[53]

One industry whose value to society is being questioned by the environmental community is the bottled water industry. The World Wide Fund for Nature, an organization with 5.2 million members, released a study in 2001 urging consumers in industrial countries to forgo bottled water, observing that it was no safer or healthier than tap water, even though it can cost 1,000 times as much.[54]

WWF notes that in the United States and Europe there are more standards regulating the quality of tap water than of bottled water. Although clever marketing in industrial countries has convinced many consumers that bottled water is healthier, the WWF study could not find any support for this claim. For those living where water is unsafe, as in some Third World cities, it is far cheaper to boil or filter water than to buy it in bottles.[55]

Phasing out the use of bottled water would eliminate the need for billions of plastic bottles and the fleets of trucks that haul and distribute the water. This in turn would eliminate the traffic congestion, air pollution, and rising carbon dioxide levels from operating the trucks.[56]

A brief review of the environmental effects of gold mining raises doubts about whether the industry is a net benefit to society. In addition to the extensive release of mercury and cyanide into the environment, annual gold production of 2,500 tons requires the processing of 750 million tons of ore—second only to the 2.5 billion tons of ore processed to produce 1 billion tons of raw steel.[57]

Over 80 percent of all the gold mined each year is used to produce jewelry that is often worn as a status symbol, a way of displaying wealth by a tiny affluent minority of the world's people. Birsel Lemke, a widely respected Turkish environmentalist, questions the future of gold mining, wondering whether it is worth turning large areas into what she calls "a lunar landscape." She is not against gold per se, but against the deadly chemicals—cyanide and mercury—that are released in processing the gold ore.[58]

To get an honest market price for gold means imposing a tax on it that would cover the cost of cleaning up the mercury and cyanide pollution from mining plus the costs of landscape restoration in mining regions. Such a tax, which would enable the price of this precious metal to reflect its full cost to society, would likely raise its price severalfold.

Another option for reducing the use of raw materials would be to eliminate subsidies that encourage their use. Nowhere are these greater than in the aluminum industry. For example, a study by the Australia Institute reports that smelters in Australia buy electricity at an astoundingly low subsidized rate of 0.7–1.4¢ per kilowatt-hour, while other industries pay 2.6–3.1¢. Without this huge subsidy, we might not have nonrefillable aluminum beverage containers. This subsidy to aluminum indirectly subsidizes both airlines and automobiles, thus encouraging travel, an energy-intensive activity.[59]

The most pervasive policy initiative to dematerialize the economy is the proposed tax on the burning of fossil fuels, a tax that would reflect the full cost to society of mining coal and pumping oil, of the air pollution associated with their use, and of climate disruption. A carbon tax will lead to a more realistic energy price, one that will permeate the energy-intensive materials economy and reduce materials use.

The challenge in building an eco-economy materials sector is to ensure that the market is sending honest signals. In the words of Ernst von Weizsäcker, "The challenge is to get the market to tell the *ecological* truth." To help the market to tell the truth, we need not only a carbon tax, but also a landfill tax so that those generating garbage pay the full cost of getting rid of it.[60]

New Industries, New Jobs

Describing the eco-economy is obviously speculative, but less so than it might seem simply because its broad outlines are defined by the principles of ecology. The specific trends and shifts described here are not projections of what will happen, though the term "will" is often used for the sake of efficiency. No one knows if these shifts "will" in fact occur, but it will take something similar to this if we are to build an eco-economy.

Building a new economy involves phasing out old industries, restructuring existing ones, and creating new ones. For example, coal use is being phased out, replaced by efficiency gains in many countries, but also by natural gas, as in the United Kingdom, and by wind power, as in Denmark and Germany.[61]

The world automobile industry faces a modest restructuring as it shifts from the gasoline-powered internal combustion engine to the gas-electric hybrid, the diesel-electric hybrid, or the high-efficiency diesel that is so popular in Europe. This will require both a retooling of engine plants and the retraining of automotive engineers and automobile mechanics.

The new economy will also bring major new industries, ones that either do not yet exist or are just beginning. Wind electricity generation is one such industry, incorporating three subsidiary industries: turbine manufacturing, installation, and maintenance. Now in its embryonic stage, this promises to become the foundation of the new energy economy. Millions of turbines soon will be converting wind into cheap electricity, becoming part of the landscape, generating income and jobs in rural communities throughout the world.

As wind power emerges as a mainstream low-cost source of electricity, it will spawn another industry—hydrogen production. Once wind turbines are in wide use, there will be a large, unused capacity during the night when electricity use drops. With this essentially free electricity, turbine owners can turn on the hydrogen generators, converting the wind power into hydrogen. This can then be used to run power plants now fueled with natural gas, as gas becomes too costly or is no longer available. The wind turbine will replace the coal mine, the oil well, and the gas field.

Among the many changes in the world food economy will be the continuing shift to fish farming. Aquaculture, the fastest growing subsector of the world food economy, has expanded at 9 percent a year since 1990. The farming of fish, particularly omnivorous species such as carp, catfish, and tilapia, is likely to continue expanding rapidly simply because these fish convert grain into animal protein so efficiently. With this aquacultural growth comes the need for a rapidly expanding aquafeed industry, one where feeds are formulated by fish nutritionists, much as they are for the poultry industry today.[62]

Bicycle manufacturing and servicing is a growth industry. As recently as 1965, world production of cars and bikes was essentially the same, with each at nearly 20 million, but as of 2003 bike production had climbed to over 100 million per year compared with 42 million cars. This growth in bicycle sales reflects growth in the ranks of those reaching the bicycle level of affluence, principally in Asia. Among industrial countries, the urban transport model being pioneered in the Netherlands and Denmark, where bikes are featured prominently, gives a sense of the bicycle's future role worldwide.[63]

As bicycle use expands, interest in battery-assisted bikes is also growing. Similar to existing bicycles, except for a tiny battery-powered electric motor that can either power the bicycle entirely or assist elderly riders or those living in hilly terrain, its soaring sales are expected to continue climbing.[64]

Yet another growth industry is raising water productivity. Just as the last half-century was devoted to raising land productivity, this half-century will be focused on raising water productivity. Irrigation technologies will become more efficient. The continuous recycling of urban water supplies, already started in some cities, will become common, replacing the "flush and forget" system.

As oil prices rise, teleconferencing gains appeal. To save fuel and time, individuals will be "attending" conferences electronically with both audio and visual connections. One day there will likely be literally thousands of firms organizing electronic conferences.

Other promising growth industries are solar cell manufacturing, light rail construction, and tree planting. For the 1.7 billion people living in developing countries and villages that lack electricity, the mass production of solar cells is the best bet for electrification. As people tire of traffic congestion and pollution, cities throughout the world are restricting car use and turning to light rail to provide mobility. As efforts to reforest the earth gain momentum, and as tree plantations expand, tree planting will emerge as a leading economic activity.[65]

Restructuring the global economy will create not only new industries, but also new jobs—indeed, whole new professions and new specialties within professions. Turning to wind in a big way will require thousands of wind meteorologists to analyze potential wind sites, identifying the best sites for wind farms. The role of wind meteorologists in the new economy will be comparable to that of petroleum geologists in the old economy.

There is a growing demand for environmental architects who can design buildings that are energy- and materials-efficient and that maximize natural heating, cooling, and lighting. In a future of water scarcity, watershed hydrologists will be needed to study the local hydrological cycle, including the movement of underground water, and to determine the sustainable yield of aquifers. They will be at the center of watershed management regimes.

As the world shifts from a throwaway economy, engineers will be needed to design products that can be recycled—from cars to computers. Once products are designed to be disassembled quickly and easily into component parts and materials, comprehensive recycling is relatively easy. These engineers will be responsible for closing the materials loop, converting the linear flow-through (throwaway) economy into a recycling economy.

In countries with a wealth of geothermal energy, it will be up to geothermal geologists to locate the best sites either for power plants or for tapping this underground energy directly to heat buildings. Retraining petroleum geologists to master geothermal technologies is one way of satisfying the likely surge in demand for geothermal geologists.

Another pressing need, particularly in developing countries, is for sanitary engineers who can design sewage systems using waterless, odorless, composting toilets, a trend that is already under way in some water-scarce communities. Yet another growing demand will be for agronomists who specialize in multiple cropping and intercropping. This requires an expertise both in the breeding and selection of crops that can fit together in a tight rotation in various locales and in agricultural practices that facilitate multiple cropping.

Corporations will obviously be challenged by economic restructuring, but so too will universities. Economic restructuring means a demand for new professions such as wind meteorologists, energy architects, and recycling engineers and thus for courses to train tomorrow's professionals.

The Environmental Revolution

Restructuring the global economy according to the principles of ecology represents the greatest investment opportunity in history. In scale, the Environmental Revolution is comparable to the Agricultural and Industrial Revolutions that preceded it.

The Agricultural Revolution involved restructuring the food economy, shifting from a nomadic life-style based on hunting and gathering to a settled life-style based on tilling the soil. Although agriculture started as a supplement to hunting and gathering, it eventually replaced it almost entirely. The Agricultural Revolution eventually cleared one tenth of the earth's land surface of either grass or trees so it could be plowed and planted to crops. Unlike the hunter-gatherer culture that had little effect on the earth, this new farming culture literally transformed the earth's surface.[66]

The Industrial Revolution has been under way for two centuries, although in some countries it is still in its early stages. At its foundation was a shift from wood to fossil fuels, a shift that set the stage for a massive expansion in economic activity. Indeed, its distinguishing feature is the harnessing of vast amounts of solar energy stored beneath the earth's surface as fossil fuels. While the Agricultural Revolution transformed the earth's surface, the Industrial Revolution is transforming the earth's atmosphere.

The additional productivity that the Industrial Revolution made possible unleashed enormous creative energies. It also gave birth to new life-styles and to the most environmentally destructive era in human history, setting the world firmly on a course of eventual economic decline.

The Environmental Revolution resembles the Industrial Revolution in that each is dependent on the shift to a new energy source. And like both earlier revolutions, the Environmental Revolution will affect the entire world.

There are differences in scale, timing, and origin among the three revolutions. Unlike the first two, the Environmental Revolution must be compressed into a matter of decades. The other revolutions were driven by new discoveries, by advances in technology, whereas this revolution, while it will be facilitated by new technologies, is being driven by our need to make peace with nature.

As noted earlier, there has not been an investment situation like this before. The $1.7 trillion that the world spends now each year on oil, the leading source of energy, provides some insight into how much it could spend on energy in the eco-economy. One difference between the investments in fossil fuels and those in wind power, solar cells, and geothermal energy is that the latter are not depletable.[67]

For developing countries dependent on imported oil, the new energy sources promise to free up capital for investment in domestic energy sources. Not many countries have their own oil fields, but all have wind and solar energy waiting to be harnessed. In terms of economic expansion and job generation, these new energy technologies are a godsend. Investments in energy efficiency will grow rapidly simply because they are profitable. In virtually all countries, saved energy is the cheapest source of new energy.

There are also abundant investment opportunities in the food economy. It is likely that the world demand for seafood, for example, will increase at least by half over the next 50 years, and perhaps much more. If so, fish farming output—now 42 million tons a year—will roughly need to double, as will the investments in fish farming. Although aquaculture's growth is likely to slow from the 9 percent a year of the last decade, it nonetheless presents a promising investment opportunity.[68]

A similar situation exists for tree plantations. As of 2000, tree plantations covered some 187 million hectares. An expansion of these by at least half will be needed both to satisfy future demand and to reduce pressures on natural forests.[69]

No sector of the global economy will be untouched by the Environmental Revolution. In this new economy, some companies will be winners and some will be losers. Those who participate in building the new economy will be the winners. Those who cling to the past risk becoming part of it.

13

Plan B: Building a New Future

As we look to the future, two questions loom large. Is civilizational decline under way? And how can we tell? Among the early social signs of possible decline are a widespread drop in life expectancy, growing numbers of hungry people, and a lengthening list of failed and failing states. For the first time in the modern era, life expectancy for a large segment of humanity—the 750 million people living in sub-Saharan Africa—has dropped precipitously, falling from 61 years to 48 years as a result of the HIV/AIDS epidemic.[1]

Over the last half-century, the number of people suffering from hunger was declining, but recently this progress was reversed as the number rose from 826 million in 1998 to 852 million in 2002. With business as usual, the number of hungry will likely continue to rise, reinforcing concerns about food security. And now we have a new wildcard in the food security deck, the fast-growing conversion of foodstuffs, such as wheat, corn, soybeans, and sugarcane, into automotive fuel. As the number of ethanol distilleries and biodiesel refineries multiplies, this threat will expand. Could food supply be the weak link in our modern civilization, as it was for the Sumerians, the Mayans, and the Easter Islanders?[2]

Perhaps the most disturbing recent development is the growing list of failed states. The *Foreign Policy* article discussed in Chapter 6 lists some 60 countries that have failed, are failing, or are at risk of failing. Governments are being overwhelmed by demographic and environmental forces. After decades of rapid population growth, many governments are suffering from demographic fatigue. With leaders unable to cope with ever-growing populations, environmental life-support systems are disintegrating and social services are breaking down.[3]

How many states have to fail before our global civilization fails? Each additional failed state further weakens the capacity of the international community to maintain stability in the monetary system, to control the spread of infectious diseases, and to deal with local famine threats. At some point, as the number of failing states multiplies, global systems begin to fail.

We know that sustaining progress depends on restructuring the global economy, shifting from a fossil-fuel-based, automobile-centered, throwaway economy to one based on renewable energy sources, a diverse transportation system, and a comprehensive reuse/recycle materials system. This can be done largely by restructuring taxes and subsidies. Sustaining progress also means eradicating poverty, stabilizing population, and restoring the earth's natural systems. Securing the additional public outlays needed to reach these goals depends on reordering fiscal priorities in response to the new threats to our security.

In this mobilization, the scarcest resource of all is time. The temptation is to reset the clock, but we cannot. Nature is the timekeeper.

Listening for Wake-up Calls

We are entering a new world. Of that there can be little doubt. What we do not know is whether it will be a world of decline and collapse or a world of environmental restoration and economic progress. Can the world mobilize quickly enough? Where will the wake-up calls come from? What form will they take? Will we hear them?

In the eyes of many, Hurricane Katrina was just such a wake-up call. Until recently, the most costly weather-related events on record were Hurricane Andrew, which struck Florida in 1992, and the flooding in China's Yangtze River basin in 1998, each causing an estimated $30 billion in damage. When Hurricane Katrina hit the U.S. Gulf Coast in late summer 2005, devastating New Orleans, its estimated cost was $200 billion—nearly seven times the previous record. Higher surface water temperatures helped make Katrina one of the most powerful storms ever to make landfall in the United States.[4]

In 1995, an intense heat wave in Chicago claimed more than 700 lives, focusing U.S. attention on climate change, but it was a minor event compared with the record 2003 heat wave in Europe that claimed 49,000 lives. France reported 14,800 deaths; Italy more than 18,000. Unfortunately this tragic loss of life was never adequately reported simply because the death toll numbers dribbled out over several months and at different times for each country. Just as the destruction from Hurricane Katrina was several times the previous record, so too the fatalities from this heat wave broke all previous fatality records by severalfold.[5]

Could a wake-up call take the form of a flood of environmental refugees? As noted earlier, political leaders in sub-Saharan Africa are talking about planting a 5-kilometer-wide and 7,000-kilometer-long belt of trees across the continent in front of the desert in an effort to stop its advance. Whether the African countries can establish a Great Green Wall, and do it quickly enough to halt the desert's advance, remains to be seen. If they fail, we are looking at millions of refugees as productive land turns to desert.[6]

In September 2005, scientists reported that the melting of ice in the Arctic may have reached a "tipping point." We may have unknowingly crossed one of nature's thresholds. According to one article, the team of scientists "believe global warming is melting Arctic ice so rapidly that the region is beginning to absorb more heat from the sun, causing the ice to melt still further and so reinforcing a vicious cycle of melting and heating." If the ice in the Arctic Sea melts and the region's climate continues to warm, the ice sheet covering Greenland, in some places a mile and a half thick, will eventually disappear. It would raise sea level by 23 feet, inundating many of the world's coastal cities and rice-growing river floodplains.[7]

If it becomes clear that we have set in motion a rise in sea level that we cannot arrest or reverse, how will this affect the way we think about ourselves as individuals and as a society? Will we face a social fracturing between generations, between those who caused the rise in sea level and those who must deal with its consequences?

Climate change, whether it is natural or human-induced, is a source of social stress. Jared Diamond notes that drought figured prominently in the collapse and disappearance of the 600-year-old Anasazi civilization in the southwestern United States shortly after 1150. A shrinking food supply led to conflict and cannibalism in this earlier New World civilization. Three centuries later, the Norse settlement in Greenland collapsed and disappeared during a period of extreme cold. For our modern civilization, it is the rise in temperature that is generating social stress in the form of crop-shrinking heat waves, ice melting, rising seas, and more-destructive storms.[8]

Is the record price of oil in late 2005 an aberration or does it reflect something more fundamental—a failure to plan for the depletion of the world's oil reserves? Is it a result of system failure? If so, can the international community pull itself together to stabilize oil prices and avoid both a possible oil-based global economic depression and spreading conflict over access to remaining oil reserves?[9]

Are these wake-up calls? If so, they have not yet awakened us. Have we pushed the snooze button so we can sleep a while longer? Or are these issues just too complicated to comprehend? Are we being overwhelmed by complexity, as Joseph Tainter postulates in his book, *The Collapse of Complex Societies,* that some earlier civilizations were?[10]

This chapter is frustratingly difficult to write because it is not about what we need to do or how to do it, but rather about how to mobilize support to do it. How do we convince ourselves of the gravity and urgency of the situation we face? It is partly a matter of overcoming vested interests and social inertia, and partly a matter of raising public understanding of the threats facing civilization.

Facing many threats simultaneously means setting priorities. Terrorism is one of those threats. No question. But it is not even close to being the top threat facing our early twenty-first century civilization. Population growth, climate change, poverty, spreading water shortages, rising oil prices, and a potential rise in food prices that could lead to unprecedented political instability are the leading threats.

New threats call for new priorities and new responses. Old priorities are hopelessly outmoded. Heavy investments in military power and sophisticated weapons systems, for instance, are of little use in dealing even with terrorism, much less climate change or aquifer depletion. Historically, it was aggressor nations building and concentrating military power that threatened the rest of the world. In contrast, today it is failing states, those that are disintegrating internally, that threaten future progress and stability.

In our new world, we need political leaders who can see the big picture, who understand the relationship between the economy and its environmental support systems. And since the principal advisors to governments are economists, we need economists who can think like ecologists. Unfortunately they are rare. Ray Anderson, founder and chairman of Atlanta-based Interface, a leading world manufacturer of industrial carpet, is especially critical of economics as it is being taught in many universities, noting that "we continue to teach economics students to trust the 'invisible hand' of the market, when the invisible hand is clearly blind to the externalities, and treats massive subsidies, such as a war to protect oil for the oil companies, as if the subsidies were deserved. Can we really trust a *blind* invisible hand to allocate resources rationally?"[11]

Some point out that neo-classical economics does recognize external costs as something to be avoided. True. But do economics instructors tabulate those costs and analyze their effects on the earth's ecosystem and its capacity to sustain the economy? For example, how many economic courses teach that our fossil-fuel-based, automobile-centered, throwaway economy is simply not a viable economic model for the world? And that the biggest challenge the world faces is to build a new economy that will sustain economic progress?

A Wartime Mobilization

As we contemplate mobilizing to rescue a planet under stress and a civilization in trouble, we see both similarities and contrasts with the mobilization for World War II. In this earlier mobilization, there was an economic restructuring, but it was temporary. Mobilizing to save civilization, in contrast, requires a permanent economic restructuring.

The U.S. entry into World War II is a fascinating case study in rapid mobilization. Initially, the United States resisted involvement in the war and responded only after it was directly attacked at Pearl Harbor on December 7, 1941. But respond it did. After an all-out commitment, the U.S. engagement helped turn the tide of war, leading the Allied Forces to victory within three-and-a-half years.[12]

In his State of the Union address on January 6, 1942, one month after the bombing of Pearl Harbor, President Roosevelt announced the country's arms production goals. The United States, he said, was planning to produce 45,000 tanks, 60,000 planes, 20,000 anti-aircraft guns, and 6 million tons of merchant shipping. He added, "Let no man say it cannot be done."[13]

No one had ever seen such huge arms production numbers. But Roosevelt and his colleagues realized that the largest concentration of industrial power in the world at that time was in the U.S. automobile industry. Even during the Depression, the United States was producing 3 million or more cars a year. After his State of the Union address, Roosevelt met with automobile industry leaders and told them that the country would rely heavily on them to reach these arms production goals. Initially they wanted to continue making cars and simply add on the production of armaments. What they did not yet know was that the sale of private automobiles would soon be banned. From the beginning of April 1942 through the end of 1944, nearly three years, there were essentially no cars produced in the United States.[14]

In addition to a ban on the production and sale of cars for private use, residential and highway construction was halted, and driving for pleasure was banned. A rationing program was also introduced. Strategic goods—including tires, gasoline, fuel oil, and sugar—were rationed beginning in 1942. Cutting back on consumption of these goods freed up material resources to support the war effort.[15]

The year 1942 witnessed the greatest expansion of industrial output in the nation's history—all for military use. Wartime aircraft needs were enormous. They included not only fighters, bombers, and reconnaissance planes, but also the troop and cargo transports needed to fight a war on two distant fronts. From the beginning of 1942 through 1944, the United States far exceeded the initial goal of 60,000 planes, turning out 229,600 aircraft, a fleet so vast it is hard even today to visualize it. Equally impressive, by the end of the war more than 5,000 ships were added to the 1,000 or so that made up the American Merchant Fleet in 1939.[16]

In her book *No Ordinary Time,* Doris Kearns Goodwin describes how various firms converted. A sparkplug factory was among the first to switch to the production of machine guns. Soon a manufacturer of stoves was producing lifeboats. A merry-go-round factory was making gun mounts; a toy company was turning out compasses; a corset manufacturer was producing grenade belts; and a pinball machine plant began to make armor-piercing shells.[17]

In retrospect, the speed of this conversion from a peacetime to a wartime economy is stunning. The harnessing of U.S. industrial power tipped the scales decisively toward the Allied Forces, reversing the tide of war. Germany and Japan, already fully extended, could not counter this effort. Winston Churchill often quoted his foreign secretary, Sir Edward Grey: "The United States is like a giant boiler. Once the fire is lighted under it, there is no limit to the power it can generate."[18]

This mobilization of resources within a matter of months demonstrates that a country and, indeed, the world can restructure the economy quickly if it is convinced of the need to do so. Many people—although not yet the majority—are already convinced of the need for a wholesale economic restructuring. The purpose of this book is to convince more people of this need, helping to tip the balance toward the forces of change and hope.

Mobilizing to Save Civilization

Mobilizing to save civilization means restructuring the economy, restoring the economy's natural support systems, eradicating poverty, and stabilizing population. We have the technologies, economic instruments, and financial resources to do this. The United States, the wealthiest society that has ever existed, has the resources to lead this effort. Jeffrey Sachs of Columbia University's Earth Institute sums it up well: "The tragic irony of this moment is that the rich countries are so rich and the poor so poor that a few added tenths of one percent of GNP from the rich ones ramped up over the coming decades could do what was never before possible in human history: ensure that the basic needs of health and education are met for all impoverished children in this world. How many more tragedies will we suffer in this country before we wake up to our capacity to help make the world a safer and more prosperous place not only through military might, but through the gift of life itself?"[19]

It is not possible to put a precise price tag on the changes needed to move our twenty-first century civilization off the overshoot-and-collapse path and onto a path that will sustain economic progress. What we can do, however, is provide some rough estimates of the scale of effort needed.

As discussed in Chapter 7, the additional external funding needed to achieve universal primary education in the more than 80 developing countries that require help, for instance, is conservatively estimated by the World Bank at $12 billion per year. Funding for an adult literacy program based largely on volunteers will take an estimated additional $4 billion annually. Providing for the most basic health care in developing countries is estimated at $33 billion by the World Health Organization. The additional funding needed to provide reproductive health care and family planning services to all women in developing countries is less than $7 billion a year.[20]

Closing the condom gap by providing the additional 9.5 billion condoms needed to control the spread of HIV in the developing world and Eastern Europe requires $2 billion—$285 million for condoms and $1.7 billion for AIDS prevention education and condom distribution. The cost of extending school lunch programs to the 44 poorest countries is $6 billion. An estimated $4 billion per year would cover the cost of assistance to preschool children and pregnant women in these countries. Altogether, the cost of reaching basic social goals comes to $68 billion a year.[21]

As noted in Chapter 8, a poverty eradication effort that is not accompanied by an earth restoration effort is doomed to fail. Protecting topsoil, reforesting the earth, restoring oceanic fisheries, and other needed measures will cost an estimated $93 billion of additional expenditures per year. The most costly activities, protecting biological diversity at $31 billion and conserving soil on cropland at $24 billion, account for over half of the earth restoration annual outlay.

Combining social goals and earth restoration components into a Plan B budget yields an additional annual expenditure of $161 billion, roughly one third of the current U.S. military budget or one sixth of the global military budget. (See Table 13–1.)[22]

Unfortunately, the United States continues to focus on building an ever-stronger military, largely ignoring the threats posed by continuing environmental deterioration, poverty, and popu-

Table 13–1. *Plan B Budget: Additional Annual Expenditures Needed to Meet Social Goals and to Restore the Earth*

Goals	Funding (billion dollars)
Basic Social Goals	
Universal primary education	12
Adult literacy	4
School lunch programs for 44 poorest countries	6
Assistance to preschool children and pregnant women in 44 poorest countries	4
Reproductive health and family planning	7
Universal basic health care	33
Closing the condom gap	2
Total	68
Earth Restoration Goals	
Reforesting the earth	6
Protecting topsoil on cropland	24
Restoring rangelands	9
Stabilizing water tables	10
Restoring fisheries	13
Protecting biological diversity	31
Total	93
Grand Total	161

Source: See endnote 22.

lation growth. Its proposed defense budget for 2006, including $50 billion for the military operations in Iraq and Afghanistan, brings the U.S. projected military expenditure to $492 billion. (See Table 13–2.) Other North Atlantic Treaty Organization members spend $209 billion a year on the military. Russia spends about $65 billion, and China, $56 billion. U.S. military spending is now roughly equal to that of all other countries combined. As the late Eugene Carroll, Jr., a retired admiral, astutely observed, "For forty-five years of the Cold War we were in an arms race with the Soviet Union. Now it appears we are in an arms race with ourselves."[23]

It is decision time. Like earlier civilizations that got into environmental trouble, we can decide to stay with business as usual and watch our modern economy decline and eventually

Table 13–2. *Comparison of Military Budgets by Country and for the World with Plan B Budget*

Country	Budget (billion dollars)
United States	492
Russia	65
China	56
United Kingdom	49
Japan	45
France	40
Germany	30
Saudi Arabia	19
India	19
Italy	18
All other	142
World Military Expenditure	975
Plan B Budget	161

Note: The U.S. number is the budget estimate for FY2006 (including the $50 billion for military operations in Iraq and Afghanistan); Russia and China data are for 2003.
Source: See endnote 23.

collapse, or we can consciously move onto a new path, one that will sustain economic progress. In this situation, no action is actually a decision to stay on the decline-and-collapse path.

It is hard to find the words to convey the gravity of our situation and the momentous nature of the decision we are about to make. How can we convey the urgency of this moment in history? Will tomorrow be too late? Do enough of us care deeply enough to turn the tide now?

Will someone somewhere one day erect a tombstone for our civilization? If so, how will it read? It cannot say we did not understand. We do understand. It cannot say we did not have the resources. We do have the resources. It can only say we were too slow to respond to the forces undermining our civilization. Time ran out.

No one can argue today that we do not have the resources to eradicate poverty, stabilize population, and protect the earth's natural resource base. We can get rid of hunger, illiteracy, disease, and poverty, and we can restore the earth's soils, forests, and fisheries. Shifting one sixth of the world military budget to the Plan B budget would be more than adequate to move the world onto a path that would sustain progress. We can build a global community where the basic needs of all the earth's people are satisfied—a world that will allow us to think of ourselves as civilized.

This economic restructuring depends on tax restructuring, on getting the market to be ecologically honest. The benchmark of political leadership in all countries will be whether or not leaders succeed in restructuring the tax system as, for example, Germany and Sweden have done. This is the key to restructuring the energy economy—both to stabilize climate and to make the transition to the post-petroleum world.[24]

It is easy to spend hundreds of billions in response to terrorist threats, but the reality is that the resources needed to disrupt a modern economy are small, and a U.S. Department of Homeland Security, however heavily funded, provides only minimal protection from suicidal terrorists. The challenge is not to provide a high-tech military response to terrorism, but to build a global society that is environmentally sustainable and equitable—one that restores hope for everyone. Such an effort would more effectively undermine the support for terrorism than any increase in military expenditures, than any new weapons systems, however advanced.

As we look at the environmentally destructive trends that are undermining our future, the world is desperately in need of visible evidence that we can indeed turn things around at the global level. Fortunately, the steps to reverse destructive trends or to initiate constructive new trends are often mutually reinforcing or win-win solutions. For example, efficiency gains that reduce oil use also reduce carbon emissions and air pollution. Steps to eradicate poverty simultaneously help eradicate hunger and stabilize population. Reforestation fixes carbon, increases aquifer recharge, and reduces soil erosion. Once we get enough trends headed in the right direction, they will often reinforce each other.

What the world needs now is a major success story in reducing carbon emissions and dependence on oil to bolster hope in the future. If the United States, for instance, were to decide to replace the existing fleet of inefficient gasoline-burning vehicles with super-efficient gas/electric hybrids over the next 10 years, gasoline use could easily be cut in half. Beyond this, a gas/electric hybrid with an additional storage battery and a plug-in capacity sets the stage for using electricity for short distance driving, such as the daily commute or grocery shopping. Then, as suggested in Chapter 10, if we invest in thousands of wind farms, Americans could do most of their short-distance driving essentially with wind energy, dramatically reducing pressures on the world's oil supplies.[25]

With many U.S. automobile assembly lines idled, it would be a relatively simple matter to retool some of them to produce wind turbines, enabling the country to quickly harness its vast wind energy potential. This would be a rather modest initiative compared with the World War II restructuring, but it would help the world to see that restructuring an economy is entirely doable and that it can be done quickly, profitably, and in a way that enhances national security by reducing dependence on vulnerable oil supplies. Globally, it would help slow the potentially disruptive rise in oil prices. Beyond this, it would reduce carbon emissions, helping to stabilize climate. And, most important, it would restore public confidence in government.

A Call to Greatness

History judges political leaders by whether or not they respond to the great issues of their time. For today's leaders, that issue is how to move the global economy onto an environmentally sound path. We need a national political leader to step forward, an environmental Churchill, to rally the world around this mobilization.

Following the terrorist attacks on the World Trade Center and the Pentagon on September 11, 2001, several world leaders suggested a twenty-first century variation of the Marshall Plan to deal with poverty and its symptoms, arguing that in an increasingly integrated world, abject poverty and great wealth cannot coexist. Gordon Brown, U.K. Chancellor of the Exchequer, notes that, "Like peace, prosperity was indivisible and to be sustained, it had to be shared." Brown sees a Marshall Plan–like initiative not as aid in the traditional sense, but as an investment in the future.[26]

French President Jacques Chirac, a political conservative, told the Earth Summit in Johannesburg in September 2002 that "the world needed an international tax to fight world poverty." He suggested a tax on airplane tickets, carbon emissions, or international currency trading. To illustrate his commitment, Chirac announced that over the next five years France would double its development aid, reaching the internationally agreed upon goal of devoting 0.7 percent of gross domestic product to aid. Going beyond economic issues, he also suggested the creation of a world environment organization to coordinate efforts to build an environmentally sustainable economy.[27]

The urgency of the situation we are now in means that individual countries will simply have to take initiatives on such things as reducing carbon emissions without waiting for a new international agreement to be negotiated. It took the better part of a decade to negotiate the grossly inadequate Kyoto Protocol. We no longer have time for prolonged negotiations.[28]

In 1999, when the German government decided to launch a tax restructuring that would raise taxes on energy use and reduce those on income, a step designed to both reduce carbon emissions and increase employment, its leaders did not insist that the rest of the world or even other European countries agree to do it. They did it because they thought it was the right thing to do for Germany. If countries take strong steps to reverse the trends undermining our future, other countries are certain to follow. At this point in history, the best way to lead is by doing.[29]

Similarly, when Sweden decided on an even more basic environmentally guided restructuring of its tax system, it did not insist that others also do so. It acted on its own and decisively, providing an example for other countries.[30]

In the United States, frustration with Washington's decision to ignore the Kyoto Protocol has led mayors of more than 180 cities to band together to honor the Protocol's goals of cutting carbon emissions 7 percent below the 1990 level over the next decade. In early June 2005, Fred Pearce wrote in the *New Scientist*, "Last month, in the boldest repudiation of a national government yet, a group of American mayors swept aside the Bush administration's refusal to cut carbon emissions." Among the cities were some of the country's largest: Los Angeles, Denver, and New York. Initiatives to achieve the carbon cutting goals are numerous and vary widely among cities. In Salt Lake City, the city authority is buying wind-generated electricity. New York City is converting its municipal motor fleet to gas-electric hybrid vehicles.[31]

A revolt is also under way at the state level. Nine states in the northeastern United States are negotiating a pact to reduce carbon emissions from power plants. State legislatures elsewhere in the country are adopting renewable portfolio standards, which establish a minimal amount of future electricity that must come from renewable energy sources. Among them are California, Colorado, Iowa, Minnesota, New York, Pennsylvania, Texas, and Wisconsin.[32]

Paralleling the need for political leadership is the need for media leadership. Given the urgency of action, and of mobilizing support for these actions, the world faces an unprecedented public education challenge. Turning the tide depends on the communications media rising to the occasion to raise public awareness about the gravity of our situation and the urgency of responding to it. Only the communications media can disseminate information on the scale needed and in the time available. No other institution has this capacity.

This position of the media industry is remarkably similar to that of the U.S. automobile industry in World War II. Like the auto industry some 60 years ago, this is not a responsibility that publishers and editors have asked for or, indeed, that they necessarily want to assume. But there is no alternative. If the communications media worldwide do not take the lead in raising public environmental understanding, the current mobilization will likely fail. We are facing a situation totally different from any faced before, one that requires an entirely new response.

On January 1, 2005, the *New York Times* took a step in this direction when it devoted four fifths of its op-ed page to a piece by Jared Diamond, based on his book *Collapse: How Societies Choose to Fail or Succeed*. Diamond reflected on the lessons we could draw from earlier civilizations that, like ours, had moved onto an economic path that was environmentally unsustainable.[33]

What Diamond learned in researching this book was that moving off the decline-and-collapse path back onto an economic path that is environmentally sustainable is not always easy. Some civilizations are able to read the warning signs and change course quickly. Others fail to do so and collapse.[34]

This research makes it clear that environmental mismanagement, if it continues long enough, leads to civilizational collapse. Diamond's article helped launch a public dialogue about the environmental parallels between our contemporary global civilization and the earlier civilizations discussed in the book.

Nongovernmental environmental groups are also answering the call. By selecting Wangari Maathai for the 2004 Peace Prize, the Nobel Peace Prize committee was recognizing grassroots environmental leadership at its best. Nearly 30 years ago, Maathai founded the Green Belt Movement, an organization that mobilized people at the local level to plant some 30 million trees in Kenya. As Geoffrey Dabelko wrote in *Grist*, the movement mobilized thousands of women, offering them empowerment, education, and even family planning. In 2002, Maathai was elected to Parliament and was shortly thereafter appointed Deputy Minister of Environment by the new government.[35]

Corporate leaders are also getting involved. Ted Turner, founder of CNN, broke new ground for individual philanthropy when he announced in 1997 a gift of $1 billion to the United Nations to support population stabilization, environmental protection, and the provision of health care. He created the UN

Foundation to serve as a vehicle through which the resources could be transferred. Turner could have waited until his death to leave a bequest for the earth, but given the urgency of the situation the world was facing, he argued that billionaires needed to respond now before problems become unmanageable.[36]

Turner undoubtedly influenced Bill Gates, founder of Microsoft, as well as other newly minted billionaires. Channeling his wealth as the world's richest individual into a foundation and allocating it to improve health in developing countries, including initiatives ranging from massive childhood vaccinations to curbing the HIV epidemic, Gates is saving millions of lives.[37]

There is a growing sense among the more thoughtful political leaders that business as usual is no longer a viable option, that unless we respond to the environmental threats to our twenty-first century civilization, we are in trouble. The prospect of failing states is growing as mega-threats such as the HIV epidemic, hydrological poverty, and land hunger threaten to overwhelm countries on the lower rungs of the global economic ladder.

You and Me

One of the questions I am frequently asked when I am speaking in various countries is, Given the environmental problems that the world is facing, can we make it? That is, can we avoid economic decline and civilizational collapse? My answer is always the same: it depends on you and me, on what you and I do to reverse these trends. It means becoming politically active. Saving our civilization is not a spectator sport.

We have moved into this new world so rapidly that we have not yet fully grasped the meaning of what is happening. Traditionally, concern for our children has translated into ensuring their health care and getting them the best education possible. But if we do not act quickly to reverse the deterioration of the earth's environmental systems, eradicate poverty, and stabilize population, their world will be declining economically and disintegrating politically. Today, securing our children's future means not only investing in their education and health care, but also investing in a program to reverse the trends that are undermining their future.

As individuals, we should continue our memberships in environmental and population organizations. We need to improve local recycling programs. We need to vote with our pocketbooks. For example, buying Green Power certificates helps drive investment in renewable energy. We need to do all the things we are now doing to protect the environment. But they are not enough. We have been doing these things for the last 35 years. We have won a lot of local battles, but we are losing the war.

The two overriding challenges are to restructure taxes and reorder fiscal priorities. Saving civilization means restructuring the economy—and at wartime speed. It means restructuring taxes to get the market to tell the ecological truth. And it means reordering fiscal priorities to get the resources needed to restore the earth, eradicate poverty, and stabilize population. Write or e-mail your elected representative about the need for tax restructuring to create an honest market. Remind him or her that corporations that left costs off the books appeared to prosper in the short run, only to collapse in the longer run.

Or better yet, meet with your elected representatives to discuss why we need to raise environmental taxes and reduce income taxes. Work with like-minded friends and associates toward this goal. Put together a delegation to meet with your elected representative. Feel free to download the information on tax restructuring in the preceding chapter from our Web site to use in these efforts. If we cannot restructure the tax system to enable the market to tell the truth, we almost certainly will not make it.

Let your political representatives know that a world spending nearly $1 trillion a year for military purposes is simply out of sync with reality in a situation where the future of civilization is in question. Ask them if $161 billion per year is an unreasonable expenditure to save civilization. Ask them if diverting one sixth of the global military budget to saving civilization is too costly.

If you like to write, try your hand at an op-ed piece for your local newspaper on the need to raise taxes on environmentally destructive activities and offset this with a lowering of income taxes. Try a letter to the editor. Organize a letter writing campaign, urging people to contact their elected representatives and local media outlets on this issue.

Push for the inclusion of poverty eradication, family planning, and reforestation in international assistance programs. Lobby for an increase in these appropriations and a cut in military appropriations, pointing out that advanced weapons systems are useless in dealing with the new threats to our civilization. Someone needs to speak on behalf of our children and grandchildren because it is their world and their futures that are at stake.

Educate yourself on environmental issues and on what happened to earlier civilizations that also found themselves in environmental trouble—and help your friends to become better informed. On this subject I recommend *Collapse* by Jared Diamond and *A Short History of Progress* by Ronald Wright. To understand the case for eradicating poverty, read "Can Extreme Poverty Be Eliminated?" by Jeffrey Sachs in the September 2005 issue of *Scientific American*. To gain a sense of the enormous potential for boosting energy efficiency, read "More Profit with Less Carbon" by Amory Lovins in the same issue.[38]

Remember, challenging though the situation may be, there are signs of the new economy emerging all over the world. We see them in the wind farms of Europe, the fast-growing U.S. fleet of gas-electric hybrid cars, the reforested hills of South Korea, the family planning program of Iran, the massive eradication of poverty in China, and the solar rooftops of Japan.

What we need to do is doable. Sit down and map out your own personal plan and timetable for what you want to do to move the world from a path headed toward economic decline to one of sustained economic progress. Sketch out a plan for the next year of the things you want to do, how you hope to do them, and whom you can work with to achieve the only goal that really counts—the preservation of civilization. What could be more rewarding?

The choice is ours—yours and mine. We can stay with business as usual and preside over an economy that continues to destroy its natural support systems until it destroys itself, or we can adopt Plan B and be the generation that changes direction, moving the world onto a path of sustained progress. The choice will be made by our generation, but it will affect life on earth for all generations to come.

Additional Resources

More information on the topics covered in Plan B 2.0 *can be found in the references listed here. Additional data and an expanded list of resources are available on the Earth Policy Institute Web site at www.earthpolicy.org/books/PB2/resources.htm.*

Chapter 1

Diamond, Jared, *Collapse: How Societies Choose to Fail or Succeed* (New York: Penguin Group, 2005).

Tainter, Joseph, *The Collapse of Complex Societies* (Cambridge, U.K.: Cambridge University Press, 1988).

United Nations Environment Programme (UNEP), www.unep.org.

United Nations Statistics Division, unstats.un.org/unsd.

Chapter 2

Association for the Study of Peak Oil and Gas, www.peakoil.net.

BP, *Putting Energy in the Spotlight: BP Statistical Review of World Energy 2005* (London: June 2005), www.bp.com/genericsection.do?categoryId=92&contentId=7005893.

Deffeyes, Kenneth S., *Beyond Oil: The View from Hubbert's Peak* (New York: Hill and Wang, 2005).

Ethanol Marketplace, www.ethanolmarketplace.com.

Heinberg, Richard, *Power Down: Options and Actions for a Post-Carbon World* (Gabriola Island, BC, Canada: New Society Publishers, 2004).

Renewable Fuels Association, *Homegrown for the Homeland: Ethanol Industry Outlook 2005* (Washington, DC: 2005).

Chapter 3

Gleick, Peter H., *The World's Water: The Biennial Report on Freshwater Resources* (Washington, DC: Island Press, various years), www.worldwater.org.

International Water Management Institute, www.iwmi.cgiar.org.

LakeNet: World Lakes Network, www.worldlakes.org.

Postel, Sandra, *Pillar of Sand: Can the Irrigation Miracle Last?* (New York: W.W. Norton & Company, 1999).

World Bank, *China: Agenda for Water Sector Strategy for North China* (Washington, DC: April 2001), lnweb18.worldbank.org/eap/eap.nsf/Attachments/WaterSectorReport.

Chapter 4

Arctic Climate Impact Assessment, *Impacts of a Warming Arctic* (Cambridge, U.K.: Cambridge University Press, 2004), www.acia.uaf.edu.

Hadley Centre for Climate Prediction and Research, www.met-office.gov.uk/research/hadleycentre/index.html.

Intergovernmental Panel on Climate Change, *Climate Change 2001. Contributions of Working Groups I, II, and III to the Third Assessment Report of the Intergovernmental Panel on Climate Change* (New York: Cambridge University Press, 2001), www.ipcc.ch.

Munich Re, *Topics Annual Review: Natural Catastrophes 2004* (Munich, Germany: 2004), www.munichre.com.

NASA, Goddard Institute for Space Studies, Surface Temperature Analysis, data.giss.nasa.gov/gistemp.

National Snow and Ice Data Center, www.nsidc.org.

Chapter 5

BirdLife International, www.birdlife.org.

Global Forest Watch, www.globalforestwatch.org.

Millennium Ecosystem Assessment, *Ecosystems and Human Well-being: Synthesis* and four technical volumes (Washington, DC: Island Press, 2005), www.millenniumassessment.org.

Species Survival Commission, *IUCN Red List of Threatened Species* (Gland, Switzerland, and Cambridge, U.K.: World Conservation Union–IUCN, various years), www.redlist.org.

U.N. Food and Agriculture Organization, *The State of World Fisheries and Aquaculture* (Rome: various years), www.fao.org/fi.

Chapter 6

Clear the Air, www.cta.policy.net.

Colborn, Theo, Dianne Dumanoski, and John Peterson Myers, *Our*

Stolen Future: Are We Threatening Our Fertility, Intelligence and Survival: A Scientific Detective Story (New York: Dutton Publishing, 1996), www.ourstolenfuture.org.

Environmental Change and Security Program at the Woodrow Wilson International Center for Scholars, www.wilsoncenter.org/index.cfm?fuseaction=topics.home&topic_id=1413.

Fund for Peace and the Carnegie Endowment for International Peace, "The Failed States Index," *Foreign Policy*, July/August 2005, www.foreignpolicy.com.

Joint United Nations Programme on HIV/AIDS (UNAIDS), *Report on the Global HIV/AIDS Epidemic* (Geneva: various years), www.unaids.org.

Sperling, Gene B., "Toward Universal Education," *Foreign Affairs*, September/October 2001, www.foreignaffairs.org.

United Nations, *World Population Prospects: The 2004 Revision* (New York: February 2005), www.un.org/esa/population/unpop.htm.

United Nations Population Database, esa.un.org/unpp.

Chapter 7

Chaya, Nada, and Sarah Haddock, *Condoms Count: Meeting the Need in the Era of HIV/AIDS, 2004 with Data Update* (Washington, DC: Population Action International, 2004), www.populationaction.org.

Global Fund to Fight AIDS, Tuberculosis, and Malaria, www.theglobalfund.org.

Global Polio Eradication Initiative, www.polioeradication.org.

Population Reference Bureau, www.prb.org.

Sachs, Jeffrey D., and the Commission on Macroeconomics and Health, *Macroeconomics and Health: Investing in Health for Economic Development* (Geneva: World Health Organization, 2001), www.paho.org/English/DPM/SHD/HP/Sachs.pdf.

United Nations, *The Millennium Development Goals Report 2005* (New York: 2005), unstats.un.org/unsd/mi/pdf/MDG%20Book.pdf.

U.N. Millennium Development Goals, www.un.org/millenniumgoals.

U.N. Population Fund (UNFPA), *The State of World Population* (New York: various years), www.unfpa.org.

Chapter 8

Balmford, Andrew, et al., "The Worldwide Costs of Marine Protected Areas," *Proceedings of the National Academy of Sciences* (vol. 101, no. 26) 29 June 2004, pp. 9,694–97, www.pnas.org.

Conservation International, Biodiversity Hotspots, www.biodiversity-hotspots.org.

Dregne, H.E., and Nan-Ting Chou, "Global Desertification Dimensions and Costs," in *Degradation and Restoration of Arid Lands* (Lubbock, TX: Texas Tech. University, 1992).

Postel, Sandra, and Brian Richter, *Rivers for Life: Managing Water for People and Nature* (Washington, DC: Island Press, 2003).

U.N. Convention to Combat Desertification, www.unccd.int.

UNEP, *Status of Desertification and Implementation of the United Nations Plan of Action to Combat Desertification* (Nairobi: 1991), www.unep.org.

U.N. Food and Agriculture Organization, Global Forest Resources Assessment, www.fao.org/forestry/fo/fra.

Chapter 9

International Food Policy Research Institute, www.ifpri.org.

International Rice Research Institute, www.irri.org.

U.N. Food and Agriculture Organization, *FISHSTAT Plus*, electronic database, www.fao.org/fi/statist/FISOFT/FISHPLUS.asp.

U.N. Food and Agriculture Organization, *The State of Food Insecurity in the World* (Rome: various years), www.fao.org/sof/sofi.

U.S. Department of Agriculture, Economic Research Service, Natural Resources and Environment Division, *Agricultural Resources and Environmental Indicators* (Washington, DC: various years), www.ers.usda.gov/publications/arei.

U.S. Department of Agriculture, *Production, Supply, & Distribution*, electronic database, Washington, DC, www.fas.usda.gov/psd.

Chapter 10

Alliance to Save Energy, www.ase.org.

American Solar Energy Society, www.ases.org.

American Wind Energy Association, www.awea.org.

Bailie, Alison, et al., *The Path to Carbon-Dioxide-Free Power: Switching to Clean Energy in the Utility Sector*, A Study for the World Wildlife Fund (Washington, DC: Tellus Institute and The Center for Energy and Climate Solutions, April 2003), www.worldwildlife.org/climate/publications/power_switch.pdf.

European Wind Energy Association, www.ewea.org.

Global Wind Energy Council, *Wind Force 12: A Blueprint to Achieve 12% of the World's Electricity from Wind Power by 2020* (Belgium: European Wind Energy Association and Greenpeace, 2005), www.gwec.net/fileadmin/documents/Publications/wf12-2005.pdf.

Lovins, Amory B., et al., *Winning the Oil Endgame: Innovation for Profits, Jobs, and Security* (Snowmass, CO: Rocky Mountain Institute, 2004), www.oilendgame.com.

Maycock, Paul, *Photovoltaic News*, www.pvenergy.com/news.html.

Chapter 11

Crawford, J. H., *Carfree Cities* (Utrecht, The Netherlands: International Books, July 2000), www.carfree.com.

Institute for Transportation and Development Policy, www.itdp.org.

O'Meara, Molly, *Reinventing Cities for People and the Planet*, Worldwatch Paper 147 (Washington, DC: Worldwatch Institute, June 1999), www.worldwatch.org.

Schrank, David, and Tim Lomax, *2005 Urban Mobility Study* (College Station, TX: Texas Transportation Institute, May 2005), mobility.tamu.edu/ums/report.

Surface Transportation Policy Project, www.transact.org.

Chapter 12

Forest Stewardship Council, www.fsc.org.

International Center for Technology Assessment, *The Real Price of Gasoline*, Report No. 3 (Washington, DC: November 1998), www.icta.org.

Marine Stewardship Council, www.msc.org.

McDonough, William, and Michael Braungart, *Cradle to Cradle: Remaking the Way We Make Things* (New York: North Point Press, 2002), www.mcdonough.com/cradle_to_cradle.htm.

Pica, Erich, ed., *Running On Empty: How Environmentally Harmful Energy Subsidies Siphon Billions from Taxpayers*, A Green Scissors Report (Washington, DC: Friends of the Earth, 2002), www.green scissors.org.

Redefining Progress: Accurate Prices Program, www.rprogress.org/programs/accurateprices.

U.S. Energy Star Program, www.energystar.gov.

Chapter 13

Anderson, Ray, "A Call for Systemic Change," speech delivered at the National Conference on Science, Policy, & the Environment: Education for a Secure and Sustainable Future, Washington, DC, 31 January 2003, www.ncseonline.org/NCSEconference/2003conference/page.cfm?FID=2504.

Center for Arms Control and Non-Proliferation, www.armscontrolcenter.org.

Diamond, Jared, "The Ends of the World as We Know Them," *New York Times*, 1 January 2005.

Sachs, Jeffrey, *The End of Poverty: Economic Possibilities for Our Time* (New York: Penguin Press, 2005).

Walton, Francis, *Miracle of World War II: How American Industry Made Victory Possible* (Macmillan: New York, 1956).

Wright, Ronald, *A Short History of Progress* (New York: Carroll and Graf Publishers, 2005).

Notes

Chapter 1. Entering A New World

1. Jared Diamond, *Collapse: How Societies Choose to Fail or Succeed* (New York: Penguin Group, 2005).
2. Mathis Wackernagel et al., "Tracking the Ecological Overshoot of the Human Economy," *Proceedings of the National Academy of Sciences,* vol. 99, no. 14 (9 July 2002), pp. 9,266–71.
3. Paul B. MacCready, AeroVironment Inc., letter to author, 19 April 2005.
4. Ned Rozell and Dan Chay, "St. Matthew Island: Overshoot & Collapse," *Energy Bulletin*, 23 November 2003.
5. Diamond, op. cit. note 1, pp. 90, 248–76; "Población Total, Por Sexo E Indice de Masculinidad, Según División Político Administrativa y Area Urbana-Rural," table in Chile Instituto Nacional de Estadísticas, *Resultados Generales Censo 2002* (Santiago, Chile: 2003).
6. United Nations, *World Population Prospects: The 2004 Revision* (New York: 2005); Population Reference Bureau, *2005 World Population Data Sheet*, wall chart (Washington, DC: August 2005); Population Reference Bureau, *2004 World Population Data Sheet*, wall chart (Washington, DC: August 2004).
7. United Nations, op. cit. note 6.
8. See Chapter 2 for further discussion of peak oil.
9. Car fleet includes passenger cars and commercial vehicles, many of which are light trucks and sport utility vehicles used for personal use, from Ward's Communications, *Ward's World Motor Vehicle Data 2004* (Southfield, MI: 2004), p. 238; population living on less than $1 a day in World Bank, *World Development Report 2005* (New York: Oxford University Press, 2004).
10. Diamond, op. cit. note 1, pp. 90, 248–76.

11. The New Road Map Foundation, "All-Consuming Passion: Waking up from the American Dream," factsheet, *EcoFuture*, updated 17 January 2002.

12. U.S. Department of Agriculture (USDA), *Production, Supply, & Distribution*, electronic database, at www.fas.usda.gov/psd, updated 13 September 2005.

13. U.N. Food and Agriculture Organization (FAO), *FAOSTAT Statistics Database*, at apps.fao.org, updated 14 July 2005.

14. U.S. Department of Energy (DOE), Energy Information Administration (EIA), "World Oil Demand," *International Petroleum Monthly*, December 2004.

15. British Petroleum (BP), *Statistical Review of World Energy 2005* (London: Group Media & Publishing, 2005).

16. International Iron and Steel Institute, *Steel Statistical Yearbook 2004* (Brussels, 2004); data for 1990–93 from Phil Hunt, Iron and Steel Statistics Bureau, e-mail to Viviana Jiménez, Earth Policy Institute, 24 January 2005.

17. *UNStats Statistics Database*, at unstats.un.org/unsd, viewed 14 February 2005; International Telecommunication Union (ITU), *Telecommunication Statistics* at www.itu.int/ITU-D/ict/statistics/at_glance/cellular03.pdf, 15 March 2005; ITU, *Telecommunication Statistics* at www.itu.int/ITU-D/ict/statistics/at_glance/internet03.pdf, 15 March 2005; Ward's Communications, op. cit. note 9.

18. Chinese economic growth from International Monetary Fund (IMF), *World Economic Outlook Database*, at www.imf.org/external/pubs/ft/weo, updated April 2005; population from United Nations, op. cit. note 6.

19. Grain from USDA, op. cit. note 12; paper includes coated papers, household and sanitary paper, newsprint, other papers, packaging, printing and writing paper, and wrapping papers, based on data from FAO, op. cit. note 13; oil from BP, op. cit. note 15; all per capita calculations based on population from United Nations, op. cit. note 6.

20. Ward's Communications, op. cit. note 9.

21. United Nations, op. cit. note 6.

22. Diamond, op. cit. note 1; Garrett Hardin, "The Tragedy of the Commons," *Science*, vol. 162 (13 December 1968).

23. Sandra Postel, *Pillar of Sand* (New York: W.W. Norton & Company, 1999), pp. 13–21.

24. Ibid.

25. Ibid.

26. Robert McC. Adams quoted in Joseph Tainter, *The Collapse of Complex Societies* (Cambridge, U.K.: Cambridge University Press, 1988), p. 1.

27. "Maya," *Encyclopaedia Britannica*, online encyclopedia, viewed 7 August 2000.

28. Ibid.

29. Jared Diamond, "Easter's End," *Discover*, August 1995, pp. 63–69.

30. Ibid.

31. United Nations, op. cit. note 6.

32. USDA, op. cit. note 12.

33. United Nations, op. cit. note 6; U.S. Census Bureau, *Foreign Trade Statistics*, "Trade: Imports, Exports and Trade Balance with China," at www.census.gov/foreign-trade/balance/c5700.html, updated June 2005; Peter Goodman, "China Tells Congress to Back Off Business," *Washington Post*, 5 July 2005.

34. Munich Re, *Topics Annual Review: Natural Catastrophes 2001* (Munich, Germany: 2002), pp. 16–17; value of China's wheat and rice harvests from USDA, op. cit. note 12, using prices from IMF, *International Financial Statistics*, electronic database, at ifs.apdi.net/imf.

35. "Forestry Cuts Down on Logging," *China Daily*, 26 May 1998; Erik Eckholm, "Chinese Leaders Vow to Mend Ecological Ways," *New York Times*, 30 August 1998; Erik Eckholm, "China Admits Ecological Sins Played Role in Flood Disaster," *New York Times*, 26 August 1998; Erik Eckholm, "Stunned by Floods, China Hastens Logging Curbs," *New York Times*, 27 February 1998.

36. Gasoline prices from DOE, EIA, *This Week in Petroleum* (Washington, DC: various issues).

37. Andrew Kimbrell et al., *The Real Price of Gasoline* (Washington, DC: International Center for Technology Assessment, 1998), p. 39.

38. James Brooke, "Japan Squeezes to Get the Most of Costly Fuel," *New York Times*, 4 June 2005; DOE and U.S. Environmental Protection Agency, *Fuel Economy Guide* (Washington, DC: 2005); Marv Balousek, "Hybrid Cars Are Catching On," *Wisconsin State Journal*, 10 August 2005.

39. Danish Wind Industry Association, "Did You Know?" fact sheet, at www.windpower.org; Colin Woodard, "Fair Winds in Denmark," *E: The Environmental Magazine*, July 2001; Marla Dickerson, "Homegrown Fuel Supply Helps Brazil Breathe Easy," *Los Angeles Times*, 15 June 2005.

40. USDA, op. cit. note 12, updated 7 September 2005; FAO, op. cit. note 13, updated 17 January 2005.

41. FAO, *FISHSTAT Plus,* electronic database, at www.fao.org/fi/statist/FISOFT/FISHPLUS.asp, updated March 2005.

42. Se-Kyung Chong, "Anmyeon-do Recreation Forest: A Millennium of Management," in Patrick B. Durst et al., *In Search of Excellence: Exemplary Forest Management in Asia and the Pacific*, Asia-Pacific

Forestry Commission (Bangkok: FAO Regional Office for Asia and the Pacific, 2005), pp. 251–59.

43. Mark Smith, "Land Retirement," in USDA, *Agricultural Resources and Environmental Indicators 2003* (Washington, DC: 2003), section 6.2 updated in December 2000, p. 14; USDA, Economic Research Service, *Agri-Environmental Policy at the Crossroads: Guideposts on a Changing Landscape*, Agricultural Economic Report No. 794 (Washington, DC: January 2001).

44. Molly O'Meara Sheehan, *City Limits: Putting the Breaks on Sprawl*, Worldwatch Paper 156 (Washington, DC: Worldwatch Institute, June 2001), p. 11.

45. Lester R. Brown, "The Short Path to Oil Independence: Gas-Electric Hybrids and Wind Power Offer Winning Combination," *Eco-Economy Update* (Washington, DC: Earth Policy Institute), 13 October 2004; Senator Joseph Lieberman, remarks prepared for the Loewy Lecture, Georgetown University (Washington, DC: 7 October 2005).

Chapter 2. Beyond the Oil Peak

1. U.S. Department of Energy (DOE), Energy Information Administration (EIA), "Select Crude Oil Spot Prices," at www.eia.doe.gov/emeu/international/crude1.html, updated 28 July 2005; John Vidal, "The End of Oil Is Closer Than You Think," *Guardian* (London), 21 April 2005; Alfred J. Cavallo, "Oil: Caveat Empty," *Bulletin of the Atomic Scientists*, vol. 61, no. 3 (May/June 2005), pp. 16–18.

2. Vidal, op. cit. note 1; Jeffrey Ball, "Dire Prophesy: As Prices Soar, Doomsayers Provoke Debate on Oil's Future—In a 1970s Echo, Dr. Campbell Warns Supply Is Drying Up, But Industry Isn't Worried—Charges of 'Malthusian Bias,'" *Wall Street Journal*, 21 September 2004.

3. DOE, EIA, "Table 11.5 World Crude Oil Production, 1960–2004," *International Energy Outlook 2004* (Washington, DC: 2004), at www.eia.doe.gov/emeu/aer/txt/ptb1105.html; Vidal, op. cit. note 1; International Energy Agency (IEA), *IEA Data Services*, at data.iea.org, updated August 2004.

4. Neil Chatterjee, "'Peak Oil' Gathering Sees $100 Crude This Decade," *Reuters*, 26 April 2005; Javier Blas and Isabel Gorst, "Oil Production in Russia Stagnates," *Financial Times*, 2 June 2005; Justin Blum, "Alaska Oil Field's Falling Production Reflects U.S. Trend," *Washington Post*, 7 June 2005; DOE, EIA, "Table 2.2 World Crude Oil Production, 1980–2003," *International Energy Annual 2003* (Washington, DC: 2005); Heather Timmons, "Shell Makes Another Cut in Reserves; Profit Jumps," *New York Times*, 4 February 2005; DOE, EIA, "Kazakhstan," *EIA Country Analysis Briefs*, (Washington, DC: updated July 2005); IEA, op. cit. note 3.

5. DOE, EIA, "Saudi Arabia," *EIA Country Analysis Briefs* (Washington, DC: updated January 2005); Chatterjee, op. cit. note 4; Adam Porter,

"Expert Says Saudi Oil May Have Peaked," *Al Jazeera*, 20 February 2005.

6. DOE, EIA, op. cit. note 3; IEA, op. cit. note 3.
7. Michael T. Klare, "The Energy Crunch to Come," *TomDispatch*, 22 March 2005; Jad Mouawad, "Big Oil's Burden of Too Much Cash," *New York Times*, 12 February 2005; Timmons, op. cit. note 4.
8. Mouawad, op. cit. note 7; Mark Williams, "The End of Oil?" *Technology Review*, February 2005; Vidal, op. cit. note 1.
9. Klare, op. cit. note 7; Timmons, op. cit. note 4; Walter Youngquist, letter to author, 29 April 2005.
10. James Picerno, "If We Really Have the Oil," *Bloomberg Wealth Manager*, September 2002, p. 45; Klare, op. cit. note 7; Kenneth S. Deffeyes, *Beyond Oil: The View from Hubbert's Peak* (New York: Hill and Wang, 2005); Richard C. Duncan and Walter Youngquist, "Encircling the Peak of World Oil Production," *Natural Resources Research*, vol. 12, no. 4 (December 2003), p. 222; A. M. Samsan Bakhtiari, "World Oil Production Capacity Model Suggests Output Peak by 2006–07," *Oil & Gas Journal*, 26 April 2004, pp. 18–20.
11. Peter Maass, "The Breaking Point," *New York Times Magazine*, 21 August 2005.
12. Robert Collier, "Canadian Oil Sands: Vast Reserves Second to Saudi Arabia Will Keep America Moving, But at a Steep Environmental Cost," *San Francisco Chronicle*, 22 May 2005; Vidal, op. cit. note 1; Walter Youngquist, "Survey of Energy Resources: Oil Shale," *Energy Bulletin*, 24 April 2005; William Brown, DOE, EIA, discussion with author, 9 August 2005.
13. "US: Caution Warranted on Oil Shale" (editorial), *Denver Post*, 18 April 2005; Gargi Chakrabarty, "Shale's New Hope," *Rocky Mountain News*, 18 October 2004; Walter Youngquist, "Alternative Energy Sources," in Lee C. Gerhard, Patrick Leahy, and Victor Yannacone, eds., *Sustainability of Energy and Water through the 21st Century*, Proceedings of the Arbor Day Farm Conference, 8–11 October 2000 (Lawrence, KS: Kansas Geological Survey, 2002), p. 65; Cavallo, op. cit. note 1.
14. DOE, EIA, "United States," *EIA Country Analysis Briefs* (Washington, DC: updated January 2005); Collier, op. cit. note 12; Thomas J. Quinn, "Turning Tar Sands into Oil," *Cleveland Plain Dealer*, 17 July 2005; "Exxon Says N. America Gas Production Has Peaked," *Reuters*, 21 June 2005.
15. Judith Crosson, "Oil Prices Prompt Another Look At Shale," *MSNBC*, 23 November 2004; Youngquist, op. cit. note 12; Youngquist, op. cit. note 13, p. 64; Vidal, op. cit. note 1.
16. Danielle Murray, "Oil and Food: A Rising Security Challenge," *Eco-Economy Update* (Washington, DC: Earth Policy Institute, 9 May 2005), p. 2 and data charts; irrigation data sources include U.S.

Department of Agriculture (USDA), "Chapter 5: Energy Use in Agriculture," *U.S. Agriculture and Forestry Greenhouse Gas Inventory: 1990–2001,* Technical Bulletin No. 1907 (Washington, DC: Global Change Program Office, Office of the Chief Economist, 2004), p. 94.

17. James Duffield, USDA, e-mail to Danielle Murray, Earth Policy Institute, 31 March 2005; USDA, *Production, Supply & Distribution,* electronic database, at www.fas.usda.gov/psd, updated 13 September 2005.

18. Conservation Technology Information Center (CTIC), "Conservation Tillage and Other Tillage Types in the United States—1990–2004," *2004 National Crop Residue Management Survey* (West Lafayette, IN: Purdue University, 2004); CTIC, "Top Ten Benefits of Conservation Tillage," at www.ctic.purdue.edu/Core4/CT/CT Survey/10Benefits.html, viewed 27 July 2005; Rolf Derpsch, "Extent of No-Tillage Adoption Worldwide," to be presented at the III World Congress on Conservation Agriculture, Nairobi, Kenya, 3–7 October 2005, e-mail to Danielle Murray, Earth Policy Institute, 9 August 2005.

19. Duffield, op. cit. note 17; tractor use and horse stocks from U.N. Food and Agriculture Organization (FAO), *FAOSTAT Statistics Database*, at apps.fao.org, updated 4 April 2005.

20. Fertilizer energy use data from Duffield, op. cit. note 17; DOE, EIA, *Annual Energy Outlook 2003* (Washington, DC: 2004); John Miranowski, "Energy Demand and Capacity to Adjust in U.S. Agricultural Production," presentation at Agricultural Outlook Forum 2005, Arlington, VA, 24 February 2005; fertilizer-to-grain ratios from USDA, op. cit. note 17; Patrick Heffer, *Short Term Prospects for World Agriculture and Fertilizer Demand 2003/04–2004/05* (Paris: International Fertilizer Industry Association (IFA), 2005); IFA Secretariat and IFA Fertilizer Demand Working Group, *Fertilizer Consumption Report* (Brussels: 2001).

21. U.S. grain production data from USDA, op. cit. note 17.

22. Brian Halweil, *Eat Here* (New York: W.W. Norton & Company, 2004), p. 29; USDA, op. cit. note 17.

23. Compiled by Earth Policy Institute from Duffield, op. cit. note 17; DOE, EIA, op. cit. note 20; USDA, National Agricultural Statistics Service, "Table 20: Energy Expenses for On-Farm Pumping of Irrigation Water by Water Source and Type of Energy: 2003 and 1998," *2003 Farm & Ranch Irrigation Survey, Census of Agriculture* (Washington, DC: 2004); irrigation and land use data from FAO, op. cit. note 19.

24. Data for 1950 from Sandra Postel, "Water for Food Production: Will There Be Enough in 2025?" *BioScience*, August 1998; irrigation and land use data from FAO, op. cit. note 19; Mark Rosengrant, Ximing Cai, and Sarah Cline, *World Water and Food to 2025: Dealing with Scarcity* (Washington, DC, and Battaramulla, Sri Lanka: International Food Policy Research Institute and International Water Management Institute, 2002), p. 155.

25. Murray, op. cit. note 16.

26. Ibid., p. 3; M. Heller and G. Keoleian, *Life-Cycle Based Sustainability Indicators for Assessment of the U.S. Food System* (Ann Arbor, MI: Center for Sustainable Systems, University of Michigan, 2000), p. 42.

27. Halweil, op. cit. note 22, p. 37; Stacy Davis and Susan Diegel, "Chapter 2: Energy," *Transportation Energy Data Book: 24th Edition* (Washington, DC: DOE, Energy Efficiency and Renewable Energy, 2004), pp. 2–17; DOE, EIA, "Chapter 5: Transportation Sector," *Measuring Energy Efficiency in the United States Economy: A Beginning* (Washington, DC: 1995), p. 31; U.S. Department of Transportation, Bureau of Transportation Statistics (BTS), *Freight Shipments in America* (Washington, DC: 2004), pp. 9–10; Andy Jones, *Eating Oil—Food in a Changing Climate* (London: Sustain and Elm Farm Research Centre, 2001), p. 2 of summary.

28. Jones, op. cit. note 27, pp. 1–2 of summary; Charlie Pye-Smith, "The Long Haul," *Race to the Top* Web site, www.racetothetop.org/case/case4.htm (London: International Institute for Environment and Development, 25 July 2002).

29. BTS and U.S. Census Bureau, "Table 14. Shipment Characteristics by Three-Digit Commodity and Mode of Transportation: 2002," *2002 Commodity Flow Survey* (Washington, DC: December 2004); Jones, op. cit. note 27; James Howard Kunstler, author of *Geography of Nowhere*, in *The End of Suburbia: Oil Depletion and the Collapse of The American Dream*, documentary film (Toronto, ON: The Electric Wallpaper Co., 2004).

30. Heller and Keoleian, op. cit. note 26, p. 42; food energy content and packaging content calculated by Danielle Murray, Earth Policy Institute, using USDA nutritional information and packaging energy costs from David Pimentel and Marcia Pimentel, *Food, Energy and Society* (Boulder, CO: University Press of Colorado, 1996), cited in Manuel Fuentes, "Alternative Energy Report," Oxford Brookes University and the Millennium Debate, 1997; Leo Horrigan, Robert S. Lawrence, and Polly Walker, "How Sustainable Agriculture Can Address the Environmental and Human Health Harms of Industrial Agriculture," *Environmental Health Perspectives*, vol. 110, no. 5 (May 2002), p. 448.

31. Murray, op. cit. note 16, pp. 1, 3; Duffield, op. cit. note 17; DOE, EIA, op. cit. note 20; USDA, op. cit. note 23; Miranowski, op. cit. note 20, p. 11.

32. Data for Table 2–1 compiled by Earth Policy Institute from International Monetary Fund (IMF), *International Financial Statistics*, online database, ifs.apdi.net, updated September 2005; IMF, *International Financial Statistics* (Washington, DC: 2005).

33. IMF, on-line database, op. cit. note 32; IMF, *International Financial Statistics*, op. cit. note 32.

34. U.S. Census Bureau, "U.S. Trade in Goods and Services," at www.census.gov/foreign-trade/statistics/historical/gands.pdf, updated 10 June 2005; IMF, on-line database, op. cit. note 32.

35. Value of grain exports and oil imports from U.S. Census Bureau, Foreign Trade Statistics, "US Imports of Crude Oil," at www.census.gov/foreign-trade/statistics/historical/petr.pdf, viewed 29 July 2005; USDA, Foreign Agricultural Service (FAS), U.S. Trade Internet System, online database, at www.fas.usda.gov/ustrade, updated July 2005; U.S. oil production and consumption from BP, *Statistical Review of World Energy 2005* (London: 2005).

36. Value of grain exports and oil imports from U.S. Census Bureau, op. cit. note 35; USDA, op. cit. note 35; U.S. oil production and consumption from BP, op. cit. note 35.

37. Figure 2–1 compiled by Earth Policy Institute from F.O. Licht, "Too Much Too Soon?—World Ethanol Production to Break Another Record in 2005," *World Ethanol and Biofuels Report*, vol. 3, no. 20 (21 June 2005), pp. 429–35, and from historical series in Molly Aeck, "Biofuel Use Growing Rapidly," in Worldwatch Institute, *Vital Signs 2005* (New York: W.W. Norton & Company, 2005), pp. 38–39; biodiesel production estimates for 2004–05 are based on preliminary data from F.O. Licht, op. cit. this note, assuming continued annual growth of 30 percent; USDA, FAS, Production Estimates and Crop Assessment Division, *EU: Biodiesel Industry Expanding Use of Oilseeds* (Washington, DC: 2003).

38. F.O. Licht, op. cit. note 37; Aeck, op. cit. note 37; biodiesel production estimates for 2004–05 are based on preliminary data from F.O. Licht, op. cit. note 37, assuming continued annual growth of 30 percent.

39. Marla Dickerson, "Homegrown Fuel Supply Helps Brazil Breathe Easy," *Los Angeles Times*, 15 June 2005; Renewable Fuels Association, *Homegrown Homeland for the Ethanol Industry Outlook 2005* (Washington, DC: 2005), pp. 2, 14–15; gasoline use from BP, op. cit. note 35, p. 12; F.O. Licht, op. cit. note 37; Karin Bendz, *EU-25—Oilseeds and Products—Biofuels Situation in the European Union—2005* (Washington, DC: USDA, FAS, 2005), p. 6.

40. DOE, EIA, op. cit. note 1; Jim Landers, "Ethanol's Sweet Allure," *Dallas Morning News*, 10 June 2005; Sergio Barros, *Brazil—Sugar—Annual Report—2005*, GAIN Report BR5008 (Washington, DC: USDA, FAS, 2005), p. 6; USDA, *Brazilian Sugar* (Washington, DC: October 2003), p. 1; Todd Benson, "In Brazil, Sugar Cane Growers Become Fuel Farmers," *New York Times*, 24 May 2005; Brazil sugarcane production and land needs calculated by Earth Policy Institute from São Paulo Sugar Cane Agroindustry Union, "Brazil as a Strategic Supplier of Fuel Ethanol," presentation, São Paulo, Brazil, January 2005.

41. USDA, *Brazilian Sugar*, op. cit. note 40, p. 4; Benson, op. cit. note 40; Emma Ross-Thomas, "Brazil Ethanol Industry Sees Japan Move in 2 Years," *Reuters*, 19 May 2005; Steve Thompson, "Great Expectations: Ethanol Is Hot, But What Is The Long-Term Outlook For Biofuel?" *Rural Cooperatives* (USDA), vol. 71, no. 3 (July–August 2004).

42. Dan Morgan, "Brazil's Biofuel Strategy Pays Off as Gas Prices Soar," *Washington Post*, 18 June 2005; Otto Doering, "U.S. Ethanol Policy: Is

It the Best Energy Alternative?" *Current Agriculture, Food and Resource Issues*, no. 5, 2004, pp. 204–05; Steve Raabe, "Drivers' Increasing Demand for Lower-cost Ethanol is Behind Plans for Three Plants on the Eastern Plains—Fill 'er Up on Corn," *Denver Post*, 19 July 2005; Suzy Fraser Dominy, "The Onward March of Ethanol," *World Grain*, 1 June 2005; Renewable Fuels Association, op. cit. note 39, pp. 8–9.

43. Christoph Berg, *World Fuel Ethanol Analysis and Outlook* (Ratzeburg, Germany: F.O. Licht, April 2004); F.O. Licht, op. cit. note 37.

44. F.O. Licht, op. cit. note 37.

45. Bendz, op. cit. note 39, p. 6; Sabine Lieberz, *Germany—Oilseeds and Products—Biofuels in Germany—2004*, GAIN Report GM4048 (Washington, DC: USDA, FAS, 2005), pp. 4, 9.

46. "France Opens Second Phase of Biofuel Plan," *Reuters*, 20 May 2005; Bendz, op. cit. note 39, pp. 1, 6; Marie-Cécile Hénard, *France—Oilseeds and Products—Biodiesel Demand Boosts Rapeseed Production—2005*, GAIN Report FR5018 (Washington, DC: USDA, FAS, 2005), p. 3; Berg, op. cit. note 43.

47. Matthew Wilde, "Soybean Farmers Could Reap Benefits from Biodiesel Industry's Rapid Growth," *Knight Ridder*, 18 July 2005; American Soybean Association, "Soybeans... The Miracle Crop," "U.S. Soybean Meal Production 1979-2004," and "U.S. Soybean Oil Production 1979-2004," *Soy Stats Online, 2005 edition*, at www.soystats.com/2005.

48. "Brazil's Fledgling Biodiesel Industry Takes Off," *Environment News Service*, 29 April 2005; Raymond Hoh, "Malaysia—Oilseeds and Products—June Update—2005," GAIN Report MY5027 (Washington, DC: USDA, FAS, 2005), p. 3; Chris Rittgers and Niniek S. Alam, "Indonesia—Oilseeds and Products—Annual—2005," GAIN Report ID5002 (Washington, DC: USDA, FAS, 2005), p. 4; Elizabeth Mello, "Brazil—Oilseeds and Products—Annual—2005," GAIN Report BR5613 (Washington, DC: USDA, FAS, 2005), p. 33; "Biofuels Take Off in Some Countries," *Reuters*, 9 June 2005; Dickerson, op. cit. note 39.

49. Table 2–2 compiled by Earth Policy Institute from FAO, op. cit. note 19, updated 14 July 2005; Manitoba Department of Energy, Science, and Technology, "Ethanol FAQ," Energy Development Initiative Web site, www.gov.mb.ca/est/energy/ethanol/ethanolfaq.html, viewed 5 August 2005; Renewable Fuels Association, op. cit. note 39; Nandini Nimbkar and Anil Rajvanshi, "Sweet Sorghum Ideal for Biofuel," *Seed World*, vol. 14, no. 8 (November 2003); Boma S. Anga, "Investment Opportunities in the Up & Down Stream Sectors of the Nigerian Cassava Industry," Cassava Agro Industries Services, www.cbc globelink.org; Ellen I. Burnes et al., *Ethanol in California: A Feasibility Framework* (Modesto, CA: Great Valley Center, 2004), p. 18; Berg, op. cit. note 43; DOE, *Biofuels from Switchgrass: Greener Energy Pastures* (Oak Ridge, TN: Oak Ridge National Laboratory, 1998); "Oil Yields and Characteristics," Journey to Forever Web site,

www.journeytoforever.org/biodiesel_yield.html, viewed 15 July 2005; soybean yield is author's estimate.

50. "Oil Yields and Characteristics," op. cit. note 49; soybean yield is author's estimate.

51. Berg, op. cit. note 43; Morgan, op. cit. note 42; Benson, op. cit. note 40; Thompson, op. cit. note 41; F.O. Licht, cited in Alfred Szwarc, "Use of Bio-Fuels in Brazil," presentation at *In-Session Workshop on Mitigation, SBSTA 21 / COP 10*, Buenos Aires: Ministry of Science and Technology, 9 December 2004; Hosein Shapouri, James A. Duffield, and Michael Wang, *The Energy Balance of Corn Ethanol: An Update*, Agricultural Economic Report No. 814 (Washington, DC: USDA, 2002), pp. 9, 11.

52. Berg, op. cit. note 43; corn-based ethanol energy balance is author's estimate, based on various sources, including F.O. Licht, cited in Szwarc, op. cit. note 51, and Shapouri, Duffield, and Wang, op. cit. note 51.

53. Aeck, op. cit. note 37, p. 38; DOE, op. cit. note 49, p. 3; David Bransby, "Switchgrass Profile," DOE Feedstock Development Program, Oak Ridge National Laboratory Web site, at bioenergy.ornl.gov/papers/misc/switchgrass-profile.html, viewed 21 June 2005.

54. DOE, op. cit. note 49, p. 2; Berg, op. cit. note 43; R. Samson et al., *The Use Of Switchgrass Biofuel Pellets as a Greenhouse Gas Offset Strategy* (Sainte Anne de Bellevue, PQ, Canada: Resource Efficient Agricultural Production-Canada (REAP), 2000).

55. Renewable Fuels Association, op. cit. note 39, pp. 2, 10; FAO and U.S. Bureau of the Census, cited in Brian Halweil, "Grain Harvest and Hunger Both Grow," in Worldwatch Institute, op. cit. note 37, p. 23; USDA, FAS, *Grain: World Markets and Trade* (Washington, DC: July 2005), pp. 8, 13.

56. Population from United Nations, *World Population Prospects: The 2004 Revision* (New York: 2005).

57. Population Reference Bureau, "Largest Urban Agglomerations, 1950, 2000, 2015," *Human Population: Fundamentals of Growth—Patterns of World Urbanization* (Washington, DC: 2005); U.N. Human Settlements Programme (UN-HABITAT), *The State of the World's Cities 2004/2005: Globalization and Urban Culture* (London: 2004), pp. 24–25; United Nations, *World Urbanization Prospects, The 2003 Revision: Data Tables and Highlights* (New York: 2004), p. 7; U.N. Department of Economic and Social Affairs, Population Division, *Urban Agglomerations 2003*, wall chart (New York: March 2004).

58. Thomas Wheeler, "It's the End of the World as We Know It: A Review of The End of Suburbia—Oil Depletion and the Collapse of the American Dream," *Alternative Press Review*, 28 July 2004.

59. Jad Mouawad, "Production Trends Point to Reliance on Imported Oil," *New York Times*, 3 January 2005; Ball, op. cit. note 2; Vidal, op. cit. note 1; Klare, op. cit. note 7.

60. BTS, "Table 1–12: U.S. Sales or Deliveries of New Aircraft, Vehicles, Vessels, and Other Conveyances," *National Transportation Statistics 2005* (Washington, DC: 2005).

61. Oliver Prichard, "SUV Drivers Reconsider," *Philadelphia Inquirer*, 1 June 2005; Danny Hakim and Jonathan Fuerbringer, "Fitch Cuts G.M. to Junk, Citing Poor S.U.V. Sales," *New York Times*, 24 May 2005; Danny Hakim, "G.M. Will Reduce Hourly Workers In U.S. By 25,000," *New York Times*, 8 June 2005.

62. Micheline Maynard, "Surging Fuel Prices Catch Most Airlines Unprepared, Adding to the Industry's Gloom," *New York Times*, 26 April 2005; "Revealed: The Real Cost of Air Travel," *The Independent* (London), 29 May 2005; Federal Aviation Administration (FAA), "Commercial Forecast Reports Eighth Consecutive Year of Aviation Growth—'Aviation Enjoyed One of its Best, If Not the Best, Decade Ever,'" press release (Washington, DC: 7 March 2000); FAA, "FAA Forecasts Passenger Levels to Top One Billion in the Next Decade," press release (Washington, DC: 17 March 2005); U.S. Department of Transportation and FAA, *FAA Aerospace Forecasts—Fiscal Years 2005–2016* (Washington, DC: 2005), p. I-25.

63. BTS, "Table 1–4: Public Road and Street Mileage in the United States by Type of Surface," *National Transportation Statistics 2005* (Washington, DC: 2005); U.S. Department of Transportation, Federal Highway Administration, *Highway Statistics* (Washington, DC: Annual Issues).

64. Nicholas Lenssen, "Nuclear Power Inches Up," in Worldwatch Institute, *Vital Signs 2001* (New York: W.W. Norton & Company, 2001), pp. 42–43.

Chapter 3. Emerging Water Shortages

1. M.T. Coe and J.A. Foley, "Human and Natural Impacts on the Water Resources of the Lake Chad Basin," *Journal of Geophysical Research (Atmospheres)*, vol. 106, no. D4 (2001), pp. 3349–56; Lynn Chandler, "Africa's Lake Chad Shrinks by 20 Times Due to Irrigation Demands, Climate Change," press release (Greenbelt, MD: NASA, Goddard Space Flight Center, 27 February 2001); population information from United Nations, *World Population Prospects: The 2004 Revision* (New York: February 2005).

2. World Bank, *China: Agenda for Water Sector Strategy for North China* (Washington, DC: April 2001); Christopher Ward, *The Political Economy of Irrigation Water Pricing in Yemen* (Sana'a, Yemen: World Bank, November 1998); U.S. Department of Agriculture (USDA), *Agricultural Resources and Environmental Indicators 2000* (Washington, DC: February 2000).

3. Water use tripling from I.A. Shiklomanov, "Assessment of Water Resources and Water Availability in the World," *Report for the Comprehensive Assessment of the Freshwater Resources of the World* (St. Petersburg, Russia: State Hydrological Institute, 1998), cited in

Peter H. Gleick, *The World's Water 2000–2001* (Washington, DC: Island Press, 2000), p. 52.

4. Jacob W. Kijne, *Unlocking the Water Potential of Agriculture* (Rome: FAO, 2003), p. 26; water use from Shiklomanov, op. cit. note 3, p. 53.
5. Grain production from USDA, *Production, Supply, & Distribution*, electronic database, www.fas.usda.gov/psd/psdselection.asp, updated 13 September 2005; Table 3–1 compiled by Earth Policy Institute from United Nations, op. cit. note 1.
6. Michael Ma, "Northern Cities Sinking as Water Table Falls," *South China Morning Post*, 11 August 2001; share of China's grain harvest from the North China Plain based on Hong Yang and Alexander Zehnder, "China's Regional Water Scarcity and Implications for Grain Supply and Trade," *Environment and Planning A*, vol. 33 (2001), and on USDA, op. cit. note 5.
7. Ma, op. cit. note 6.
8. World Bank, op. cit. note 2, pp. vii, xi.
9. John Wade, Adam Branson, and Xiang Qing, *China Grain and Feed Annual Report 2002* (Beijing: USDA, 21 February 2002).
10. Grain production from USDA, op. cit. note 5.
11. Wade, Branson, and Xiang, op. cit. note 9; grain production from USDA, op. cit. note 5.
12. Grain production from USDA, op. cit. note 5.
13. World Bank, op. cit. note 2, p. viii; calculations by Earth Policy Institute based on 1,000 tons of water to produce 1 ton of grain in U.N. Food and Agriculture Organization (FAO), *Yield Response to Water* (Rome: 1979).
14. Irrigated area from FAO, *FAOSTAT Statistics Database*, at apps.fao.org, updated 4 April 2005; grain harvest from USDA, op. cit. note 5.
15. Fred Pearce, "Asian Farmers Sucking the Continent Dry," *New Scientist*, 25 August 2004.
16. Ibid.; Tamil Nadu population from 2001 census, "Tamil Nadu at a Glance: Area and Population" at www.tn.gov.in.
17. Pearce, op. cit. note 15.
18. Tushaar Shah et al., *The Global Groundwater Situation: Overview of Opportunities and Challenges* (Colombo, Sri Lanka: International Water Management Institute, 2000).
19. USDA, op. cit. note 2, Chapter 2.1, p. 6; irrigated share calculated from FAO, op. cit. note 14; harvest from USDA, op. cit. note 5.
20. Population from United Nations, op. cit. note 1; fall in water table from "Pakistan: Focus on Water Crisis," *U.N. Integrated Regional Information Networks*, 17 May 2002.

21. "Pakistan: Focus on Water Crisis," op. cit. note 20; Garstang quoted in "Water Crisis Threatens Pakistan: Experts," *Agence France-Presse*, 26 January 2001.

22. Sardar Riaz A. Khan, "Declining Land Resource Base," *Dawn* (Pakistan), 27 September 2004.

23. USDA, op. cit. note 5.

24. Population from United Nations, op. cit. note 1; overpumping from Chenaran Agricultural Center, Ministry of Agriculture, according to Hamid Taravati, publisher, Iran, e-mail to author, 25 June 2002.

25. Craig S. Smith, "Saudis Worry as They Waste Their Scarce Water," *New York Times*, 26 January 2003; grain production from USDA, op. cit. note 5.

26. Smith, op. cit. note 25.

27. Ibid.

28. Population from United Nations, op. cit. note 1; Yemen's water situation from Christopher Ward, "Yemen's Water Crisis," based on a lecture to the British Yemeni Society in September 2000, July 2001; Ward, op. cit. note 2.

29. Marcus Moench, "Groundwater: Potential and Constraints," in Ruth S. Meinzen-Dick and Mark W. Rosegrant, eds., *Overcoming Water Scarcity and Quality Constraints* (Washington, DC: International Food Policy Research Institute, October 2001).

30. Population from United Nations, op. cit. note 1; Yemen's water situation from Ward, op. cit. note 2.

31. Deborah Camiel, "Israel, Palestinian Water Resources Down the Drain," *Reuters*, 12 July 2000.

32. Population from United Nations, op. cit. note 1; water table fall from Shah et al., op. cit. note 18; percentage of water extracted from underground from Karin Kemper, "Groundwater Management in Mexico: Legal and Institutional Issues," in Salman M.A. Salman, ed., *Groundwater: Legal and Policy Perspectives, Proceedings of a World Bank Seminar* (Washington, DC: World Bank, 1999), p. 117.

33. Colorado, Ganges, Indus, and Nile rivers from Sandra Postel, *Pillar of Sand* (New York: W.W. Norton & Company, 1999), pp. 59, 71–73, 94, 261–62; Yellow River from Lester R. Brown and Brian Halweil, "China's Water Shortages Could Shake World Food Security," *World Watch*, July/August 1998, p. 11.

34. Gleick, op. cit. note 3, p. 52.

35. Sandra Postel, *Last Oasis* (New York: W.W. Norton & Company, 1997), pp. 38–39; World Commission on Dams, *Dams and Development: A New Framework for Decision-Making* (London: Island Press, 2000), p. 8.

36. Postel, op. cit. note 33, pp. 261–62; Jim Carrier, "The Colorado: A River

Drained Dry," *National Geographic*, June 1991, pp. 4–32.

37. U.N. Environment Programme (UNEP), *Afghanistan: Post-Conflict Environmental Assessment* (Geneva: 2003), p. 60.

38. Brown and Halweil, op. cit. note 33.

39. Postel, op. cit. note 33, pp. 71, 146.

40. Ibid., pp. 56–58; population from United Nations, op. cit. note 1.

41. Moench, op. cit. note 29; population from United Nations, op. cit. note 1.

42. UNEP, "'Garden of Eden' in Southern Iraq Likely to Disappear Completely in Five Years Unless Urgent Action Taken," news release (Nairobi: 22 March 2003); Hassan Partow, *The Mesopotamian Marshlands: Demise of an Ecosystem*, Early Warning and Assessment Technical Report (Nairobi: Division of Early Warning and Assessment, UNEP, 2001).

43. Janet Larsen, "Disappearing Lakes, Shrinking Seas," *Eco-Economy Update* (Washington, DC: Earth Policy Institute, 7 April 2005).

44. David Maisel, "Lake and Bake: Photos of the Once-Mighty, Now-Drained Owens Lake," *Grist Magazine*, 19 January 2005.

45. Larsen, op. cit. note 43; "Statistics: The Measurements of the Mono Basin," Mono Lake Web site, www.monolake.org, updated 4 January 2005.

46. Megan Goldin, "Israel's Shrinking Sea of Galilee Needs Miracle," *Reuters*, 14 August 2001; Jordan River diminishing from Annette Young, "Middle East Conflict Killing the Holy Water," *The Scotsman*, 12 September 2004.

47. Caroline Hawley, "Dead Sea 'to Disappear by 2050,'" *BBC*, 3 August 2001; Gidon Bromberg, "Water and Peace," *World Watch*, July/August 2004, pp. 24–30.

48. Quirin Schiermeier, "Ecologists Plot to Turn the Tide for Shrinking Lake," *Nature*, vol. 412 (23 August 2001), p. 756.

49. "Sea to Disappear within 15 Years," *News 24*, 22 July 2003; "Kazakh Dam Condemns Most of the Shrunken Aral Sea to Oblivion," *Guardian* (London), 29 October 2003; Nikolai Mikhalchuk, "The Dying Aral Sea," *The Green Cross Optimist*, spring 2004, pp. 37–39; Fred Pearce, "Poisoned Waters," *New Scientist*, October 1995, pp. 29–33; Caroline Williams, "Long Time No Sea," *New Scientist*, 4 January 2003, pp. 34–37.

50. Larsen, op. cit. note 43; NASA, Earth Observatory, "Aral Sea," at earthobservatory.nasa.gov/Newsroom/NewImages/images.php3?img_id=16277, viewed 25 January 2005; Alex Kirby, "Kazakhs 'to Save North Aral Sea,'" *BBC*, 29 October 2003.

51. "Kazakh Dam Condemns Most of the Shrunken Aral Sea to Oblivion," op. cit. note 49.

52. Lester R. Brown, "Worsening Water Shortages Threaten China's Food

Security," *Eco-Economy Update* (Washington, DC: Earth Policy Institute, 4 October 2001); Li Heng, "20 Natural Lakes Disappear Each Year in China," *People's Daily*, 21 October 2002; Xinhua, "Glaciers Receding, Wetlands Shrinking in River Fountainhead Area," *China Daily*, 7 January 2004.

53. Prakriiti Gupta, "Last SOS for Dal Lake," *People & the Planet*, 8 June 2004; Hilal Bhat, "Silenced Springs," *Down to Earth*, vol. 13, no. 18 (5 February 2005).

54. Jim Carlton, "Shrinking Lake in Mexico Threatens Future of Region," *Wall Street Journal*, 3 September 2003; population from United Nations, op. cit. note 1.

55. Water to make steel from Postel, op. cit. note 35, p. 137; price of steel as of June 2005 from Michael Fenton, Iron & Steel Commodity Specialist, U.S. Geological Survey, e-mail to Erin Greenfield, Earth Policy Institute, 21 July 2005; 1,000 tons of water for 1 ton of grain from FAO, *Yield Response to Water* (Rome: 1979); price of wheat from International Monetary Fund, *International Financial Statistics*, http://ifs.apdi.net, July 2005.

56. Noel Gollehon and William Quinby, "Irrigation in the American West: Area, Water and Economic Activity," *Water Resources Development*, vol. 16, no. 2 (2000), pp. 187–95; Postel, op. cit. note 35, p. 137.

57. Gershon Feder and Andrew Keck, *Increasing Competition for Land and Water Resources: A Global Perspective* (Washington, DC: World Bank, March 1995), pp. 28–29; population projections from United Nations, op. cit. note 1; China water demand from World Bank, op. cit. note 8; Brown and Halweil, op. cit. note 33.

58. Postel, op. cit. note 33, pp. 65–66.

59. Brown and Halweil, op. cit. note 33.

60. Shah et al., op. cit. note 18.

61. Gollehon and Quinby, op. cit. note 56, pp. 187–95; *The Water Strategist*, various issues at www.waterstrategist.com.

62. Arkansas River basin from Joey Bunch, "Water Projects Forecast to Fall Short of Needs: Study Predicts 10% Deficit in State," *Denver Post*, 22 July 2004.

63. Dean Murphy, "Pact in West Will Send Farms' Water to Cities," *New York Times*, 17 October 2003; Tim Molloy, "California Water District Approves Plan to Pay Farmers for Irrigation Water," *Associated Press*, 13 May 2004.

64. "China Politics: Growing Tensions Over Scarce Water," *The Economist*, 21 June 2004.

65. Population from United Nations, op. cit. note 1.

66. FAO, op. cit. note 13.

67. Grain production from USDA, op. cit. note 5; Jonathan Watts, "No

Longer Self-Sufficient in Food, the Country Today Has to Buy Abroad, Raising Global Prices: China's Farmers Cannot Feed Hungry Cities," *Guardian*, 26 August 2004; Peter Goodman, "A New Use for Good Earth: Chinese Farmers Pay Price in Drive to Build Golf Centers," *Washington Post*, 13 April 2004; Jim Yardley, "China Races to Reverse Falling Grain Production, " *New York Times*, 2 May 2004; population from United Nations, op. cit. note 1.

68. Grain from USDA, Foreign Agricultural Service, *Grain: World Markets and Trade* (Washington, DC: various years).

69. Population from United Nations, op. cit. note 1; grain production from USDA, op. cit. note 5.

70. Population from United Nations, op. cit. note 1; grain production from USDA, op. cit. note 5.

71. Nile River flow from Postel, op. cit. note 33, p. 77; grain imports from USDA, op. cit. note 5; calculation based on 1,000 tons of water for 1 ton of grain from FAO, op. cit. note 13.

72. Population from United Nations, op. cit. note 1; grain production from USDA, op. cit. note 5.

73. Andrew Keller, R. Sakthivadivel, and David Seckler, *Water Scarcity and the Role of Storage in Development*, Research Report 39 (Colombo, Sri Lanka: International Water Management Institute, 2000), p. 5.

74. USDA, op. cit. note 2, p. 7; USDA, National Agricultural Statistics Service, *Agricultural Statistics 2003* (Washington, DC: U.S. Government Printing Office, 2003), pp. I-6 – I-42.

75. David Seckler, David Molden, and Randolph Barker, "Water Scarcity in the Twenty-First Century," *Water Brief 1* (Colombo, Sri Lanka: International Water Management Institute, 1999), p. 2; United Nations, op. cit. note 1.

76. USDA, op. cit. note 5; FAO, op. cit. note 13.

Chapter 4. Rising Temperatures and Rising Seas

1. Sir David King, "Global Warming: The Science of Climate Change—the Imperatives for Action," presented as the 3rd Greenpeace Business Lecture (London: 12 October 2004); Paul Brown, "Melting Ice: The Threat to London's Future," *The Guardian* (London), 14 July 2004; ice core study in EPICA Community Members, "Eight Glacial Cycles from an Antarctic Ice Core," *Nature*, vol. 429 (10 June 2004), pp. 623–28; Jerry F. McManus, "A Great Grand-Daddy of Ice Cores," *Nature*, vol. 429 (10 June 2004), pp. 611–12; Gabrielle Walker, "Frozen Time," *Nature*, vol. 429 (10 June 2004), pp. 596–97.

2. EPICA Community Members, op. cit. note 1; current carbon dioxide level from C. D. Keeling and T. P. Whorf, "Atmospheric CO_2 Records from Sites in the SIO Air Sampling Network," in *Trends: A Compendium of Data on Global Change* (Oak Ridge, TN: Carbon

Dioxide Information Analysis Center, Oak Ridge National Laboratory, May 2005); Brown, op. cit. note 1; Quirin Schiermeier, "A Rising Tide," *Nature*, vol. 428 (11 March 2004), pp. 114–15.

3. U.S. Department of Agriculture (USDA), *Production, Supply, & Distribution*, electronic database, at www.fas.usda.gov/psd, updated 13 September 2005; Janet Larsen, "Record Heat Wave in Europe Takes 35,000 Lives," *Eco-Economy Update* (Washington, DC: Earth Policy Institute, 9 October 2003); USDA National Agricultural Statistics Service (NASS), "Crop Production," news release (Washington, DC: 12 August 2005).

4. Cindy Schreuder and Sharman Stein, "Heat's Toll Worse Than Believed, Study Says at Least 200 More Died," *Chicago Tribune*, 21 September 1995; "India Heat Wave Toll Tops 1,000," *CNN*, 22 May 2002; "India's Heatwave Toll 1,200, No Respite in Sight," *Agence France-Presse*, 23 May 2002.

5. Centers for Disease Control and Prevention (CDC), "Heat-Related Deaths—Chicago, Illinois, 1996–2001, and United States, 1979–1999," *Morbidity and Mortality Weekly Report,* 4 July, 2003; estimate of deaths across Europe compiled in Larsen, op. cit. note 3, updated with Istituto Nazionale di Statistica, *Bilancio Demografico Nazionale: Anno 2003* (Rome: 15 July 2004); death toll from the 11 September 2001 attacks from National Commission on Terrorist Attacks Upon the United States, *The 9/11 Commission Report* (Washington, DC: U.S. Government Printing Office, 2004).

6. Andrew Simms, "Farewell Tuvalu," *The Guardian* (London), 29 October 2001; Jacopo Pasotti, "Maldives Experience That Sinking Feeling," *Science Now*, 17 June 2005; Brown, op. cit. note 1; Stuart R. Gaffin, *High Water Blues: Impacts of Sea Level Rise on Selected Coasts and Islands* (Washington, DC: Environmental Defense Fund, 1997), p. 27.

7. "Awful Weather We're Having," *The Economist,* 2 October 2004; Richard Milne, "Hurricanes Cost Munich Re Reinsurance," *Financial Times*, 6 November 2004.

8. J. Hansen, NASA's Goddard Institute for Space Studies (GISS), "Global Temperature Anomalies in 0.1 C," at http://data.giss.nasa.gov/gistemp/tabledata/GLB.Ts.txt, updated September 2005; climate monitoring stations from Reto A. Ruedy, GISS, e-mail to Janet Larsen, Earth Policy Institute, 14 May 2003.

9. Figure 4–1 from Hansen, op. cit. note 8; crops from USDA, op. cit. note 3; USDA, *Grain: World Markets and Trade* (Washington, DC: various months).

10. Figure 4–2 from Keeling and Whorf, op. cit. note 2, with historical carbon dioxide estimate in data from Seth Dunn, "Carbon Emissions Dip," in Worldwatch Institute, *Vital Signs 1999* (New York: W.W. Norton & Company, 1999), pp. 60–61; update from Hansen, op. cit. note 8.

11. Intergovernmental Panel on Climate Change (IPCC), *Climate Change*

2001: The Scientific Basis. Contribution of Working Group I to the Third Assessment Report of the Intergovernmental Panel on Climate Change (New York: Cambridge University Press, 2001); Arctic Climate Impact Assessment (ACIA), *Impacts of a Warming Arctic* (Cambridge, U.K.: Cambridge University Press, 2004); National Snow and Ice Data Center (NSIDC), "Arctic Sea Ice Shrinking, Greenland Ice Sheet Melting, According to Study," press release, 7 December 2002; Frank Paul et al., "Rapid Disintegration of Alpine Glaciers Observed with Satellite Data," *Geophysical Research Letters*, vol. 31, L21402 (12 November 2004); Hansen, op. cit. note 8; comparison to time since Ice Age from Warren Washington, cited in Stephen Phillips, "Ignoring Climate Will Land Us in Hot Water," *Times Higher Education Supplement*, 7 February 2003.

12. IPCC, op. cit. note 11.

13. Shaobing Peng et al., "Rice Yields Decline with Higher Night Temperature from Global Warming," *Proceedings of the National Academy of Sciences*, 6 July 2004, pp. 9971–75; John Krist, "Water Issues Will Dominate California's Agenda This Year," *Environmental News Network*, 21 February 2003; Thomas R. Knutson and Robert E. Tuleya, "Impact of CO_2-Induced Warming on Simulated Hurricane Intensity and Precipitation: Sensitivity to the Choice of Climate Model and Convective Parameterization," *Journal of Climate*, vol. 17, no. 18 (15 September 2004), pp. 3477–95; Aiguo Dai, Kevin E. Trenberth, and Taotao Qian, "A Global Dataset of Palmer Drought Severity Index for 1870–2002: Relationship with Soil Moisture and Effects of Surface Warming," *Journal of Hydrometeorology*, vol. 5 (December 2004), pp. 1117–30; "Even Modest Climate Change Means More and Larger Fires," *Environment News Service*, 31 August 2004.

14. National Center for Atmospheric Research and UCAR Office of Programs, "Drought's Growing Reach: NCAR Study Points to Global Warming as Key Factor," press release (Boulder, CO: 10 January 2005); Dai, Trenberth, and Qian, op. cit. note 13.

15. Donald McKenzie et al., "Climatic Change, Wildfire, and Conservation," *Conservation Biology*, vol. 18, no. 4 (August 2004), pp. 890–902.

16. "Elizabeth Gillespie, "Global Warming May Be Making Rivers Too Hot: Cold-Water Fish Will Struggle, Report Says," *Seattle Post-Intelligencer*, 24 March 2005.

17. Camille Parmesan and Hector Galbraith, *Observed Impacts of Global Climate Change in the U.S.* (Arlington, VA: Pew Center on Global Climate Change, 2004); DeNeed L. Brown, "Signs of Thaw in a Desert of Snow, *Washington Post*, 28 May 2002.

18. Parmesan and Galbraith, op. cit. note 17; J. R. Pegg, "Global Warming Disrupting North American Wildlife," *Environment News Service*, 16 December 2004.

19. Douglas B. Inkley et al., *Global Climate Change and Wildlife in North America* (Bethesda, MD: The Wildlife Society, December 2004).

20. John E. Sheehy, International Rice Research Institute, Philippines, e-

mail to Janet Larsen, Earth Policy Institute, 1 October 2002; Pedro Sanchez, "The Climate Change–Soil Fertility–Food Security Nexus," speech, *Sustainable Food Security for All by 2020,* Bonn, Germany, 4–6 September 2002; USDA, op. cit. note 3.

21. Mohan K. Wali et al., "Assessing Terrestrial Ecosystem Sustainability," *Nature & Resources,* October-December 1999, pp. 21–33.

22. Sheehy, op. cit. note 20; Sanchez, op. cit. note 20.

23. Peng et al., op. cit. note 13.

24. Ibid.; Proceedings of the National Academy of Sciences, "Warmer Evening Temperatures Lower Rice Yields," press release (Washington, DC: 29 June 2004).

25. David B. Lobell and Gregory P. Asner, "Climate and Management Contributions to Recent Trends in U.S. Agricultural Yields," *Science*, vol. 299 (14 February 2003), p. 1032.

26. K. S. Kavi Kumar and Jyoti Parikh, "Socio-Economic Impacts of Climate Change on Indian Agriculture," *International Review for Environmental Strategies,* vol. 2, no. 2 (2001), pp. 277–93; United Nations, *World Population Prospects: The 2004 Revision* (New York: February 2005).

27. Krist, op. cit. note 13.

28. Michael J. Scott et al., "Climate Change and Adaptation in Irrigated Agriculture—A Case Study of the Yakima River," in *UCOWR/NIWR Conference, Water Allocation: Economics and the Environment* (Carbondale, IL: Universities Council on Water Resources, 2004); Pacific Northwest National Laboratory, "Global Warming to Squeeze Western Mountains Dry by 2050," press release (Richland, WA: 16 February 2004); Pacific Northwest National Laboratory, "We're Here, We're Warming, Can We Get Used to It?" press release (Richland, WA: 17 February 2005).

29. Bhawani S. Dongol et al., "Low Flows in the Middle Mountain Watersheds of the Hindu-Kush Himalayas (HKH)," paper presented at the International Conference on the Great Himalayas: Climate, Health, Ecology, Management and Conservation, Kathmandu, Nepal, January 2004; Mountain Agenda, *Mountains of the World: Water Towers for the 21st Century* (Bern: 1998); Mehrdad Khalili, "The Climate of Iran: North, South, Kavir (Desert), Mountains," *San'ate Hamlo Naql*, March 1997, pp. 48–53.

30. For more information see Evelyne Yohe, "Sizing Up the Earth's Glaciers," NASA Earth Observatory, at earthobservatory.nasa.gov/Study/GLIMS, 22 June 2004.

31. Crop harvests from USDA, op. cit. note 3.

32. Robert Marquand, "Glaciers in the Himalayas Melting at Rapid Rate," *Christian Science Monitor*, 5 November 1999.

33. Paul et al., op. cit. note 11; Lonnie G. Thompson, "Disappearing

Glaciers Evidence of a Rapidly Changing Earth," American Association for the Advancement of Science Annual Meeting, San Francisco, CA, February 2001; Juan Forero, "As Andean Glaciers Shrink, Water Worries Grow," *New York Times*, 24 November 2002; Monica Vargas, "Peru's Snowy Peaks May Vanish as Planet Heats Up," *Reuters*, 23 July 2004.

34. IPCC, op. cit. note 11.

35. University of Colorado at Boulder, "Global Sea Levels Likely to Rise Higher in 21st Century than Previous Predictions," press release (Boulder, CO: 16 February 2002).

36. "Alaska Examines Impacts of Global Warming," *National Geographic News*, 21 December 2001; Myrna H. P. Hall and Daniel B. Fagre, "Modeled Climate-Induced Glacier Change in Glacier National Park, 1850–2100," *BioScience*, February 2003, pp. 131–40.

37. American Institute of Physics, "New Research Shows Mountain Glaciers Shrinking Worldwide," press release (Boston: 30 May 2001).

38. Thompson, op. cit. note 33; Eric Hansen, "Hot Peaks," *OnEarth*, fall 2002, p. 8.

39. Hansen, op. cit. note 38.

40. Paul et al., op. cit. note 11; Ceri Radford, "Melting Swiss Glaciers Threaten Alps," *Reuters*, 16 November 2004.

41. Thompson, op. cit. note 11; "The Peak of Mt Kilimanjaro As It Has Not Been Seen for 11,000 Years," *The Guardian* (London), 14 March 2005.

42. Kargel quoted in Hansen, op. cit. note 38.

43. Jonathan Watts, "Highest Icefields Will Not Last 100 Years, Study Finds: China's Glacier Research Warns of Deserts and Floods Due to Warming," *The Guardian* (London), 24 September 2004; "China Warns of 'Ecological Catastrophe' from Tibet's Melting Glaciers," *Agence France-Presse*, 5 October 2004; "Glacier Study Reveals Chilling Prediction," *China Daily*, 23 September 2004.

44. Watts, op. cit. note 43; "China Warns of 'Ecological Catastrophe' from Tibet's Melting Glaciers," op. cit. note 43; "Glacier Study Reveals Chilling Prediction," op. cit note 43.

45. ACIA op. cit. note 11; ACIA Web site, www.acia.uaf.edu, updated 13 July 2005; "Rapid Arctic Warming Brings Sea Level Rise, Extinctions," *Environment News Service,* 8 November 2004.

46. J. R. Pegg, "The Earth is Melting, Arctic Native Leader Warns," *Environment News Service*, 16 September 2004.

47. ACIA, op. cit. note 11.

48. Erik Stokstad, "Defrosting the Carbon Freezer of the North," *Science*, vol. 304 (11 June 2004), pp. 1618–20; carbon emissions in G. Marland, T. A. Boden, and R. J. Andres, "Global, Regional, and National CO_2 Emissions," in Oak Ridge National Laboratory, op. cit. note 2.

49. R. Warrick et al., "Changes in Sea-Level," in J.T. Houghton et al., eds., *Climate Change, 1995: The Science of Climate Change* (Cambridge, U.K.: Cambridge University Press, 1995), pp. 359–405, cited in Dorthe Dahl-Jensen, "The Greenland Ice Sheet Reacts," *Science*, vol. 289 (21 July 2000), pp. 404–05.

50. IPCC, *Climate Change 2001: Impacts, Adaptation, and Vulnerability. Contribution of Working Group II to the Third Assessment Report of the Intergovernmental Panel on Climate Change* (New York: Cambridge University Press, 2001), pp. 948–51; Committee of Abrupt Climate Change, *Abrupt Climate Change: Inevitable Surprises* (Washington, DC: National Research Council, 2002).

51. Joe Friesen, "Arctic Melt May Open Up Northwest Passage: Portal Could Cut Nearly 5,000 Nautical Miles From Asia-Europe Trip Via Panama Canal," *Globe and Mail* (Toronto), 9 November 2004.

52. U.S. Department of Energy, Energy Information Administration, "Antarctica: Fact Sheet," at www.eia.doe.gov/emeu/cabs/antarctica.html, September 2000.

53. Andrew Shepherd, "Larsen Ice Sheet Has Progressively Thinned," *Science*, vol. 302 (31 October 2003), pp. 856–59; "Breakaway Bergs Disrupt Antarctic Ecosystem," *Environment News Service*, 9 May 2002; "Giant Antarctic Ice Shelves Shatter and Break Away," *Environment News Service*, 19 March 2002.

54. NSIDC, "Antarctic Ice Shelf Collapses," at nsidc.org/iceshelves/larsenb2002, 19 March 2002; "Breakaway Bergs Disrupt Antarctic Ecosystem," op. cit. note 53; "Giant Antarctic Ice Shelves Shatter and Break Away," op. cit. note 53.

55. NSIDC, op. cit. note 54; "Breakaway Bergs Disrupt Antarctic Ecosystem," op. cit. note 53; "Giant Antarctic Ice Shelves Shatter and Break Away," op. cit. note 53; Vaughan quoted in Andrew Revkin, "Large Ice Shelf in Antarctica Disintegrates at Great Speed," *New York Times*, 20 March 2002.

56. Michael Byrnes, "New Antarctic Iceberg Split No Threat," *Reuters*, 20 May 2002; Young quoted in "Giant Antarctic Ice Shelves Shatter and Break Away," op. cit. note 53.

57. Boesch cited in Bette Hileman, "Consequences of Climate Change," *Chemical & Engineering News*, 27 March 2000, pp. 18–19.

58. World Bank, *World Development Report 1999/2000* (New York: Oxford University Press, 2000), p. 100; population from United Nations, op. cit. note 26; Shanghai population from United Nations, *World Urbanization Prospects: The 2003 Revision* (New York: 2004); Shanghai from Stuart R. Gaffin, *High Water Blues: Impacts of Sea Level Rise on Selected Coasts and Islands* (Washington, DC: Environmental Defense Fund, 1997), p. 27.

59. James E. Neumann et al., *Sea-level Rise & Global Climate Change: A Review of Impacts to U.S. Coasts* (Arlington, VA: Pew Center on Global Climate Change, 2000); Gaffin, op. cit. note 58.

60. IPCC, op. cit. note 11, p. 665.

61. Knutson and Tuleya, op. cit. note 13.

62. Janet N. Abramovitz, "Averting Unnatural Disasters," in Lester R. Brown et al., *State of the World 2001* (New York: W.W. Norton & Company, 2001) pp. 123–42.

63. Storm death toll from National Climatic Data Center, National Oceanic & Atmospheric Administration, U.S. Department of Commerce, "Mitch: The Deadliest Atlantic Hurricane Since 1780," www.ncdc.noaa.gov/oa/reports/mitch/mitch.html, updated 1 July 2004; Flores quoted in Arturo Chavez et al., "After the Hurricane: Forest Sector Reconstruction in Honduras," *Forest Products Journal*, November/December 2001, pp. 18–24; gross domestic product from International Monetary Fund (IMF), *World Economic Outlook Database,* at www.imf.org/external/pubs/ft/weo, updated April 2003.

64. Michael Smith, "Bad Weather, Climate Change Cost World Record $90 Billion," *Bloomberg*, 15 December 2004; "Insurers See Hurricane Costs as High as $23 Billion," *Reuters*, 4 October 2004.

65. Lisa Rein and Dan Balz, "240,000 Evacuees Strain Capacity," *Washington Post*, 4 September 2005; National Climatic Data Center, "Climate of 2005: Summary of Hurricane Katrina," fact sheet, www.ncdc.noaa.gov/oa/climate/research/2005/katrina.html, updated 1 September 2005; P.J. Webster et al., "Changes in Tropical Cyclone Number, Duration, and Intensity in a Warming Environment," *Science*, vol. 309 (16 September 2005), pp. 1844–46; "Katrina May Cost as Much as Four Years of War: Government Certain to Pay More than $200 Billion Following Hurricane," *Associated Press,* 10 September 2005.

66. "Awful Weather We're Having," op. cit. note 7; Munich Re, *Topics Geo Annual Review: Natural Catastrophes 2004* (Munich, Germany: 2005), p. 15.

67. "Disaster and Its Shadow," *The Economist,* 14 September 2002, p. 71; "Moody's Downgrades Munich Re's Ratings to 'Aa1,'" *Insurance Journal,* 20 September 2002; Hilary Burke, "Insurers to Pay Record Disaster Damages in 2004," *Reuters*, 16 December 2004; Richard Milne, "Hurricanes Cost Munich Re Reinsurance," *Financial Times*, 6 November 2004.

68. Tim Hirsch, "Climate Change Hits Bottom Line," *BBC News*, 15 December 2004.

69. Munich Re, "Natural Disasters: Billion-$ Insurance Losses," in Louis Perroy, "Impacts of Climate Change on Financial Institutions' Medium to Long Term Assets and Liabilities," paper presented to the Staple Inn Actuarial Society, 14 June 2005.

70. Munich Re, *Topics Annual Review: Natural Catastrophes 2001* (Munich, Germany: 2002), pp. 16–17; value of China's wheat and rice harvests from USDA, op. cit. note 3, using prices from IMF, *International Financial Statistics,* electronic database, at ifs.apdi.net/imf.

71 Munich Re, op. cit. note 69.

72. Andrew Dlugolecki, "Climate Change and the Financial Services Industry," speech delivered at the opening of the UNEP Financial Services Roundtable, Frankfurt, Germany, 16 November 2000; "Climate Change Could Bankrupt Us by 2065," *Environment News Service*, 24 November 2000.

73. Bjorn Larsen, *World Fossil Fuel Subsidies and Global Carbon Emissions in a Model with Interfuel Substitution*, Policy Research Working Paper 1256 (Washington, DC: World Bank, February 1994), p. 7.

74. Contributions from the Center for Responsive Politics, "Oil and Gas: Long Term Contribution Trends," at www.opensecrets.org/industries/indus.asp?Ind=E01, updated 10 May 2005; Committee on Ways and Means, *Incentives for Domestic Oil and Gas Production and Status of the Industry*, Hearing Before the Subcommittee on Oversight of the Committee on Ways and Means, House of Representatives (Washington, DC: U.S. Government Printing Office, February 1999), p. 16.

75. Kym Anderson and Warwick J. McKibbin, "Reducing Coal Subsidies and Trade Barriers: Their Contribution to Greenhouse Gas Abatement," *Environment and Development Economics*, October 2000, pp. 457–81.

76. Military expenditures from Graham E. Fuller and Ian O. Lesser, "Persian Gulf Myths," *Foreign Affairs*, May–June 1997, pp. 42–53; value of Persian Gulf oil imports from U.S. Department of Energy, Energy Information Administration, *Annual Energy Review* (Washington, DC: 2001), p. 165.

77. Mark M. Glickman, *Beyond Gas Taxes: Linking Driving Fees to Externalities* (Oakland, CA: Redefining Progress, 2001), p. 1; number of taxpayers from Internal Revenue Service, "Number of Returns Filed, by Type of Return and State, Fiscal Year 2000," in *2000 IRS Data Book* (Washington, DC: 2001).

Chapter 5. Natural Systems Under Stress

1. Walter C. Lowdermilk, *Conquest of the Land Through 7,000 Years*, USDA Bulletin No. 99 (Washington, DC: U.S. Department of Agriculture (USDA), Natural Resources Conservation Service, 1939).

2. Ibid., p. 10.

3. U.N. Food and Agriculture Organization (FAO), "FAO/WFP Crop and Food Assessment Mission to Lesotho Special Report," at www.fao.org, viewed 29 May 2002; Michael Grunwald, "Bizarre Weather Ravages Africans' Crops," *Washington Post*, 7 January 2003.

4. Number of hungry from FAO, *The State of Food Insecurity in the World 2004* (Rome: 2004).

5. Species Survival Commission, *2000 IUCN Red List of Threatened Species* (Gland, Switzerland, and Cambridge, U.K.: World Conservation Union-IUCN, 2000), p. 1.

6. Teresa Cerojano, "Decades of Illegal Logging Blamed for High Death Toll in Philippine Storm," *Associated Press*, 1 December, 2004; Thailand from Patrick B. Durst et al., *Forests Out of Bounds: Impacts and Effectiveness of Logging Bans in Natural Forests in Asia-Pacific* (Bangkok: FAO, Asia-Pacific Forestry Commission, 2001); Munich Re, "Munich Re's Review of Natural Catastrophes in 1998," press release (Munich: 19 December 1998); Harry Doran, "Human Activities Aid Force of Nature: Massive Destruction Has Worsened the Floods Which Have Struck Throughout History, But Lessons Are Being Learned," *South China Morning Post*, 24 July 2003; John Pomfret, "China's Lumbering Economy Ravages Border Forests," *Washington Post*, 26 March 2001.
7. World forested area from FAO, "Table 1.2. Forest Area by Region 2000," Forest Resources Assessment (FRA) 2000 (Rome: 2001).
8. FAO, *Agriculture: Towards 2015/30*, Technical Interim Report (Rome: 2000).
9. Forest Frontiers Initiative, *The Last Frontier Forests: Ecosystems and Economies on the Edge* (Washington, DC: WRI, 1997).
10. FAO, *FAOSTAT Statistics Database*, apps.fao.org, updated 21 January 2005.
11. Alain Marcoux, "Population and Deforestation," in *Population and the Environment* (Rome: FAO, 2000); March Turnbull, "Life in the Extreme," *Africa Geographic Online*, at www.africa-geographic.com, 4 April 2005.
12. Nigel Sizer and Dominiek Plouvier, *Increased Investment and Trade by Transnational Logging Companies in Africa, the Caribbean, and the Pacific* (Belgium: World Wide Fund for Nature (WWF) and WRI Forest Frontiers Initiative, 2000), pp. 21–35; Lester R. Brown, "Nature's Limits," in Lester R. Brown et al., *State of the World 1995* (New York: W.W. Norton & Company: 1995), p. 9.
13. Maria Pia Palermo, "Brazil Losing Fight to Save the Amazon," *Reuters*, 22 May 2005; Steve Kingstone, "Amazon Destruction Accelerating," *BBC News*, 19 May 2005.
14. Mario Rautner, Martin Hardiono, and Raymond J. Alfred, *Borneo: Treasure Island at Risk* (Frankfurt: WWF Germany, June 2005), p. 7.
15. "Haitian Storm Deaths Blamed on Deforestation," Environment News Service, 27 September 2004; "Haiti Floods Due to Deforestation," *CBSNews.com*, 23 September 2004.
16. Mozambique flooding from "Aid Agencies Gear Up in Mozambique Flood Rescue Effort," *CNN*, 1 March 2000; loss of forest cover from Carmen Revenga et al., *Watersheds of the World* (Washington, DC: WRI and Worldwatch Institute, 1998).
17. "Madagascar's Rainforest Faces Destruction," *Guardian* (London), 29 June 2003.
18. Eneas Salati and Peter B. Vose, "Amazon Basin: A System in Equilibrium," *Science*, vol. 225 (13 July 1984), pp. 129–38.

19. Philip Fearnside quoted in Barbara J. Fraser, "Putting a Price on the Forest," LatinamericaPress.org, 10 November 2002; Philip M. Fearnside, "The Main Resources of Amazonia," paper for presentation at the Latin American Studies Association XX International Congress, Guadalajara, Mexico, 17–19 April 1997.
20. Charles Mkoka, "Unchecked Deforestation Endangers Malawi Ecosystems," *Environment News Service*, 16 November 2004.
21. Anscombe quoted in Mkoka, op. cit. note 20.
22. Durst et al., op. cit. note 6; Zhu Chunquan, Rodney Taylor, and Feng Guoqiang, *China's Wood Market, Trade and Environment* (Monmouth Junction, NJ, and Beijing: Science Press USA Inc. and WWF International, 2004).
23. One third is author's estimate; Lester R. Brown, *Building a Sustainable Society* (New York: W.W. Norton & Company, 1981), p. 3.
24. Yang Youlin, Victor Squires, and Lu Qi, eds., *Global Alarm: Dust and Sandstorms from the World's Drylands* (Bangkok: Secretariat of the U.N. Convention to Combat Desertification, 2002), pp. 15–28.
25. John Steinbeck, *The Grapes of Wrath* (New York: Viking Penguin, Inc., 1939).
26. FAO, *The State of Food and Agriculture 1995* (Rome: 1995), p. 175.
27. Ibid.; USDA, *Production, Supply, & Distribution*, electronic database, at www.fas.usda.gov/psd, updated 13 September 2005; FAO, op. cit. note 10, updated 14 July 2005.
28. U.N. Environment Programme (UNEP), *Mongolia: State of the Environment 2002* (Pathumthani, Thailand: Regional Resource Centre for Asia and the Pacific, 2001), pp. 3–7; USDA, op. cit. note 27; population from United Nations, *World Population Prospects: The 2004 Revision* (New York: February 2005).
29. National Aeronautics and Space Administration (NASA) Earth Observatory, "Dust Storm off Western Sahara Coast," at earthobservatory.nasa.gov/NaturalHazards/natural_hazards_v2.php3?img_id=12664, viewed 9 January 2005.
30. Paul Brown, "4x4s Replace the Desert Camel and Whip Up a Worldwide Dust Storm," *Guardian* (London), 20 August 2004.
31. Ibid.
32. Hong Yang and Xiubin Li, "Cultivated Land and Food Supply in China," *Land Use Policy*, vol. 17, no. 2 (2000), p. 5.
33. Asif Farrukh, *Pakistan Grain and Feed Annual Report 2002* (Islamabad, Pakistan: USDA Foreign Agricultural Service (FAS), 2003).
34. Lester R. Brown and Edward C. Wolf, *Soil Erosion: Quiet Crisis in the World Economy*, Worldwatch Paper 60 (Washington, DC: Worldwatch Institute, 1984), p. 20.
35. Land area estimate from Stanley Wood, Kate Sebastian, and Sara J.

Scherr, *Pilot Analysis of Global Ecosystems: Agroecosystems* (Washington, DC: International Food Policy Research Institute and WRI, 2000), p. 3; livestock counts from FAO, op. cit. note 10, updated 14 July 2005.

36. Number of pastoralists from "Investing in Pastoralism," *Agriculture Technology Notes* (Rural Development Department, World Bank), March 1998, p. 1; FAO, op. cit. note 10, updated 14 July 2005.

37. FAO, op. cit. note 10, updated 14 July 2005; United Nations, op. cit. note 28.

38. USDA, *Livestock and Poultry: World Markets and Trade* (Washington, DC: USDA FAS, March 2000); population from United Nations, op. cit. note 28.

39. Robin P. White, Siobhan Murray, and Mark Rohweder, *Pilot Analysis of Global Ecosystems: Grassland Ecosystems* (Washington, DC: WRI, 2000); FAO, op. cit. note 10, updated 14 July 2005; United Nations, op. cit. note 28; Southern African Development Coordination Conference, *SADCC Agriculture: Toward 2000* (Rome: FAO, 1984).

40. FAO, op. cit. note 10, updated 14 July 2005; United Nations, op. cit. note 28.

41. FAO, op. cit. note 10, with livestock data updated 14 July 2005.

42. B.S. Sathe, "Dairy/Milk Production," in Livestock Investment Opportunities in India, FAO Web site, www.fao.org/DOCREP/ARTICLE/AGRIPPA/657_en00.htm, viewed 9 September 2005.

43. H. Dregne et al., "A New Assessment of the World Status of Desertification," *Desertification Control Bulletin*, no. 20, 1991.

44. Population from United Nations, op. cit. note 28.

45. "Case Studies of Sand-Dust Storms in Africa and Australia," in Yang, Squires, and Lu, eds., op. cit. note 24, pp. 123–66.

46. Government of Nigeria, *Combating Desertification and Mitigating the Effects of Drought in Nigeria*, National Report on the Implementation of the United Nations Convention to Combat Desertification (Nigeria: November 1999); population from United Nations, op. cit. note 28; livestock from FAO, op. cit. note 10, updated 14 July 2005.

47. Iranian News Agency, "Official Warns of Impending Desertification Catastrophe in Southeast Iran," *BBC International Reports*, 29 September 2002.

48. UNEP, *Afghanistan: Post-Conflict Environmental Assessment* (Geneva: 2003), p. 52.

49. Wang Hongchang, *Deforestation and Desiccation in China: A Preliminary Study* (Beijing, China: Center for Environment and Development, Chinese Academy of Social Sciences, 1999).

50. Wang Tao, Cold and Arid Regions Environmental and Engineering Research Institute (CAREERI), Chinese Academy of Sciences, e-mail to

author, 4 April 2004; Wang Tao, "The Process and Its Control of Sandy Desertification in Northern China," CAREERI, Chinese Academy of Sciences, seminar on desertification, held in Lanzhou, China, May 2002.

51. Ann Schrader, "Latest Import From China: Haze," *Denver Post*, 18 April 2001; Brown, op. cit. note 30.

52. Howard W. French, "China's Growing Deserts Are Suffocating Korea," *New York Times*, 14 April 2002.

53. See Table 1–1 in Lester R. Brown, Janet Larsen, and Bernie Fischlowitz-Roberts, *The Earth Policy Reader* (New York: W.W. Norton & Company, 2002), p. 13.

54. U.S. Embassy, "Desert Mergers and Acquisitions," *Beijing Environment, Science, and Technology Update* (Beijing: 19 July 2002), p. 2.

55. See Table 5–2 in Lester Brown, *Outgrowing the Earth* (New York: W.W. Norton & Company, 2005), pp. 86–87.

56. Calculations by Earth Policy Institute from FAO, *FISHSTAT Plus*, electronic database, at www.fao.org/fi/statist/FISOFT/FISHPLUS.asp, updated March 2005; United Nations, op. cit. note 28.

57. FAO, *The State of World Fisheries and Aquaculture 2004* (Rome: 2004), pp. 24, 32.

58. Ransom A. Myers and Boris Worm, "Rapid Worldwide Depletion of Predatory Fish Communities," *Nature*, vol. 432 (15 May 2003), pp. 280–83; Charles Crosby, "'Blue Frontier' is Decimated," *Dalhousie News*, 11 June 2003.

59. Myers and Worm, op. cit. note 58; Crosby, op. cit. note 58.

60. Myers and Worm, op. cit. note 58.

61. Andrew Revkin, "Tracking the Imperiled Bluefin from Ocean to Sushi Platter," *New York Times*, 3 May 2005; Ted Williams, "The Last Bluefin Hunt," in Valerie Harms et al., *The National Audubon Society Almanac of the Environment* (New York: Grosset/Putnam, 1994), p. 185; Konstantin Volkov, "The Caviar Game Rules," *Reuters-IUCN*, 2001.

62. Lauretta Burke et al., *Pilot Analysis of Global Ecosystems: Coastal Ecosystems* (Washington, DC: WRI, 2000), pp. 19, 51; coastal wetland loss in Italy from Brown and Kane, op. cit. note 43, p. 82.

63. Clive Wilkinson, ed., *Status of Coral Reefs of the World: 2004* (Townsville, Australia: Global Coral Reef Monitoring Network, 2004), p. 9.

64. Organisation for Economic Cooperation and Development, *OECD Environmental Outlook* (Paris: 2001), pp. 109–20.

65. J. A. Gulland, ed., *Fish Resources of the Ocean* (Surrey, U.K.: Fishing News Ltd., 1971), an FAO-sponsored publication that estimated that oceanic fisheries would not be able to sustain an annual yield of more than 100 million tons.

66. Caroline Southey, "EU Puts New Curbs on Fishing," *Financial Times*, 16 April 1997.

67. Dan Bilefsky, "North Sea's Cod Grounds to be Closed for 12 Weeks," *Financial Times*, 25 January 2001; Paul Brown and Andrew Osborn, "Ban on North Sea Cod Fishing," *Guardian* (London), 25 January 2001; Alex Kirby, "UK Cod Fishing 'Could be Halted,'" *BBC News*, 6 November 2000; "Reforming the Common Fisheries Policy," European Union Web site, at europa.eu.int/comm/fisheries/reform /index_en.htm, viewed 8 October 2003.

68. Diadie Ba, "Senegal, EU Prepare for Fisheries Deal Tussle," *Reuters*, 28 May 2001; Charles Clover, *The End of the Line: How Overfishing is Changing the World and What We Eat* (London: Ebury Press, 2004), pp. 37–46.

69. Clover, op. cit. note 68, p. 38.

70. FAO, op. cit. note 56; United Nations, op. cit. note 28.

71. David Quammen, "Planet of Weeds," *Harper's Magazine*, October 1998.

72. Species Survival Commission, *2004 IUCN Red List of Threatened Species* (Gland, Switzerland, and Cambridge, U.K.: World Conservation Union-IUCN, 2004).

73. Ibid., p. 11.

74. Ibid.; TRAFFIC, *Food for Thought: The Utilization of Wild Meat in Eastern and Southern Africa* (Cambridge, U.K.: 2000).

75. Danna Harman, "Bonobos' Threat to Hungry Humans," *Christian Science Monitor*, 7 June 2001.

76. Species Survival Commission, op. cit. note 72; "Birds on the IUCN Red List," *Bird Life International*, 2005 update, at www.birdlife.org; "Great Indian Bustard Facing Extinction," *India Abroad Daily*, 12 February 2001; Çagan Sekercioglu, Gretchen C. Daily, and Paul R. Ehrlich, "Ecosystem Consequences of Bird Declines," *Proceedings of the National Academy of Sciences*, vol. 101, no. 52 (28 December 2004).

77. Michael McCarthy, "Mystery of the Silent Woodlands: Scientists Are Baffled as Bird Numbers Plummet," *Independent* (London), 25 February 2005; British Trust for Ornithology, "Tough Time for Woodland Birds," press release (Thetford, Norfolk, U.K.: 25 February 2005); J. A. Thomas et al., "Comparative Losses of British Butterflies, Birds, and Plants and the Global Extinction Crisis," *Science*, vol. 303, 19 March 2004, pp. 1,879–81; Dan Vergano, "1 in 10 Bird Species Could Vanish Within 100 Years," *USA Today*, 14 December 2004.

78. Janet N. Abramovitz, *Imperiled Waters, Impoverished Future: The Decline of Freshwater Ecosystems*, Worldwatch Paper 128 (Washington, DC: Worldwatch Institute, March 1996), p. 59; Species Survival Commission, op. cit. note 72, p. 89.

79. James R. Spotila et al., "Pacific Leatherback Turtles Face Extinction,"

Nature, vol. 405 (1 June 2000), pp. 529–30; "Leatherback Turtles Threatened," *Washington Post*, 5 June 2000.

80. Lauretta Burke and Jonathan Maidens, *Reefs at Risk in the Caribbean* (Washington DC: WRI, 2004), pp. 12–14, 27–31.

81. Mohammed Kotb et al., "Status of Coral Reefs in the Red Sea and Gulf of Aden in 2004," in Wilkinson, op. cit. note 63, pp. 137–39.

82. David Kaimowitz et al., *Hamburger Connection Fuels Amazon Destruction* (Jakarta, Indonesia: Center for International Forestry Research, 2004).

83. Conservation International, "The Brazilian Cerrado," at www.biodiversityhotspots.org, viewed 10 September 2004.

84. Species Survival Commission, op. cit. note 72, p. 92; Species Survival Commission, op. cit. note 5, p. 28.

Chapter 6. Early Signs of Decline

1. United Nations, *World Population Prospects: The 2004 Revision* (New York: 2005); Joint United Nations Programme on HIV/AIDS (UNAIDS), *2004 Report on the Global AIDS Epidemic* (Geneva: 2004), p. 191.

2. United Nations, op. cit. note 1; UNAIDS, op. cit. note 1.

3. United Nations, op. cit. note 1; health insurance from U.S. Census Bureau News, "Income Stable, Poverty Up, Numbers of Americans With and Without Health Insurance Rise, Census Bureau Reports," press release (Washington, DC: 26 August 2004).

4. World Health Organization (WHO) cited in Gary Gardner and Brian Halweil, *Underfed and Overfed: The Global Epidemic of Malnutrition*, Worldwatch Paper 150 (Washington, DC: Worldwatch Institute, 2000), p. 7.

5. WHO and UNICEF, *Global Water Supply and Sanitation Assessment 2000 Report* (New York: 2000), pp. v, 2; Gardner and Halweil, op. cit. note 4, p. 7.

6. "Trends in Educational Attainment of the 25- to 34-Year-Old Population (1991–2002)," in Organization for Economic Cooperation and Development (OECD), *Education at a Glance 2004* (Paris: 2004); UNICEF, *Progress for Children: A Report Card on Gender Parity and Primary Education* (New York: 2005), p. 3; The Education for All (EFA) Global Monitoring Report Team, *EFA Global Monitoring Report 2005: The Quality Imperative* (Paris: UNESCO, 2004).

7. Population growth rates from Population Reference Bureau (PRB), *2005 World Population Data Sheet*, wall chart (Washington, DC: August 2005); Hilaire A. Mputu, *Literacy and Non-Formal Education in the E-9 Countries* (Paris: UNESCO, 2001), pp. 5–13; UNESCO Institute for Statistics, "Youth (15–24) and Adult (15+) Literacy Rates by Country and by Gender for 2000–2004," at www.uis.unesco.org, May 2005.

8. Gene B. Sperling, "Toward Universal Education," *Foreign Affairs*, September/October 2001, pp. 7–13.

9. WHO and UNICEF, op. cit. note 5; Peter H. Gleick, *Dirty Water: Estimated Deaths from Water-Related Disease 2000–2020* (Oakland, CA: Pacific Institute, 2002); United Nations, op. cit. note 1.

10. Hunger as a risk factor for disease in WHO, *World Health Report 2002* (Geneva: 2002), and in Majid Ezzati et al., "Selected Major Risk Factors and Global and Regional Burden of Disease," *The Lancet*, 30 October 2002, pp. 1–14.

11. U.N. Food and Agriculture Organization (FAO), *The State of Food Insecurity in the World 2004* (Rome: 2004).

12. FAO, *The State of Food Insecurity in the World 2002* (Rome: 2002); population from United Nations, op. cit. note 1.

13. FAO, op. cit. note 11.

14. Ibid.

15. Gary Gardner and Brian Halweil, "Nourishing the Underfed and Overfed," in Lester R. Brown et al., *State of the World 2000* (New York: W.W. Norton & Company, 2000), pp. 70–73.

16. WHO/UNICEF, *World Malaria Report 2005* (Geneva: 2005); Anne Platt McGinn, "Malaria's Lethal Grip Tightens," in Worldwatch Institute, *Vital Signs 2001* (New York: W.W. Norton & Company, 2001), pp. 134–35; Sachs from Center for International Development at Harvard University and London School of Hygiene and Tropical Medicine, "Executive Summary for *Economics of Malaria*," www.rbm.who.int/docs/abuja_sachs2.htm, viewed 3 August 2005; malaria deaths calculated from United Nations, op. cit. note 1, and WHO/UNICEF, op. cit. this note.

17. More deaths from AIDS than wars from Lawrence K. Altman, "U.N. Forecasts Big Increase in AIDS Death Toll," *New York Times*, 3 July 2002.

18. UNAIDS, *AIDS Epidemic Update* (Geneva: December 2004), p. 1; UNAIDS, op. cit. note 1, pp. 189–207; total deaths and historical estimates calculated using UNAIDS statistics in Worldwatch Institute, *Signposts 2004*, CD-Rom (Washington, DC: 2004); anti-retroviral treatment in sub-Saharan Africa from WHO, "Access to HIV Treatment Continues to Accelerate in Developing Countries, but Bottlenecks Persist, Says WHO/UNAIDS Report," press release (Geneva: 29 June 2005).

19. UNAIDS, op. cit. note 1, pp. 39–66, 191.

20. AIDS and food security in UNAIDS, op. cit. note 1, pp. 39–66; FAO, *The Impact of HIV/AIDS on Food Security*, 27th Session of the Committee on World Food Security, Rome, 28 May–1 June 2001.

21. "Strategic Caring: Firms Strategize About AIDS," *The Economist*, 5 October 2002; UNAIDS, op. cit. note 1, pp. 39–66.

22. EFA Global Monitoring Report Team, op. cit. note 6; UNAIDS, op. cit. note 1, pp. 39–66; Prega Govender, "Shock AIDS Test Result at Varsity," *Sunday Times* (Johannesburg), 25 April 1999; "South Africa: University Finds 25 Percent of Students Infected," *Kaiser Daily HIV/AIDS Report,* 27 April 1999.

23. UNAIDS, op. cit. note 1, pp. 39–66.

24. UNAIDS, UNICEF, and U.S. Agency for International Development (USAID), *Children on the Brink 2004: A Joint Report on New Orphan Estimates and a Framework for Action* (Washington, DC: July 2004), p. 29; Michael Grunwald, "Sowing Harvests of Hunger in Africa," *Washington Post*, 17 November 2002.

25. Stephen Lewis, press briefing (New York: 8 January 2003); Edith M. Lederer, "Lack of Funding for HIV/AIDS is Mass Murder by Complacency, Says U.N. Envoy," *Associated Press*, 9 January 2003.

26. Alex de Waal, "What AIDS Means in a Famine," *New York Times*, 19 November 2002.

27. Sarah Janssen, Gina Solomon, and Ted Schettler, *Chemical Contaminants and Human Disease: A Summary of Evidence* (Boston: Alliance for a Healthy Tomorrow, 2004); Geoffrey Lean, "US Study Links More than 200 Diseases to Pollution," *Independent News* (London), 14 November 2004.

28. Jane Houlihan et al., *Body Burden: The Pollution in Newborns* (Washington, DC: Environmental Working Group, 2005).

29. Bernie Fischlowitz-Roberts, "Air Pollution Fatalities Now Exceed Traffic Fatalities by 3 to 1," *Eco-Economy Update* (Washington, DC: Earth Policy Institute, September 2002), citing WHO, "Air Pollution," Fact Sheet 187 (Geneva: revised September 2000); N. Künzli et al., "Public-Health Impact of Outdoor and Traffic-related Air Pollution: A European Assessment," *Lancet*, 2 September 2000, p. 795; traffic accident deaths from British Red Cross, "May 8 Spotlight on the Millions Injured and Disabled by Road Accidents," press release (London: 9 May 2001); 70,000 American deaths from Joel Schwartz, quoted in Harvard School of Public Health, "Air Pollution Deadlier Than Previously Thought," press release (Cambridge, MA: 2 March 2000).

30. C. Pritchard, D. Baldwin, and A. Mayers, "Changing Patterns of Adult (45–74 years) Neurological Deaths in the Major Western World Countries 1979–1987," *Public Health*, vol. 118, issue 4 (June 2004), pp. 268–83; Juliette Jowit, "Pollutants Cause Huge Rise in Brain Diseases: Scientists Alarmed as Number of Cases Triples in 20 Years," *The Observer* (London), 15 August 2004.

31. Sharon LaFraniere, "Mother Russia's Poisoned Land," *Washington Post,* 22 June 1999.

32. "Mercury Poisoning Disease Hits Amazon Villages," *Reuters*, 4 February 1999; mercury emissions from U.S. coal plants in U.S. Environmental Protection Agency (EPA), Office of Air Quality Planning and Standards and Office of Research and Development, *Mercury Study*

Report to Congress Volume II (Washington, DC: December 1997), p. ES-4; Patricia Glick, *The Toll from Coal* (Washington, DC: National Wildlife Federation, 2000), p. 9; EPA, "EPA Decides Mercury Emissions from Power Plants Must Be Reduced," press release (Washington, DC: 15 December 2000); Ilan Levin and Eric Schaeffer, *Dirty Kilowatts: America's Most Polluting Power Plants* (Washington, DC: Environmental Integrity Project, 2005).

33. EPA, Office of Water, "2004 National Listing of Fish Advisories," *EPA Fact Sheet* (Washington, DC: September 2005); Kathryn Mahaffey, EPA, *Methylmercury: Epidemiology Update*, presentation at The National Forum on Contaminants in Fish, San Diego: January 2004, at www.epa.gov/waterscience/fish/forum/2004/presentations/monday/mahaffey.pdf; Leonardo Trasande, Philip J. Landrigan, and Clyde Schechter, "Public Health and Economic Consequences of Methyl Mercury Toxicity to the Developing Brain," *Environmental Health Perspectives,* vol. 13, no. 5 (May 2005).

34. Anne Platt McGinn, *Why Poison Ourselves? A Precautionary Approach to Synthetic Chemicals,* Worldwatch Paper 153 (Washington, DC: Worldwatch Institute, November 2000), p. 7; 200 chemicals in body from Pete Myers, plenary discussion on Emerging Environmental Issues, at USAID Environmental Officers Training Workshop, "Meeting the Environmental Challenges of the 21st Century," Airlie Center, Warrenton, VA, 26 July 1999.

35. EPA, "Toxics Release Inventory (TRI) Program," fact sheet, at www.epa.gov/tri, updated 17 May 2005; EPA, "EPA Issues New Toxics Report, Improves Means of Reporting," press release (Washington, DC: 11 April 2001).

36. Rachel Carson, *Silent Spring* (Boston: Houghton Mifflin Company, 2002); Theo Colborn, Dianne Dumanoski, and John Peterson Myers, *Our Stolen Future* (New York: Dutton Publishing, 1996).

37. Helen Spiegelman and Bill Sheehan, "Products, Waste, and the End of the Throwaway Society," in Carolyn Raffensperger and Nancy Myers, eds., *The Networker: The Newsletter of the Science and Environmental Health Network,* electronic newsletter, vol. 10, no. 2 (May 2005).

38. Calculated by Earth Policy Institute from United States Geological Survey, *Mineral Commodity Summaries 2005* (Washington, DC: U.S. Government Printing Office, 2005).

39. Eric Lipton, "The Long and Winding Road Now Followed by New York City's Trash," *New York Times*, 24 March 2001.

40. Lester R. Brown, "New York: Garbage Capital of the World," *Eco-Economy Update* (Washington, DC: Earth Policy Institute, April 2002); calculations by author, updated with The City of New York Department of Sanitation, "DSNY–Fact Sheet," at www.nyc.gov/html/dos/html/dosfact.html, updated 27 October 2003; Kirk Johnson, "To City's Burden, Add 11,000 Tons of Daily Trash," *New York Times,* 24 March 2001; Lhota quoted in Eric Lipton, "The Long and Winding Road Now Followed by New York City's Trash," *New York Times*, 24 March 2001.

41. Gilmore quoted in Lipton, op. cit. note 40.

42. Joel Kurth, "N.J. Piles Demolition Trash on Michigan," *Detroit News*, 28 December 2004; Lipton, op. cit. note 40.

43. Günther Baechler, "Why Environmental Transformation Causes Violence: A Synthesis," *Environmental Change and Security Project Report,* Issue 4 (spring 1998), pp. 24–44.

44. Grainland area in 1950 from U.S. Department of Agriculture (USDA), *Production, Supply, and Distribution Country Reports,* October 1990; 2004 grainland area from USDA, *Production, Supply, & Distribution,* electronic database, at www.fas.usda.gov/psd, updated 13 September 2005; population from United Nations, op. cit. note 1.

45. Editorial Desk, "Time for Action on Sudan," *New York Times,* 18 June 2004.

46. Sudan update from Coalition for International Justice (CIJ), "New Analysis Claims Darfur Deaths Near 400,000: Experts Estimate 500 People a Day Are Dying," press release (Washington, DC: 21 April 2005); CIJ, Table: "Estimates from Retrospective Mortality Surveys in Darfur and Chad Displacement Camps, Circa February 2003—April 2005," at www.cij.org, April 2005; "Sudan," in U.S. Central Intelligence Agency, *World Fact Book*, at www.cia.gov/cia/publications/factbook, updated 30 August 2005.

47. Somini Sengupta, "Where the Land is a Tinderbox, the Killing Is a Frenzy," *New York Times*, 16 June 2004; Nigeria population data from United Nations, op. cit. note 1; Government of Nigeria, *Combating Desertification and Mitigating the Effects of Drought in Nigeria,* National Report on the Implementation of the United Nations Convention to Combat Desertification (Nigeria: November 1999).

48. Sengupta, op. cit. note 47.

49. Ibid.

50. James Gasana, "Remember Rwanda?" *World Watch*, September/October 2002, pp. 24–32.

51. Ibid.

52. Population from United States Census Bureau, Population Division, International Programs Center, *International Database*, at www.census.gov/ipc/www/idbacc.html, updated 26 April 2005; demand for firewood from Gasana, op. cit. note 50.

53. Gasana, op. cit. note 50; Emily Wax, "At the Heart of Rwanda's Horror: General's History Offers Clues to the Roots of Genocide," *Washington Post*, 21 September 2002.

54. United Nations, op. cit. note 1.

55. Ibid.; Gasana, op. cit. note 50.

56. United Nations, op. cit. note 1; Nile River discussed in Sandra Postel, *Pillar of Sand* (New York: W.W. Norton & Company, 1999), pp. 141–49.

57. Population from United Nations, op. cit. note 1; income per person calculated from gross domestic product based on purchasing-power-parity in International Monetary Fund, *World Economic Outlook Database,* Washington, DC, updated April 2005; Postel, op. cit. note 56.

58. Postel, op. cit. note 56; United Nations, op. cit. note 1.

59. O'Hara quoted in Michael Wines, "Grand Soviet Scheme for Sharing Water in Central Asia is Foundering," *New York Times*, 9 December 2002.

60. Alan Cowell, "Migrants Found off Italy Boat Piled With Dead," *International Herald Tribune,* 21 October 2003, cited in Lester R. Brown, "Troubling New Flows of Environmental Refugees," *Eco-Economy Update* (Washington, DC: Earth Policy Institute, January 2004).

61. Ibid.

62. "Mexico's Immigration Problem: The Kamikazes of Poverty," *The Economist*, 31 January 2004; calculation by author.

63. "Mexico's Immigration Problem," op. cit. note 62.

64. Norman Myers, "Environmental Refugees: A Growing Phenomenon of the 21st Century," *Philosophical Transactions: Biological Sciences,* 29 April 2002, pp. 609–13, cited in Brown, op. cit. note 60.

65. Frank Bruni, "Off Sicily, Tide of Bodies Roils Immigrant Debate," *New York Times*, 23 September 2002; "Boat Sinks Off Coast of Turkey: One Survivor and 7 Bodies Found," *Agence France-Presse*, 22 December 2003; Flora Botsford, "Spain Recovers Drowned Migrants," *BBC News*, 25 April 2002; Mary Jordan and Kevin Sullivan, "Trade Brings Riches, But Not to Mexico's Poor," *Washington Post,* 22 March 2003; Robert McLeman and Barry Smit, "Climate Change, Migration and Security," *Commentary No. 86* (Ottawa: Canadian Security Intelligence Service, 2 March 2004); Arizona Desert deaths from "Humane Approach to Border," *Denver Post*, 24 April 2003; Ralph Blumenthal, "Citing Violence, 2 Border States Declare a Crisis," *New York Times*, 17 August 2005.

66. U.S. Dust Bowl from Yang Youlin, Victor Squires, and Lu Qi, eds., *Global Alarm: Dust and Sandstorms from the World's Drylands* (Bangkok: Secretariat of the U.N. Convention to Combat Desertification, 2002), pp. 109–22.

67. Jonathan Shaw, "The Great Global Experiment," *Harvard Magazine*, November–December 2002, p. 35; Tomas Alex Tizon, "Can One Man Turn the Tide? As Erosion Eats Away at Tiny Newtok, Alaska, the Relocation of Its Yupik Eskimo Villagers and Their Homes Has Fallen to the Local Grocer," *New York Times*, 28 October 2004.

68. Abandoned villages in India from Tushaar Shah et al., *The Global Groundwater Situation: Overview of Opportunities and Challenges* (Colombo, Sri Lanka: International Water Management Institute, 2000); population from United Nations, op. cit. note 1.

69. Wang Tao, Cold and Arid Regions Environmental and Engineering Research Institute (CAREERI), Chinese Academy of Sciences, e-mail to author, 4 April 2004; Wang Tao, "The Process and Its Control of Sandy Desertification in Northern China," CAREERI, Chinese Academy of Sciences, seminar on desertification, held in Lanzhou, China, May 2002.

70. Iranian News Agency, "Official Warns of Impending Desertification Catastrophe in Southeast Iran," *BBC International Reports*, 29 September 2002; Government of Nigeria, op. cit. note 47, p. 6.

71. Intergovernmental Panel on Climate Change (IPCC), *Climate Change 2001: The Scientific Basis. Contributions of Working Group I to the Third Assessment Report of the Intergovernmental Panel on Climate Change* (New York: Cambridge University Press, 2001); World Bank, *World Development Report 1999/2000* (New York: Oxford University Press, 2000), p. 100; population from United Nations, op. cit. note 1.

72. Fund for Peace and the Carnegie Endowment for International Peace, "The Failed States Index," *Foreign Policy*, July/August 2005, p. 56–65.

73. Ibid.

74. Ibid.

75. Ibid.

76. Ibid.

77. Ibid.

78. Ibid.

79. Richard Cincotta, Robert Engelman, and Daniele Anastasion, *The Security Demographic: Population and Civil Conflict After the Cold War* (Washington, DC: Population Action International, 2003).

80. Ed Stoddard, "Environment Looms as Major Security Threat," *Reuters*, 1 March 2004.

81. Ginger Thompson, "A New Scourge Afflicts Haiti: Kidnappings," *New York Times*, 6 July 2005; Madeleine K. Albright and Robin Cook, "The World Needs to Step It Up in Afghanistan," *International Herald Tribune*, 5 October 2004; Desmond Butler, "5-Year Hunt Fails to Net Qaeda Suspect in Africa," *New York Times*, 14 June 2003.

82. Abraham McLaughlin, "Can Africa Solve African Problems?" *Christian Science Monitor*, 4 January 2005; Marc Lacey, "Beyond the Bullets and Blades," *New York Times*, 20 March 2005.

83. UNAIDS, op. cit. note 1, p. 191; AIDS orphans from *Children on the Brink 2004: A Joint Report on New Orphan Estimates and a Framework for Action* (Washington, DC: UNAIDS, UNICEF, and USAID, 2004), p. 29.

84. "Afghanistan: The Ignored War," in Christy Harvey, Judd Legum and Jonathan Baskin, *The Progress Report* (Washington, DC: American Progress Action Fund, 2005); McLaughlin, op. cit. note 82; "A Failing

State: The Himalayan Kingdom Is a Gathering Menace," *The Economist*, 4 December 2004.

85. United Nations, "United Nations Peacekeeping Operations," background note, at www.un.org/Depts/dpko/dpko/bnote.htm, 30 June 2005; Marc Lacey, "Congo Tribal Killings Create a New Wave of Refugees," *New York Times*, 6 March 2005.

86. United Nations World Food Programme (WFP), "New Operation Provides WFP Food Aid to 550,000 Haitians," news release (Rome: 5 May 2005); WFP, "India Helps WFP Feed Afghan Schoolchildren," news release (Rome: 17 May 2005).

Chapter 7. Eradicating Poverty, Stabilizing Population

1. United Nations General Assembly, "United Nations Millennium Declaration," resolution adopted by the General Assembly, 8 September 2000; United Nations, *The Millennium Development Goals Report 2005* (New York: 2005); "More or Less Equal? Is Economic Inequality Around the World Getting Better or Worse?" *The Economist*, 13 March 2004; International Monetary Fund, *World Economic Outlook*, electronic database, www.imf.org, updated September 2005.

2. World Bank, *World Development Report 2005* (New York: Oxford University Press, 2004); Jeffrey D. Sachs, "India Takes the Lead," *Korea Herald*, 4 August 2004.

3. United Nations, "Poverty, Percentage of Population Below $1 (1993 PPP) Per Day Consumption (World Bank)," Millennium Development Goals Indicators Database, updated 26 August 2005.

4. United Nations, *World Population Prospects: The 2004 Revision* (New York: 2005); G8 leaders, "Gleneagles Communiqué on Africa, Climate Change, Energy and Sustainable Development," document from G8 Summit, Gleneagles, Scotland, July 2005.

5. United Nations General Assembly, op. cit. note 1.

6. United Nations, op. cit. note 1; UNICEF, *Children Under Threat: The State of the World's Children 2005* (New York: 2004).

7. UNICEF, *Progress for Children: A Report Card on Gender Parity and Primary Education* (New York: 2005), p. 3; Hilaire A. Mputu, *Literacy and Non-Formal Education in the E-9 Countries* (Paris: UNESCO, 2001), p. 5; Paul Blustein, "Global Education Plan Gains Backing," *Washington Post*, 22 April 2002; Gene Sperling, "Educate Them All," *Washington Post*, 20 April 2002; Polly Curtis, "Lack of Education 'a Greater Threat than Terrorism': Sen," *The Guardian* (London), 28 October 2003.

8. United Nations General Assembly, op. cit. note 1; Blustein, op. cit. note 7; Sperling, op. cit. note 7; World Bank, "World Bank Announces First Group of Countries for 'Education For All' Fast Track," press release (Washington, DC: 12 June 2002); World Bank, "Education for All the World's Children: Donors Have Agreed to Help First Group of

Countries on Education Fast-Track," press release (Washington, DC: 27 November 2002); Gene Sperling, "The G-8—Send 104 Million Friends to School," *Bloomberg News*, 20 June 2005. For more information on the Millennium Development Goals, see www.un.org/millenniumgoals; for more information on the World Bank's and the international community's involvement in the Education For All program, see www1.worldbank.org/education/efa.asp.

9. See education chapter in World Bank, *Poverty Reduction Strategy Paper Sourcebook* (Washington, DC: 2001), pp. 2–4.

10. Gene B. Sperling, "Toward Universal Education," *Foreign Affairs*, September/October 2001, pp. 7–13.

11. Sperling, op. cit. note 7.

12. The Education for All (EFA) Global Monitoring Report Team, *EFA Global Monitoring Report 2005: The Quality Imperative* (Paris: UNESCO, 2004), p. 21; U.N. Commission on Population and Development, Thirty-sixth Session, *Population, Education, and Development*, press releases, 31 March–4 April 2003; UNESCO, "Winners of UNESCO Literacy Prizes 2003," press release, 27 May 2003.

13. Blustein, op. cit. note 7; United Nations, "Progress Towards the Millennium Development Goals, 1990–2005," New York, 13 June 2005.

14. George McGovern, "Yes We CAN Feed the World's Hungry," *Parade*, 16 December 2001; George McGovern, *The Third Freedom: Ending Hunger in Our Time* (New York: Simon & Schuster: 2001), chapter 1.

15. Jeffrey Sachs, "A New Map of the World," *The Economist*, 22 June 2000; McGovern, "Yes We CAN Feed the World's Hungry," op. cit. note 14.

16. McGovern, "Yes We CAN Feed the World's Hungry," op. cit. note 14.

17. Ibid.

18. Ibid.

19. Population from United Nations, op. cit. note 4; Population Reference Bureau (PRB), *2004 World Population Data Sheet*, wall chart (Washington, DC: August 2004).

20. United Nations, op. cit. note 4; PRB, *2005 World Population Data Sheet*, wall chart (Washington, DC: August 2005).

21. United Nations, op. cit. note 4.

22. U.N. Population Fund (UNFPA), *The State of World Population 2004* (New York: 2004), p. 39; the 201 million women who want to limit their family size but lack access to a choice of effective contraception consist of some 137 million women with an unmet need for contraception and another 64 million who are using less reliable traditional family planning methods.

23. Janet Larsen, "Iran's Birth Rate Plummeting at Record Pace," in Lester R.

Brown, Janet Larsen, and Bernie Fischlowitz-Roberts, *The Earth Policy Reader* (New York: W.W. Norton & Company, 2002), pp. 190–94; see also Homa Hoodfar and Samad Assadpour, "The Politics of Population Policy in the Islamic Republic of Iran," *Studies in Family Planning*, March 2000, pp. 19–34, and Farzaneh Roudi, "Iran's Family Planning Program: Responding to a Nation's Needs," *MENA Policy Brief*, June 2002; Iran population growth rate from United Nations, op. cit. note 4.

24. Larsen, op. cit. note 23.

25. Ibid.

26. Ibid.; population growth rates from PRB, op. cit. note 20; United Nations, op. cit. note 4.

27. Janet Larsen, "World Population Grew by 76 Million People in 2004—3 Million Added in the Industrial World and 73 Million in the Developing World," *Eco-Economy Indicator* (Washington, DC: Earth Policy Institute, 21 December 2004); UNFPA, "Meeting the Goals of the ICPD: Consequences of Resource Shortfalls up to the Year 2000," paper presented to the Executive Board of the U.N. Development Programme and the UNFPA, New York, 12–23 May 1997; UNFPA, *Population Issues Briefing Kit* (New York: Prographics, Inc., 2001), p. 23; UNFPA, op. cit. note 22, pp. 89–90.

28. UNFPA, op. cit. note 22, p. 39.

29. Honduran Ministry of Health, *Encuesta Nacional de Epidemiología y Salud Familiar* (National Survey of Epidemiology and Family Health) (Tegucigalpa: 1996), cited in George Martine and Jose Miguel Guzman, "Population, Poverty, and Vulnerability: Mitigating the Effects of Natural Disasters," in *Environmental Change and Security Project Report* (Washington, DC: Woodrow Wilson International Center for Scholars, 2002), pp. 45–68.

30. "Bangladesh: National Family Planning Program," *Family Planning Programs: Diverse Solutions for a Global Challenge* (Washington, DC: PRB, 1994).

31. John Donnelly, "U.S. Seeks Cuts in Health Programs Abroad," *Boston Globe*, 5 February 2003.

32. Jeffrey D. Sachs and the Commission on Macroeconomics and Health, *Macroeconomics and Health: Investing in Health for Economic Development* (Geneva: World Health Organization (WHO), 2001); Ruth Levine and the What Works Working Group, *Millions Saved: Proven Successes in Global Health* (Washington, DC: Center for Global Development, 2004).

33. WHO, *World Health Report 2002* (Geneva: 2002), p. 10; "The Tobacco Epidemic: A Crisis of Startling Dimensions," in *Message From the Director-General of the World Health Organization for World No-Tobacco Day 1998*, at www.who.int/archives/ntday/ntday98/ad98e_1.htm; air pollution from WHO, "Air Pollution," fact sheet 187 (Geneva: revised September 2000).

34. Alison Langley, "Anti-Smoking Treaty Is Adopted by 192 Nations," *New York Times*, 22 May 2003; information on WHO's Tobacco Free Initiative is at www5.who.int/tobacco/index.cfm.

35. Cigarette consumption from U.S. Department of Agriculture (USDA), *Production, Supply, & Distribution*, electronic database, Washington, DC, updated 31 May 2005; per capita estimates made using population from United Nations, op. cit. note 4; Daniel Yee, "Smoking Declines in U.S.—Barely," *CBS News*, 10 November 2004.

36. USDA, op. cit. note 35; per capita estimates made using population from United Nations, op. cit. note 4.

37. "Smoking Bans Around the World," *Reuters*, 10 January 2005.

38. "Bangladesh Bans Smoking in Many Public Places," *Reuters*, 15 March 2005; "New Zealand Stubs Out Smoking in Bars, Restaurants," *Reuters*, 13 December 2004; Bernard Wysocki, Jr., "Companies Get Tough With Smokers, Obese to Trim Costs," *Wall Street Journal*, 12 October 2004.

39. Bill and Melinda Gates Foundation, "Vaccine-Preventable Diseases," at www.gatesfoundation.org/GlobalHealth/Pri_Diseases/Vaccines/default, viewed 9 September 2005.

40. Sachs and Commission on Macroeconomics and Health, op. cit. note 32; WHO, "Smallpox," fact sheet at www.who.int, viewed 10 October 2005.

41. United Nations Foundation, "The United Nations Foundation Honors Canadian Government for Contributions in Fight Against Polio," press release, 27 January 2005; United Nations Foundation, "Donate: Polio," at www.unfoundation.org/donate/polio.asp, viewed 9 September 2005.

42. David Brown, "A Blow to Anti-Polio Campaign," *Washington Post*, 10 May 2005; Global Polio Eradication Initiative, "Polio Eradication Situation Report—September 2005," press release (Geneva: September 2005).

43. Global Polio Eradication Initiative, op. cit. note 42; Global Polio Eradication Initiative, "Global Case Count," at www.polioeradication.org/casecount.asp, 20 September 2005.

44. Sachs and Commission on Macroeconomics and Health, op. cit. note 32.

45. Nita Bhalla, "Teaching Truck Drivers About AIDS," *BBC*, 25 June 2001; Hugh Ellis, "Truck Drivers Targeted in New AIDS Offensive," *The Namibian*, 17 March 2003; C. B. S. Venkataramana and P. V. Sarada, "Extent and Speed of Spread of HIV Infection in India Through the Commercial Sex Networks: A Perspective," *Tropical Medicine and International Health*, vol. 6, no. 12 (December 2001), pp. 1,040–61, cited in "HIV Spread Via Female Sex Workers in India Set to Increase Significantly by 2005," *Reuters Health*, 26 December 2001.

46. Mark Covey, "Target Soldiers in Fight Against AIDS Says New Report," press release (London: Panos Institute, 8 July 2002); "Free Condoms for Soldiers," *South Africa Press Association*, 5 August 2001; HIV prevalence rate from Joint United Nations Programme on HIV/AIDS (UNAIDS), *2004 Report on the Global AIDS Epidemic* (Geneva: July 2004), p. 191.

47. Nada Chaya and Sarah Haddock, *Condoms Count: Meeting the Need in the Era of HIV/AIDS, 2004 Data Update* (Washington, DC: Population Action International, 2004); Nada Chaya and Kai-Ahset Amen, with Michael Fox, *Condoms Count: Meeting the Need in the Era of HIV/AIDS* (Washington, DC: Population Action International, 2002); Population Action International, "Counting Condoms: Donors Coming Up Short," press release (Washington, DC: 14 July 2004); 2 billion condoms needed for contraception based on estimates from Robert Gardner et al., *Closing the Condom Gap* (Baltimore, MD: Johns Hopkins University School of Public Health, Population Information Program, April 1999); "Who Pays for Condoms," in Chaya and Amen, with Fox, op. cit. this note, pp. 29–36; Communications Consortium Media Center, "U.N. Special Session on Children Ends in Acrimony," *PLANetWIRE.org*, 14 May 2002; Adam Clymer, "U.S. Revises Sex Information, and a Fight Goes On," *New York Times*, 27 December 2002.

48. Chaya and Amen, with Fox, op. cit. note 47.

49. "Who Pays for Condoms," op. cit. note 47, pp. 29–36; Communications Consortium Media Center, op. cit. note 47; Clymer, op. cit. note 47.

50. UNAIDS, op. cit. note 46; UNAIDS, *AIDS Epidemic Update* (Geneva: December 2004), p. 13; UNAIDS, *Report on the Global HIV/AIDS Epidemic* (Geneva: June 2000), pp. 9–11.

51. UNAIDS and WHO, *Progress on Global Access to HIV Antiretroviral Therapy: An Update on "3 by 5"* (Geneva: 2005), pp. 7, 13.

52. Clive Bell, Shantayanan Devarajan, and Hans Gersbach, "The Long-run Economic Cost of AIDS: Theory and an Application to South Africa," *Policy Research Working Paper Series* (Washington, DC: World Bank, 2003); "AIDS Summit: The Economics of Letting People Die," *Star Tribune*, 16 July 2003; Deborah Mitchell, "HIV Treatment: 2 Million Years of Life Saved," *Reuters Health*, 28 February 2005.

53. "AIDS Summit," op. cit. note 52.

54. Organisation for Economic Co-operation and Development (OECD), *Agricultural Policies in OECD Countries: Monitoring and Evaluation 2005, Highlights* (Paris: 2005); U.S. Bureau of International Information Programs (IIP), "Official Aid to Developing Countries Rose 4.6 Percent in 2004," press release, (Washington, DC: 11 April 2005); "The Hypocrisy of Farm Subsidies," *New York Times*, 1 December 2002.

55. Roger Thurow and Geoff Winestock, "Addiction to Sugar Subsidies Chokes Poor Nations' Exports," *Wall Street Journal*, 16 September 2002; Mark Turner, "African Nations 'Off Track' in Reducing Poverty," *Financial Times*, 9 July 2003.

56. OECD, op. cit. note 54; "The Hypocrisy of Farm Subsidies," op. cit. note 54.

57. U.S. IIP, op. cit. note 54; "South Africa: Weaning States Off Subsidies," *Africa News*, 19 August 2005.

58. See Chapter 2 for further discussion of oil prices and ethanol.

59. Kevin Watkins and Joachim von Braun, "Time to Stop Dumping on the World's Poor," in *Trade Policies and Food Security* (Washington, DC: International Food Policy Research Institute: 2003); population from United Nations, op. cit. note 4.

60. Elizabeth Becker, "Looming Battle Over Cotton Subsidies, *New York Times,* 24 January 2004; Elizabeth Becker, "U.S. Will Cut Farm Subsidies in Trade Deal," *New York Times*, 31 July 2004.

61. "Ending the Cycle of Debt," *New York Times*, 1 October 2004.

62. G8 Leaders, "G8 Finance Ministers' Conclusions on Development," Pre Summit Statement by G8 Finance Ministers, London, 10–11 June 2005; Oxfam International, "Gleneagles: What Really Happened at the G8 Summit?" Oxfam Briefing Note (London: 29 July 2005).

63. Abid Aslam, "18 Poor Countries to See Debt Slate Wiped Clean, Saving $10 Million Per Week," *One World US*, 26 September 2005.

64. UNFPA, op. cit. note 22, pp. 14–15.

65. Population from United Nations, op. cit. note 4; UNFPA, op. cit. note 22.

66. Costs of meeting social goals in Table 7–1 calculated by Earth Policy Institute, based on the following sources: universal primary education from World Bank, cited in Blustein, op. cit. note 7; adult literacy campaign is author's estimate; school lunch program from McGovern, "Yes We CAN Feed the World's Hungry," op. cit. note 14; assistance to preschool children and pregnant women is author's estimate of extending the U.S.'s Women, Infants, and Children program, based on ibid.; reproductive health and family planning based on the goals from and the progress since the 1994 International Conference on Population and Development (UNFPA, "Meeting the Goals of the ICPD," op. cit. note 27), combining the $5 billion shortfalls of the developing-country and industrial-country groups; universal basic health care from Sachs and the Commission on Macroeconomics and Health, op. cit. note 32; closing the condom gap estimated from Chaya and Amen, with Fox, op. cit. note 47, and from Gardner et al., op. cit. note 47.

67. Sachs and the Commission on Macroeconomics and Health, op. cit. note 32.

68. Ibid.; U.N. Development Programme, "World on Track to Meet Millennium Goal on Extreme Poverty Thanks to India: Report," press release (New York: 8 July 2003); Wu Xiaoling, "Statement of Madam. Wu Xiaoling, Deputy Governor of the People's Bank of China," speech delivered at the 39th Annual Meeting of the Board of Governors of the African Development Bank (Group), Kampala, Uganda, 25–26 May 2004.

Chapter 8. Restoring the Earth

1. Jonathan Lash, "Dealing with the Tinder As Well As the Flint," *Science*, vol. 294, no. 5548 (30 November 2001), p. 1,789.
2. Craig A. Cox, "Conservation Can Mean Life or Death," *Journal of Soil and Water Conservation*, November/December 2004.
3. Remaining forests from U.N. Food and Agriculture Organization (FAO), "Table 1.2. Forest Area by Region 2000," *Forest Resources Assessment (FRA) 2000* (Rome: 2001).
4. Janet N. Abramovitz, "Paper Recycling Remains Strong," in Lester R. Brown et al., *Vital Signs 2000* (New York: W.W. Norton & Company, 2000), pp. 132–33.
5. Ibid.; U.S. Environmental Protection Agency, *Municipal Solid Waste Generation, Recycling, and Disposal in the United States: Facts and Figures for 2003* (Washington, DC: 2003).
6. Fuelwood as a proportion of total harvested wood from FAO, *FAOSTAT Statistics Database*, at apps.fao.org, forest data updated 12 August 2005; Daniel M. Kammen, "From Energy Efficiency to Social Utility: Lessons from Cookstove Design, Dissemination, and Use," in José Goldemberg and Thomas B. Johansson, *Energy as an Instrument for Socio-Economic Development* (New York: U.N. Development Programme, 1995).
7. Solar Cooking International Volunteers, *Solar Cookers International, Creating Healthy Communities*, at www.edc-cu.org/pdf/Solar%20Cookers%20International.pdf, viewed 9 September 2005; Kevin Porter, "Final Kakuma Evaluation: Solar Cookers Filled a Critical Gap," in Solar Cookers International, *Solar Cooker Review*, November 2004; Solar Cookers International cost from "Breakthrough in Kenyan Refugee Camps," at solarcooking.org/kakuma-m.htm, viewed 27 September 2005.
8. FAO, *Agriculture: Towards 2015/30, Technical Interim Report* (Geneva: Economic and Social Department, 2000), pp. 156–57.
9. Johanna Son, "Philippines: Row Rages Over Lifting of Ban on Lumber Exports," *InterPress Service*, 17 April 1998.
10. Reed Funk, letter to author, 9 August 2005.
11. Alliance for Forest Conservation and Sustainable Use, "WWF/World Bank Forest Alliance Launches Ambitious Program to Reduce Deforestation and Curb Illegal Logging," press release (New York: World Bank/WWF, 25 May 2005); Alliance for Forest Conservation and Sustainable Use, "World Bank/WWF Alliance for Forest Conservation & Sustainable Use: Questions & Answers," fact sheet, World Bank/WWF, at lnweb18.worldbank.org/ESSD/envext.nsf/80ByDocName/WorldBankWWFAllianceQA/$FILE/QAAlliance.pdf, viewed 4 October 2005.
12. Forest Stewardship Council, *FSC Certified Forests* (Bonn, Germany: 2005), pp. 34, 40, 53; Forest Stewardship Council, "FSC Regional

Totals," www.certified.forests.org/data/global_table.htm, viewed 10 August 2005.

13. Plantation area from FAO, op. cit. note 3, p. 402; grain area from U.S. Department of Agriculture (USDA), *Production, Supply, & Distribution*, electronic database, Washington, DC, at www.fas.usda.gov/psd, updated 13 September 2005; FAO, op. cit. note 8, p. 167.

14. Chris Brown and D. J. Mead, eds., "Future Production from Forest Plantations," *Forest Plantation Thematic Paper* (Rome: FAO, 2001), p. 9; FAO, op. cit. note 6.

15. FAO, op. cit. note 8, p. 161; FAO, op. cit. note 3, updated 10 April 2001; Ashley T. Mattoon, "Paper Forests," *World Watch*, March/April 1998, p. 20.

16. Mattoon, op. cit. note 15; corn yields from USDA, op. cit. note 13.

17. FAO, op. cit. note 8, p. 185; Brown and Mead, op. cit. note 14.

18. M. Davis et al., "New England-Acadian Forests," in Taylor H. Ricketts et al., eds., *Terrestrial Ecoregions of North American: A Conservation Assessment* (Washington, DC: Island Press, 1999); David R. Foster, "Harvard Forest: Addressing Major Issues in Policy Debates and in the Understanding of Ecosystem Process and Pattern," *LTER Network News: The Newsletter of the Long-term Ecological Network*, spring/summer 1996.

19. C. Csaki, "Agricultural Reforms in Central and Eastern Europe and the Former Soviet Union: Status and Perspectives," *Agricultural Economics*, vol. 22 (2000), pp. 37–54; Igor Shvytov, *Agriculturally Induced Environmental Problems in Russia*, Discussion Paper No. 17 (Halle, Germany: Institute of Agricultural Development in Central and Eastern Europe, 1998), p. 13.

20. Se-Kyung Chong, "Anmyeon-do Recreation Forest: A Millennium of Management," in Patrick B. Durst et al., *In Search of Excellence: Exemplary Forest Management in Asia and the Pacific*, Asia-Pacific Forestry Commission (Bangkok: FAO Regional Office for Asia and the Pacific, 2005), pp. 251–59.

21. Ibid.

22. The Turkish Foundation for Combating Soil Erosion (TEMA), at english.tema.org.tr, viewed 10 August 2005.

23. "China's Great Green Wall," *BBC*, 3 March 2001; Evan Ratliff, "The Green Wall of China," *Wired*, April 2003.

24. United Nations, "The Great North American Dust Bowl: A Cautionary Tale," *Global Alarm Dust and Sandstorms from the World's Drylands* (Bangkok: Secretariat of the U.N. Convention to Combat Desertification, 2002), pp. 77–121.

25. USDA, Economic Research Service (ERS), *Agri-Environmental Policy at the Crossroads: Guideposts on a Changing Landscape*, Agricultural Economic Report No. 794 (Washington, DC: 2001); USDA, Farm Service Agency Online, "History of the CRP," in *The Conservation*

Reserve Program, at www.fsa.usda.gov/dafp/cepd/12crplogo/history.htm, viewed 28 September 2005.

26. USDA, *Agri-Environmental Policy at the Crossroads*, op. cit. note 25, p. 16; loss of topsoil from water erosion from USDA, *Summary Report: 1997 Natural Resources Inventory* (Washington, DC, and Ames, IA: Natural Resources Conservation Service and Statistical Laboratory, Iowa State University, 1999, rev. 2000), pp. 46–51.

27. R. Neil Sampson, *Farmland or Wasteland* (Emmaus, PA: Rodale Press, 1981), p. 242.

28. USDA, Natural Resources Conservation Service, *CORE4 Conservation Practices Training Guide: The Common Sense Approach to Natural Resource Conservation* (Washington, DC: August 1999); Rolf Derpsch, "Frontiers in Conservation Tillage and Advances in Conservation Practice," in D. E. Stott, R. H. Mohtar, and G. C. Steinhardt, eds., *Sustaining the Global Farm*, selected papers from the 10th International Soil Conservation Organization Meeting, at Purdue University and USDA-ARS National Soil Erosion Research Laboratory, 24–29 May 1999 (Washington, DC: 2001), pp. 248–54.

29. Conservation Technology Information Center, Purdue University, "National Tillage Trends (1990-2004)," from the 2004 National Crop Residue Management Survey Data, at www.ctic.purdue.edu/ctic/CRM2004/1990-2004data.pdf, viewed 10 August 2005; FAO, *Intensifying Crop Production with Conservation Agriculture*, at www.fao.org/ag/ags/aGSE/main.htm, viewed 20 May 2003; Rolf Derpsch and J. R. Benites, "The Extent of CA / No-tillage Adoption Worldwide" to be presented at the Third World Congress on Conservation Agriculture, Nairobi, Kenya, 3–7 October 2005, e-mail to Danielle Murray, Earth Policy Institute, 9 August 2005.

30. FAO, op. cit. note 29.

31. "Algeria to Convert Large Cereal Land to Tree-Planting," *Reuters*, 8 December 2000; Souhail Karam, "Drought-Hit North Africa Seen Hunting for Grains," *Reuters*, 15 July 2005.

32. Silvia Aloisi, "Senegal Mulls 'Green Wall' to Stop Desert Advance,' *Reuters*, 1 August 2005.

33. Ratliff, op. cit. note 23; Sun Xiufang and Ralph Bean, *China Solid Wood Products Annual Report 2002* (Beijing: USDA, 2002).

34. Author's discussion with officials of Helin County, Inner Mongolia (Nei Monggol), 17 May 2002.

35. Ibid.

36. U.S. Embassy, *Grapes of Wrath in Inner Mongolia* (Beijing: May 2001).

37. India's dairy industry from A. Banerjee, "Dairying Systems in India," *World Animal Review*, vol. 79/2 (Rome: FAO, 1994).

38. Sandra Postel and Brian Richter, *Rivers for Life: Managing Water for People and Nature* (Washington, DC: Island Press, 2003), p. 85.

39. Megan Dyson, Ger Bergkamp, and John Scanlon, eds., *Flow: The Essentials of Environmental Flows* (Gland, Switzerland, and Cambridge, U.K.: World Conservation Union–IUCN, 2003), p. 2.

40. Sandra Postel, *Pillar of Sand* (New York: W.W. Norton & Company, 1999), pp. 121–22.

41. Ibid.

42. John Tibbetts, "Making Amends: Ecological Restoration in the United States," *Environmental Health Perspectives*, vol. 108, no. 8 (August 2000), pp. A357–A361.

43. Definition of marine reserves network from "Scientific Consensus Statement on Marine Reserves and Marine Protected Areas," presented at the AAAS annual meeting, 15-20 February 2001, initial signatories include Steven Gaines, Jane Lubchenco, Stephen Palumbi, and Megan Detheir, p. 2.

44. Andrew Balmford et al., "The Worldwide Costs of Marine Protected Areas," *Proceedings of the National Academy of Sciences,* vol. 101, no. 26 (29 June 2004), pp. 9,694–97; "Costs of a Worldwide System of Marine Parks," press release (York: The University of York, 12 July 2004).

45. Balmford et al., op. cit. note 44; Tim Radford, "Marine Parks Can Solve Global Fish Crisis, Experts Say," *Guardian* (London), 15 June 2004.

46. Balmford op. cit. note 44; Radford, op. cit. note 45.

47. Radford, op. cit. note 45; Richard Black, "Protection Needed for 'Marine Serengetis,'" *BBC News*, 6 August 2003; Balmford et al., op. cit. note 44.

48. American Association for the Advancement of Science (AAAS), "Leading Marine Scientists Release New Evidence that Marine Reserves Produce Enormous Benefits within Their Boundaries and Beyond," press release (Washington, DC: 12 March 2001); "Scientific Consensus Statement," op. cit. note 43.

49. AAAS, op. cit. note 48; "Scientific Consensus Statement," op. cit. note 43.

50. R. J. Diaz, J. Nestlerode, and M. L. Diaz, "A Global Perspective on the Effects of Eutrophication and Hypoxia on Aquatic Biota," in G. L. Rupp and M. D. White, eds., *Proceedings of the 7th Annual Symposium on Fish Physiology, Toxicology and Water Quality*, Estonia, 12–15 May 2003 (Athens, GA: U.S. Environmental Protection Agency, Ecosystems Research Division, 2004); United Nations Environment Programme (UNEP), *GEO Yearbook 2003* (Nairobi: 2004).

51. Diaz, Nestlerode, and Diaz, op. cit. note 50; UNEP, op. cit. note 50; Mark Peters et al., "Reducing Nitrogen Flow to the Gulf of Mexico: Strategies for Agriculture," *Agricultural Outlook*, November 1999, pp. 20–24.

52. Organisation for Economic Co-operation and Development, *Review of Fisheries in OECD Countries: Policies and Summary Statistics* (Paris: 2003), pp. 55–56; World Wildlife Fund, *Hard Facts, Hidden Problems: A Review of Current Data on Fishing Subsidies* (Washington, DC: 2001), pp. ii; Balmford et al., op. cit. note 44; Radford, op. cit. note 45.

53. United Nations, *World Population Prospects: The 2004 Revision* (New York: 2005).

54. J.R. Pegg, "Global Forces Threaten World's Parks," *Environment News Service*, 27 August 2003.

55. Conservation International, "Biodiversity Hotspots," at www.biodiversityhotspots.org/xp/Hotspots, viewed 10 August 2005; Steve Connor, "New Biodiversity Hotspots Revealed," *Independent* (London), 7 September 2005.

56. U.S. Fish and Wildlife Service, *The Endangered Species Act of 1973*, at www.fws.gov/endangered/esaall.pdf, viewed 10 August 2005.

57. Table 8–1 sources include reforesting the earth and protecting topsoil on cropland from Lester R. Brown and Edward C. Wolf, "Reclaiming the Future," in Lester R. Brown et al., *State of the World 1988* (New York: W.W. Norton & Company, 1988), p. 174, using data from FAO, *Fuelwood Supplies in the Developing Countries*, Forestry Paper 42 (Rome: 1983); restoring rangelands from UNEP, *Status of Desertification and Implementation of the United Nations Plan of Action to Combat Desertification* (Nairobi: 1991), pp. 73–92; marine reserves from Balmford et al., op. cit. note 44; protecting biological diversity from World Parks Congress, "Financial Security for Protected Areas," at www.iucn.org/themes/wcpa/wpc2003/pdfs/outputs/recommendations/approved/english/html/r07.htm; World Parks Congress, "The Durban Accord," at www.iucn.org/themes/wcpa/wpc2003/english/outputs/durban/durbanaccord.htm.

58. Chong, op. cit. note 20; TEMA, op. cit. note 22; "The Nobel Peace Prize 2004," press release, The Norwegian Nobel Committee, 8 October 2004, at www.nobelprize.org.

59. Brown and Wolf, op. cit. note 57, p. 175.

60. Runsheng Yin et al., "China's Ecological Rehabilitation: The Unprecedented Efforts and Dramatic Impacts of Reforestation and Slope Protection in Western China," in Woodrow Wilson International Center for Scholars, China Environment Forum, *China Environment Series,* Issue 7, Washington, DC, 2005, pp. 17–32.

61. Brown and Wolf, op. cit. note 57, p. 176.

62. Ibid., pp. 173–74.

63. Ibid., p. 174.

64. Ibid.

65. Ibid.

66. UNEP, op. cit. note 57, with dollar figures converted from 1990 to 2004 dollars using implicit price deflators from U.S. Department of Commerce, Bureau of Economic Analysis, "Table C.1. GDP and Other Major NIPA Aggregates," in *Survey of Current Business*, September 2005, p. D-48.

67. H. E. Dregne and Nan-Ting Chou, "Global Desertification Dimensions and Costs," in *Degradation and Restoration of Arid Lands* (Lubbock, TX: Texas Tech. University, 1992); UNEP, op. cit. note 57.

68. Balmford et al., op. cit. note 44.

69. World Parks Congress, "Financial Security for Protected Areas" and "The Durban Accord," op. cit. note 57.

70. Irrigated cropland from FAO, op. cit. note 6, land data updated 4 April 2005.

71. Jordan from Tom Gardner-Outlaw and Robert Engelman, *Sustaining Water, Easing Scarcity: A Second Update* (Washington, DC: Population Action International, 1997); Mexico from Sandra Postel, *Last Oasis* (New York: W.W. Norton & Company, 1997), pp.150–51.

72. Postel, op. cit. note 40, pp. 230–35; Postel, op. cit. note 71, pp. 167–68.

Chapter 9. Feeding Seven Billion Well

1. "Last Food Shipment Signals End of 25-Year WFP Aid to China," *Asian Economic News*, 8 April 2005; U.S. Department of Agriculture (USDA), *Production, Supply, & Distribution*, electronic database, at www.fas.usda.gov/psd, updated 13 July 2005.

2. U.N. Food and Agriculture Organization (FAO), *The State of Food Insecurity in the World 2004* (Rome: 2004), p. 6.

3. Thomas R. Sinclair, "Limits to Crop Yield," paper presented at the 1999 National Academy Colloquium, *Plants and Populations: Is There Time?* Irvine, CA, 5–6 December 1998; FAO, *FAOSTAT Statistics Database*, at apps.fao.org, with fertilizer use data updated 4 April 2005.

4. United Nations, *World Population Prospects: The 2004 Revision* (New York: 2005).

5. USDA, op. cit. note 1.

6. Ibid.

7. John Wade, Adam Branson, and Xiang Qing, *China Grain and Feed Annual Report 2002* (Beijing: USDA, 2002); USDA, op. cit. note 1.

8. Double-cropping yields from USDA, *India Grain and Feed Annual Report 2003* (New Delhi: 2003); population from United Nations, op. cit. note 4; USDA, op. cit. note 1.

9. Grain harvested area from USDA, op. cit. note 1; USDA, *Japan Grain and Feed Annual Report 2003* (Tokyo: 2003).

10. USDA, op. cit. note 1.

11. Richard Magleby, "Soil Management and Conservation," in USDA, *Agricultural Resources and Environmental Indicators 2003* (Washington, DC: February 2003), Chapter 4.2, p. 14.
12. USDA, op. cit. note 1; Randall D. Schnepf et al., *Agriculture in Brazil and Argentina* (Washington, DC: USDA Economic Research Service (ERS), 2001), pp. 8–10.
13. FAO, op. cit. note 3; USDA, op. cit. note 1.
14. Pedro Sanchez, "The Climate Change–Soil Fertility–Food Security Nexus," summary note (Bonn: International Food Policy Research Institute, 4 September 2001).
15. Water requirements for grain production from FAO, *Yield Response to Water* (Rome: 1979); water use from I.A. Shiklomanov, "Assessment of Water Resources and Water Availability in the World," Report for the Comprehensive Assessment of the Freshwater Resources of the World (St. Petersburg, Russia: State Hydrological Institute, 1998), cited in Peter H. Gleick, *The World's Water 2000–2001* (Washington, DC: Island Press, 2000), p. 53.
16. Water use for grain production from FAO, op. cit. note 15.
17. Sandra Postel and Amy Vickers, "Boosting Water Productivity," in Worldwatch Institute, *State of the World 2004* (New York: W.W. Norton & Company, 2004), pp. 51–52.
18. Ibid.
19. Wang Shucheng, private meeting with author, Beijing, May 2004.
20. FAO, *Crops and Drops* (Rome: 2002), p. 17; Alain Vidal, Aline Comeau, and Hervé Plusquellec, *Case Studies on Water Conservation in the Mediterranean Region* (Rome: FAO, 2001), p. vii.
21. FAO, op. cit. note 20; Vidal, Comeau, and Plusquellec, op. cit. note 20.
22. Postel and Vickers, op. cit. note 17, p. 53.
23. Sandra Postel et al., "Drip Irrigation for Small Farmers: A New Initiative to Alleviate Hunger and Poverty," *Water International*, March 2001, pp. 3–13.
24. Ibid.
25. For more information on water users associations, see R. Maria Saleth and Arial Dinar, *Water Challenge and Institutional Response: A Cross-Country Perspective* (Washington, DC: World Bank, 1999), p. 26.
26. Saleth and Dinar, op. cit. note 25, p. 6.
27. World Bank and Swiss Agency for Development and Cooperation, Summary Report, *Middle East and North Africa Regional Water Initiative Workshop on Sustainable Groundwater Management*, Sana'a, Yemen, 25–28 June 2000, p. 19.
28. Peter Wonacott, "To Save Water, China Lifts Price," *Wall Street Journal*, 14 June 2004.

29. USDA, op. cit. note 1.

30. Population from United Nations, op. cit. note 4; grain consumption from USDA, op. cit. note 1; water calculation based on 1,000 tons of water for 1 ton of grain from FAO, op. cit. note 15.

31. USDA, op. cit. note 1.

32. FAO, op. cit. note 3, with livestock data updated 14 July 2005; 2005 production estimates from FAO, Global Information and Early Warning System on Food and Agriculture (GIEWS), *Food Outlook*, No. 1 (Rome: April 2005).

33. Feed-to-poultry conversion ratio derived from data in Robert V. Bishop et al., *The World Poultry Market—Government Intervention and Multilateral Policy Reform* (Washington, DC: USDA, 1990); conversion ratio of grain to beef based on Allen Baker, Feed Situation and Outlook staff, ERS, USDA, discussion with author, 27 April 1992; pork data from Leland Southard, Livestock and Poultry Situation and Outlook staff, ERS, USDA, discussion with author, 27 April 1992; fish from Rosamond L. Naylor et al., "Effect of Aquaculture on World Fish Supplies," *Nature*, vol. 405 (29 June 2000), pp. 1,017–24.

34. Figure 9–1 from FAO, op. cit. note 3, with livestock data updated 14 July 2005; FAO, GIEWS, op. cit. note 32; fish data from FAO, *FISHSTAT Plus*, electronic database, at www.fao.org/fi/statist/FISOFT/FISHPLUS.asp, updated March 2005.

35. FAO, op. cit. note 34.

36. Naylor et al., op. cit. note 33; polyculture in India from W. C. Nandeesha et al., "Breeding of Carp with Oviprim," in Indian Branch, Asian Fisheries Society, India, Special Publication No. 4 (Mangalore, India: 1990), p. 1.

37. Krishen Rana, "Changing Scenarios in Aquaculture Development in China," *FAO Aquaculture Newsletter*, August 1999, p. 18.

38. Catfish requirements from Naylor et al., op. cit. note 33; U.S. catfish production data from USDA, National Agricultural Statistics Service, *Catfish Production* (Washington, DC: February 2003), p. 5.

39. FAO, op. cit. note 34; Naylor et al., op. cit. note 33; Taija-Riitta Tuominen and Maren Esmark, *Food For Thought: The Use of Marine Resources in Fish Feed* (Oslo: WWF-Norway, 2003); Rebecca Goldburg and Rosamond Naylor, "Future Seascapes, Fishing, and Fish Farming," *Frontiers in Ecology and the Environment*, vol. 3, no. 1 (February 2005), pp. 21–28.

40. FAO, op. cit. note 34; FAO, *The State of World Fisheries and Aquaculture 2004* (Rome: 2004).

41. USDA, op. cit. note 1; Suzi Fraser Dominy, "Soy's Growing Importance," *World Grain*, 13 April 2004.

42. Use of soy is from author's calculations based on USDA, op. cit. note 1, and on USDA, Foreign Agricultural Service (FAS), various agricul-

tural reports (Washington, DC: various years); growth in biodiesel discussed in more detail in Chapter 2.

43. USDA, op. cit. note 1.

44. Ibid.

45. Ibid.; David McKee, "Crushing Competition," *World Grain*, 13 April 2004; USDA, FAS, *China Oilseeds and Products Annual Report 2004* (Beijing: March 2004); Dominy, op. cit. note 41.

46. Historical statistics in Worldwatch Institute, *Signposts 2002*, CD-Rom (Washington, DC: 2002); USDA, op. cit. note 1.

47. Figure 9–2 from FAO, op. cit. note 3, updated 14 July 2005; preliminary 2005 production estimates from FAO, GIEWS, *Food Outlook*, No. 2 (Rome: June 2005).

48. S. C. Dhall and Meena Dhall, "Dairy Industry—India's Strength in Its Livestock," *Business Line*, Internet Edition of *Financial Daily* from *The Hindu* group of publications, 7 November 1997; see also Surinder Sud, "India Is Now World's Largest Milk Producer," *India Perspectives*, May 1999, pp. 25–26; A. Banerjee, "Dairying Systems in India," *World Animal Review*, vol. 79, no. 2 (1994).

49. Milk supply per person from FAO, op. cit. note 3, updated 27 August 2004; United Nations, op. cit. note 4.

50. Banerjee, op. cit. note 48; Dhall and Dhall, op. cit. note 48.

51. Wade, Branson, and Xiang, op. cit. note 7; China's crop residue production and use from Gao Tengyun, "Treatment and Utilization of Crop Straw and Stover in China," *Livestock Research for Rural Development*, February 2000.

52. USDA, ERS, "China's Beef Economy: Production, Marketing, Consumption, and Foreign Trade," *International Agriculture and Trade Reports: China* (Washington, DC: July 1998), p. 28.

53. S. F. Li, "Aquaculture Research and Its Relation to Development in China," in World Fish Center, *Agricultural Development and the Opportunities for Aquatic Resources Research in China* (Penang, Malaysia: 2001), p. 26; FAO, op. cit. note 34.

54. FAO, op. cit. note 34; FAO, op. cit. note 3, updated 14 July 2005.

55. United Nations, op. cit. note 4; FAO, op. cit. note 3, updated 14 July 2005.

56. China's economic growth from International Monetary Fund, *World Economic Outlook Database*, at www.imf.org/external/pubs/ft/weo, updated 13 April 2005.

57. Meat consumption from FAO, op. cit. note 3, updated 14 July 2005.

58. Author's calculations from USDA, op. cit. note 1; United Nations, op. cit. note 4.

59. USDA, op. cit. note 1; United Nations, op. cit. note 4.

60. USDA, op. cit. note 1; United Nations, op. cit. note 4; Organisation for Economic Co-operation and Development, "Total Health Expenditure Per Capita, US$ PPP," table, OECD Health Data, www.oecd.org, June 2005.

61. FAO, op. cit. note 3, updated 14 July 2005.

62. USDA, op. cit. note 1.

63. Poultry derived from data in Bishop et al., op. cit. note 33; catfish and carp from Naylor et al., op. cit. note 33.

64. Naylor et al., op. cit. note 33; feed-to-poultry conversion ratio derived from data in Bishop et al., op. cit. note 33.

65. Aquaculture output from FAO, op. cit. note 34.

66. Ibid.; "Mekong Delta to Become Biggest Aquatic Producer in Vietnam," *Vietnam News Agency*, 3 August 2004.

67. USDA, ERS, Natural Resources and Environment Division, *Agricultural Resources and Environmental Indicators, 1996–1997*, Agricultural Handbook No. 712 (Washington, DC: 1997).

Chapter 10. Stabilizing Climate

1. United Nations, *Kyoto Protocol to the United Nations Framework Convention on Climate Change* (New York: 1997); S. Pacala and R. Socolow, "Stabilization Wedges: Solving the Climate Problem for the Next 50 Years with Current Technologies," *Science*, 13 August 2004.

2. European Commission, "Commissioner Piebalgs: Europe Could Save 20% of Its Energy by 2020," press release (Brussels: 22 June 2005); "Europe Tries to Replace Fossil Fuels With Sustainable Energy," *Environment News Service,* 18 July 2005.

3. European Commission, op. cit. note 2; "Europe Tries to Replace Fossil Fuels With Sustainable Energy," op. cit. note 2.

4. James Brooke, "Japan Squeezes to Get the Most of Costly Fuel," *New York Times*, 4 June 2005.

5. Ibid.

6. Ralph Torrie, Richard Parfett, and Paul Steenhof, *Kyoto and Beyond: The Low-Emission Path to Innovation and Efficiency* (Ottawa: The David Suzuki Foundation and Climate Action Network Canada, October 2002); Alison Bailie et al., *The Path to Carbon-Dioxide-Free Power: Switching to Clean Energy in the Utility Sector*, A Study for the World Wildlife Fund (Washington, DC: Tellus Institute and The Center for Energy and Climate Solutions, April 2003).

7. Ontario Ministry of Energy, "McGuinty Government Unveils Bold Plan to Clean Up Ontario's Air," press release (Toronto: 15 June 2005); EIN Publishing, "Ontario Unveils Plan for Replacing Coal-fired Power Plants," *Global Warming Today,* 28 June 2005; Gibbons quoted in Martin Mittelstaedt, "Putting Out the Fires," *Globe and Mail* (Toronto), 15 March 2003.

8. Ray Anderson, writing in Torrie, Parfett, and Steenhof, op. cit. note 6, p. 2.

9. Per capita energy consumption in U.S. Department of Energy (DOE), Energy Information Administration (EIA), "France," "Germany," "Spain," "United Kingdom," "United States," *EIA Country Analysis Briefs* (Washington, DC: updated at various times between November 2004 and July 2005).

10. Bill Prindle, "How Energy Efficiency Can Turn 1300 New Power Plants Into 170," fact sheet (Washington, DC: Alliance to Save Energy, 2 May 2001).

11. Howard Geller, "Compact Fluorescent Lighting," *American Council for an Energy-Efficient Economy Technology Brief*, www.aceee.org, viewed 1 May 2003.

12. Gasoline savings based on Malcolm A. Weiss et al., *Comparative Assessment of Fuel Cell Cars* (Cambridge, MA: Massachusetts Institute of Technology, February 2003); 2004 sales estimate from "Sales Numbers and Forecasts for Hybrid Vehicles," at www.hybridcars.com, viewed 29 August 2005; 2008 sales projections from David L. Greene, K. G. Duleep, and Walter McManus, *Future Potential of Hybrid and Diesel Powertrains in the U.S. Light-Duty Vehicle Market* (Oak Ridge, Tennessee: Oak Ridge National Laboratory, 2004).

13. Figure 10–1 from Worldwatch Institute, *Signposts 2004*, CD-Rom (Washington, DC: 2004), updated by Earth Policy Institute from Global Wind Energy Council (GWEC), "Global Wind Power Continues Expansion: Pace of Installation Needs to Accelerate to Combat Climate Change," press release (Brussels: 4 March 2005); American Wind Energy Association (AWEA), *Global Wind Energy Market Report* (Washington, DC: March 2004). Oil, natural gas, coal, and nuclear power from British Petroleum (BP), *BP Statistical Review of World Energy 2005* (London: Group Media & Publishing, 2005), pp. 9, 25, 33–34.

14. Worldwatch Institute, op. cit. note 13, updated by Earth Policy Institute from GWEC, op. cit. note 13; Danish Wind Industry Association, "Did You Know?" fact sheet, at www.windpower.org, viewed 1 August 2005; BTM Consult ApS, "International Wind Energy Development: World Market Update 2004: Forecast 2005–2009," press release (Ringkøbing, Denmark: 31 March 2005).

15. GWEC, op. cit. note 13; GWEC, *Wind Force 12: A Blueprint to Achieve 12% of the World's Electricity from Wind Power by 2020* (Belgium: European Wind Energy Association and Greenpeace, 2005); European Wind Energy Association (EWEA), *Wind Power Targets for Europe: 75,000 MW by 2010* (Belgium: October 2003).

16. GWEC, op. cit. note 13; GWEC, op. cit. note 15; Garrad Hassan and Partners, *Sea Wind Europe* (London: Greenpeace, March 2004).

17. British Wind Energy Association (BWEA), "Statistics," fact sheet, www.bwea.org, viewed 8 August 2005; "Big Boost for Offshore Wind Power," *Reuters*, 19 December 2003.

18. Estimate of heat wave deaths across Europe compiled in Janet Larsen, "Record Heat Wave in Europe Takes 35,000 Lives," *Eco-Economy Update* (Washington, DC: Earth Policy Institute, 9 October 2003), updated with Istituto Nazionale di Statistica (Istat), *Bilancio Demografico Nazionale: Anno 2003* (Rome: Istituto Nazionale di Statistica, 2004); wind power from GWEC, op. cit. note 13; Les Perreaux, "Windmill Project To Push Quebec Past Alberta In Wind Energy Production," *Canadian Press*, 5 October 2004; Stephen Leahy, "Change in the Chinese Wind," *Wired News*, 4 October 2004; GWEC, op. cit. note 15.

19. D. L. Elliott, L. L. Wendell, and G. L. Gower, *An Assessment of the Available Windy Land Area and Wind Energy Potential in the Contiguous United States* (Richland, WA: Pacific Northwest Laboratory, 1991).

20. Ibid.; C. L. Archer and M. Z. Jacobson, "The Spatial and Temporal Distributions of U.S. Winds and Wind Power at 80 m Derived from Measurements," *Journal of Geophysical Research*, 16 May 2003.

21. Larry Flowers, National Renewable Energy Laboratory, "Wind Power Update," www.eren.doe.gov/windpoweringamerica/pdfs/wpa/wpa_update.pdf, viewed 19 June 2002; Glenn Hasek, "Powering the Future," *Industry Week*, 1 May 2000; 2¢ per kilowatt-hour from EWEA and Greenpeace, *Wind Force 12* (Brussels: May 2003).

22. "US Wind Power Industry Gets Tax Credit Boost," *Reuters*, 13 March 2002; "Blocked US Energy Bill Slows Wind Power Projects," *Reuters*, 12 January 2004; American Wind Energy Association, "Energy Bill Extends Wind Power Incentive Through 2007: First-Ever 'Seamless' Extension Will Spur Investment, Job Creation, and Clean Energy Production," press release (Washington, DC: 29 July 2005).

23. Internet from Molly O'Meara Sheehan, "Communications Networks Expand," in Worldwatch Institute, *Vital Signs 2003* (New York: W.W. Norton & Company, 2003), pp. 60–61.

24. Harry Braun, *The Phoenix Project: Shifting From Oil to Hydrogen with Wartime Speed*, prepared for the Renewable Hydrogen Roundtable, World Resources Institute, Washington, DC, 10–11 April 2003, pp. 3–4; ability of U.S. automobile industry to produce a million wind turbines per year is author's estimate.

25. Fossil fuel subsidies from Bjorn Larsen, *World Fossil Fuel Subsidies and Global Carbon Emissions in a Model with Interfuel Substitution*, Policy Research Working Paper 1256 (Washington, DC: World Bank, 1994), p. 7; companies involved in wind from Birgitte Dyrekilde, "Big Players to Spark Wind Power Consolidation," *Reuters*, 18 March 2002.

26. Jim Dehlsen, Clipper Wind, discussion with author, 30 May 2001; Massachusetts Institute of Technology, "MIT Hosts Hearing On Cape Wind Farm," press release (Cambridge, MA: 14 December 2004).

27. AWEA, "Wind Energy Projects," fact sheet (Washington, DC: 24 April 2005); calculation of electricity production from Tom Gray, AWEA, e-mail to author, 12 June 2002.

28. Wind royalties from Union of Concerned Scientists, "Farming the Wind: Wind Power and Agriculture," www.ucsusa.org/clean_energy/renewable_energy/page.cfm?pageID=128; for corn, calculations by author using data from John Dittrich, American Corn Growers Association, "Major Crops: A 27-Year History with Inflation Adjustments," *Key Indicators of the U.S. Farm Sector* (Washington, DC: January 2002); beef is author's estimate.

29. "Benchmark Oil Price Hits Dollars 66.50 A Barrel," *Financial Times*, 29 September 2005.

30. DOE and U.S. Environmental Protection Agency (EPA), *Fuel Economy Guide* (Washington, DC: 2005); gasoline savings based on Malcolm A. Weiss et al., *Comparative Assessment of Fuel Cell Cars* (Cambridge, MA: Massachusetts Institute of Technology, 2003).

31. DOE and EPA, op. cit. note 30; Marv Balousek, "Hybrid Cars Are Catching On," *Wisconsin State Journal*, 10 August 2005; EPA, "Emission Facts," fact sheet, www.epa.gov/otaq/consumer/f00013.htm, updated 15 July 2005.

32. John Porretto, "Ford Expands Lineup of Hybrid SUVs," *Chicago Sun-Times*, 14 April 2004; Matthew L. Wald, "Designed to Save, Hybrids Burn Gas in Drive for Power," *New York Times*, 17 July 2005; General Motors, "Hybrid Power to the People," *New York Times*, 27 September 2004.

33. Lester R. Brown, "The Short Path to Oil Independence," *Eco-Economy Update* (Washington, DC: Earth Policy Institute, 13 October 2004); Senator Joseph Lieberman, remarks prepared for Loewy Lecture, Georgetown University (Washington, DC: 7 October 2005).

34. Amory B. Lovins et al., *Winning the Oil Endgame: Innovation for Profits, Jobs, and Security* (Snowmass, CO: Rocky Mountain Institute, 2004), p. 64.

35. Associated Press, "Review Faults Electricity Grid System," *Los Angeles Times*, 30 September 2004.

36. C. L. Archer and M. Z. Jacobson, "Evaluation of Global Wind Power," *Journal of Geophysical Research*, vol. 110, no. D12110 (2005), pp. 1–20.

37. Denis Hayes, "Sunpower," in Energy Foundation, *2001 Annual Report* (San Francisco: February 2002), pp. 10–18.

38. Figure 10–2 shows cumulative solar installations with data compiled from Paul Maycock, "PV News Annual Market Survey Results," *Photovoltaic News*, April 2005; Janet L. Sawin, "Solar Energy Markets Booming," in Worldwatch Institute, *Vital Signs 2005* (New York: W.W. Norton & Company, 2005), pp. 36–37; market share from Katharine Mieszkowski, "How George Bush Lost the Sun," *Salon*, 25 October 2004; Michael Schmela, "This is a Sharp World," *Photon International*, March 2004.

39. William J. Kelly, "German Renewables Law Portends Tight California

Market," *California Energy Circuit,* 18 May 2004; Office of Energy Efficiency and Renewable Energy (EERE), DOE, "Net Metering, Tax Credits for Solar Energy Included in Energy Act," *EERE Network News*, 10 August 2005.

40. European Photovoltaic Industry Association and Greenpeace, *Solar Generation* (Brussels: September 2001); Paul Maycock, "Japanese PV Residential Dissemination Program Exceeds Goals," *Photovoltaic News*, January 2004.

41. Paul Maycock, "German 100,000 Roofs Program Tops 130 MW in 2003," *Photovoltaic News*, August 2004.

42. Kelly, op. cit. note 39.

43. "Manchester's Tallest Building Gets Europe's Largest Solar Array," *Environment News Service*, 9 November 2004.

44. "Power to the Poor," *The Economist*, 10 February 2001, pp. 21–23.

45. Bernie Fischlowitz-Roberts, "Sales of Solar Cells Take Off," *Eco-Economy Update* (Washington, DC: Earth Policy Institute, 11 June 2002); population without electricity in World Summit on Sustainable Development, Department of Public Information, Press Conference on Global Sustainable Energy Network (Johannesburg: 1 September 2002).

46. Paul Maycock, "Japanese Issue 'Roadmap to 2030,'" *Photovoltaic News*, December 2004, p. 1, and Mantik Kusjanto and Anneli Palmen, "Germany's Solar World Seeks Place in the Sun," *Reuters*, 13 January 2005, cited in Sawin, op. cit. note 38; EERE, op. cit. note 39.

47. Robert H. Williams, "Facilitating Widespread Deployment of Wind and Photovoltaic Technologies," in Energy Foundation, *2001 Annual Report* (San Francisco: February 2002), pp. 19–30.

48. Scott Sklar, "Sleepers That Are Coming to Light," *Earthscan*, 7 February 2005; EERE, "Spain to Build an 11-Megawatt Solar Power Tower," *EERE Network News*, 24 August 2005.

49. Sawin, op. cit. note 38.

50. Li Hua, "From Quantity to Quality: How China's Maturing Solar Thermal Industry Will Need to Face Up to Market Challenges," *Renewable Energy World*, January-February 2005, pp. 56–57, cited in Sawin, op. cit. note 38; Germany from David Sharrock, "Spain Makes Solar Panels Mandatory in New Buildings," *Times Online (U.K.)*, 9 November 2004.

51. Sawin, op. cit. note 38; Sharrock, op. cit. note 50.

52. Charlene Wardlow, "The Environmental Benefits of Geothermal Energy," presented at Environmental and Energy Study Institute, "Geothermal Energy: Heating Up the Renewable Energy Portfolio," briefing to United States House of Representatives, Washington, DC, 8 February 2005.

53. Japan from Hal Kane, "Geothermal Power Gains," in Lester R. Brown

et al., *Vital Signs 1993* (New York: W.W. Norton & Company, 1993), p. 54; DOE, EIA, "Japan," *EIA Country Analysis Brief* (Washington, DC: updated August 2004); other potential in World Bank, "Geothermal Energy," prepared under the PB Power and World Bank partnership program, www.worldbank.org/html/fpd/energy/geothermal, viewed 23 January 2003.

54. Mary H. Dickson and Mario Fanelli, "What is Geothermal Energy?" (Pisa, Italy: Istituto di Geoscienze e Georisorse, CNR, February 2004), online at International Geothermal Association, iga.igg.cnr.it/index.php; 1990 data from International Geothermal Association, "Electricity Generation," at iga.igg.cnr.it/index.php, updated 20 July 2005.

55. Dickson and Fanelli, op. cit. note 54; Philippines share from World Bank, op. cit. note 53; California from Alyssa Kagel, Diana Bates and Karl Gawell, *A Guide to Geothermal Energy and the Environment* (Washington, DC: Geothermal Energy Association, 22 April 2005).

56. World Bank, op. cit. note 53.

57. John W. Lund and Derek H. Freeston, "World-Wide Direct Uses of Geothermal Energy 2000," *Geothermics*, vol. 30 (2001), pp. 34, 51, 53; Ben Hirschler, "Hydrogen Puts Iceland on Road to Oil-Free Future," *Reuters*, 31 May 2002.

58. Lund and Freeston, op. cit. note 57.

59. Ibid.; California in World Bank, op. cit. note 53.

60. World Bank, op. cit. note 53.

61. Lund and Freeston, op. cit. note 57, pp. 46, 53.

62. Population from United Nations, *World Population Prospects: The 2004 Revision* (New York: February 2005); Peter Janssen, "The Too Slow Flow: Why Indonesia Could Get All Its Power From Volcanoes—But Doesn't," *Newsweek*, 20 September 2004.

63. Calculation of electricity production from Gray, op. cit. note 27; Renewable Fuels Association, "Homegrown for the Homeland: Ethanol Industry Outlook 2005" (Washington, DC: 2005).

64. Renewable Fuels Association, op. cit. note 63; average ethanol yield in Brazil calculated by Earth Policy Institute from São Paulo Sugar Cane Agroindustry Union (UNICA), cited in Alfred Szwarc, "Use of Bio-Fuels in Brazil," presentation at *In-Session Workshop on Mitigation, SBSTA 21 / COP 10*, Buenos Aires: Ministry of Science and Technology, 9 December 2004; Christoph Berg, *World Fuel Ethanol Analysis and Outlook* (Ratzeburg, Germany: F.O. Licht, April 2004); net energy yields from F.O. Licht, cited in Szwarc, op. cit. this note.

65. "Oil Yields and Characteristics," Journey to Forever, at www.journeytoforever.org/biodiesel_yield.html, viewed 15 July 2005; soybean yield is author's estimate.

66. Geothermal heat and hydrogen from Árni Ragnarsson and Thorkell Helgason, eds., *Energy In Iceland: Historical Perspective, Present*

Status, Future Outlook (Reykjavik, Iceland: National Energy Authority (Orkustofnun) and Ministries of Industry and Commerce, February 2004), pp. 21, 42; hydropower from Ragnheidur Inga Thorarinsdottir and Helga Bardadottir, eds., *Energy Statistics in Iceland* (Reykjavik, Iceland: National Energy Authority (Orkustofnun), September 2004.

67. Lovins et al., op. cit. note 34.
68. DOE, EIA, "United States," *EIA Country Analysis Brief* (Washington, DC: updated January 2005).
69. Donald W. Aitken, "Germany Launches Its Transition: How One of the Most Advanced Industrial Nations Is Moving to 100 Percent Energy from Renewable Sources," *Solar Today*, March/April 2005, pp. 26–29.
70. Janssen, op. cit. note 62; hydroelectricity from DOE, EIA, "Indonesia," *EIA Country Analysis Brief* (Washington, DC: updated July 2004).
71. GWEC, op. cit. note 15, p. 7.
72. Brazil's ethanol self-sufficiency potential calculated by Earth Policy Institute from UNICA, "Brazil as a Strategic Supplier of Fuel Ethanol," presentation for the Governors' Ethanol Coalition, January 2005.
73. Hydropower and electricity generation from DOE, EIA, "China," *EIA Country Analysis Brief* (Washington, DC: updated August 2005); wind potential from GWEC, op. cit. note 15, p. 28.
74. GWEC, op. cit. note 15, p. 43.
75. DOE, op. cit. note 68; C. Palese et al., "Wind Regime and Wind Power in North Patagonia, Argentina," *Wind Engineering*, 1 September 2000, pp. 361–77; "Clean Energy in Patagonia from Wind and Hydrogen," *Inter Press Service*, 15 May 2005.
76. Kelly, op. cit. note 39.

Chapter 11. Designing Sustainable Cities

1. United Nations, *World Urbanization Prospects: The 2003 Revision* (New York: 2004), p. 129.
2. Christopher Flavin, "Hearing on Asia's Environmental Challenges: Testimony of Christopher Flavin," Committee on International Relations, U.S. House of Representatives, Washington, DC, 22 September 2004; David Schrank and Tim Lomax, *2005 Urban Mobility Study* (College Station, TX: Texas Transportation Institute, May 2005), p. 1.
3. Susan Ives, "The Politics of Happiness," *Trust for Public Land*, 9 August 2002; Lisa Jones, "A Tale of Two Mayors: The Improbable Story of How Bogota, Colombia, Became Somewhere You Might Actually Want To Live," *Grist Magazine*, 4 April 2002.
4. Enrique Peñalosa, "Parks for Livable Cities: Lessons from a Radical Mayor," keynote address at the Urban Parks Institute's Great

Parks/Great Cities Conference (Chicago: 30 July 2001); Ives, op. cit. note 3; Jones, op. cit. note 3; Claudia Nanninga, "Energy Efficient Transport—A Solution for China," *Voices of Grassroots*, November 2004.

5. Peñalosa, op. cit. note 4.

6. Jones, op. cit. note 3; Molly O'Meara, *Reinventing Cities for People and the Planet*, Worldwatch Paper 147 (Washington, DC: Worldwatch Institute, June 1999), p. 47.

7. O'Meara, op. cit. note 6, p. 47.

8. Urban population in 1900 from Mario Polèse, "Urbanization and Development, *Development Express*, no. 4, 1997; United Nations, *World Urbanization Prospects, The 2003 Revision: Data Tables and Highlights* (New York: 2004), pp. 1, 6.

9. O'Meara, op. cit. note 6, pp. 14–15; United Nations, op. cit. note 8, p. 7; United Nations, *World Population Prospects, The 2004 Revision: Highlights* (New York: 2005), pp. 1, 28; U.N. Department of Economic and Social Affairs, Population Division, *Urban Agglomerations 2003*, wall chart (New York: March 2004).

10. World Health Organization, "The World Health Organization Warns of the Rising Threat of Heart Disease and Stroke as Overweight and Obesity Rapidly Increase: WHO Urges Healthy Diet, Physical Activity, No Tobacco Use," press release (Geneva: 22 September 2005); Jane E. Brody, "As America Gets Bigger, the World Does Too," *New York Times*, 19 April 2005.

11. Los Angeles from Sandra Postel, *Last Oasis*, rev. ed. (New York: W.W. Norton & Company, 1997), p. 20; Mexico City from Joel Simon, *Endangered Mexico* (San Francisco, CA: Sierra Club Books, 1997); "Beijing Residents to Drink Water from Yangtze," *Xinhua News Agency*, 12 May 2005.

12. U.S. Department of Agriculture, Foreign Agricultural Service, *Grain: World Markets and Trade* and *Oilseeds: World Markets and Trade* (Washington, DC: various issues).

13. "China Faces Water Shortage of 40 Billion Cubic Meters Every Year," *Agence France-Presse*, 28 December 2004.

14. Richard Register, "Losing the World, One Environmental Victory at a Time—And a Way to Solve That Problem," essay (Oakland, CA: Ecocity Builders, Inc., 31 August 2005); Richard Register, *Ecocities: Building Cities in Balance with Nature* (Berkeley, CA: Berkeley Hill Books, 2002).

15. Register, "Losing the World," op. cit. note 14.

16. Register, "Losing the World," op. cit. note 14.; 2003 population estimate from U.S. Census Bureau, "San Luis Obispo (city), California," factsheet, at quickfacts.census.gov/ qfd/states/06/0668154.html, revised 29 September 2005.

17. Register, "Losing the World," op. cit. note 14.

18. See Chapters 2 and 10 for further discussion of the energy economy.

19. Jay Walljasper, "Unjamming the Future," *Ode*, October 2005, pp. 36–41.

20. Molly O'Meara Sheehan, "Making Better Transportation Choices," in Lester R. Brown et al., *State of the World 2001* (New York: W.W. Norton & Company, 2001), p. 116.

21. William D. Eggers, Peter Samuel, and Rune Munk, *Combating Gridlock: How Pricing Road Use Can Ease Congestion* (New York: Deloitte, November 2003); Tom Miles, "London Drivers to Pay UK's First Congestion Tax," *Reuters*, 28 February 2002; Randy Kennedy, "The Day The Traffic Disappeared," *New York Times Magazine*, 20 April 2003, pp. 42–45.

22. Transport for London, *Central London Congestion Charging: Impacts Monitoring—Third Annual Report* (London: 2005), p. 1; Transport for London, *Central London Congestion Charging: Impacts Monitoring— Second Annual Report* (London: April 2004), pp. 2, 4, 13; Transport for London, *Impacts Monitoring Programme: First Annual Report* (London: 2003), p. 52; bicycles and mopeds from Transport for London data cited in Blake Shaffer and Georgina Santos, *Preliminary Results of the London Congestion Charging Scheme* (Cambridge, U.K.: 2003), p. 22.

23. "Cardiff Congestion Charge Looming," *BBC News*, 12 July 2005; Juliette Jowit, "Congestion Charging Sweeps The World—A Rash Of Cities Round The Globe Is Set To Travel The Same Road as London," *Guardian* (London), 15 February 2004; Rachel Gordon, "London's Traffic Tactic Piques Interest in S.F.—Congestion Eased by Making Drivers Pay to Traverse Busiest Areas at Peak Times," *San Francisco Chronicle*, 4 June 2005; Andy Moore and John Lamb, "Congestion Charging," *SEPA View* (Scottish Environmental Protection Agency), no. 18 (Winter 2004); Transportation Alternatives, *London Businesses Still Back Congestion Charging*, press release (New York: 4 September 2003); Jim Motavalli, "Climate for Change: England Gets Serious About Global Warming," *E: The Environmental Magazine*, May-June 2005; "Swedish Government Approves Congestion Tax for Stockholm on Trial Basis," *Associated Press*, 29 April 2005.

24. O'Meara, op. cit. note 6, p. 45.

25. China's bicycle production compiled from United Nations, *The Growth of World Industry: 1969 Edition*, vol. 1 (New York: 1970), from *Yearbook of Industrial Statistics* (New York: various years), and from *Industrial Commodity Statistics Yearbook* (New York: various years); "World Market Report," *Interbike Directory* (Laguna Beach, CA: Miller-Freeman, various years); "China's Bicycle Output to Stabilize Until 2008," *Global Sources*, 5 August 2005; 143 bicycles per 100 households in 2002 from "China Ends 'Bicycle Kingdom' As Embracing Cars," *China Daily*, 11 November 2004; 3.39 people per household

in 2002 from "Chinese Families Shrinking in Size," *China Today*, August 2005; 2002 population in China from United Nations, *World Population Prospects: The 2004 Revision*, op. cit. note 9; cars in China from Ward's Communications, *Ward's World Motor Vehicle Data 2004* (Southfield, MI: 2004), p. 16.

26. Number of police forces in Matthew Hickman and Brian A. Reaves, *Local Police Departments 1999* (Washington, DC: U.S. Department of Justice, Bureau of Justice Statistics, 2001); arrest rate from a conversation with a member of the Washington, DC, police force.

27. Glenn Collins, "Old Form of Delivery Thrives in New World of E-Commerce," *New York Times*, 24 December 1999.

28. O'Meara, op. cit. note 6, pp. 47–48.

29. Ibid.

30. Spanish Railway Foundation, "Spanish Greenways Programme," Vias Verdes Web site, at www.ffe.es/viasverdes/programme.htm, viewed 10 August 2005.

31. Walljasper, op. cit. note 19.

32. O'Meara, op. cit. note 6, pp. 47–48; Japan from author's personal observation.

33. "Farming in Urban Areas Can Boost Food Security," *FAO Newsroom*, 3 June 2005.

34. Ibid.

35. Jac Smit, "Urban Agriculture's Contribution to Sustainable Urbanisation," *Urban Agriculture*, August 2002, p. 13.

36. Ibid.

37. Ibid., p. 12.

38. "Gardening for the Poor," *FAO Newsroom*, 2004, at www.fao.org/newsroom/en/field/2004/37627/article_37647en.html, viewed 27 June 2005.

39. Ibid.

40. "Cuba: Ciudad de la Habana," *Urban Agriculture*, August 2002, p. 22; Lawrence Solomon, "Sowing the Skyline," *National Post* (Urban Renaissance Institute), 13 November 2004; Katherine H. Brown and Anne Carter, *Urban Agriculture and Community Food Security in the United States: Farming from the City Center to the Urban Fringe* (Venice, CA: Community Food Security Coalition, October 2003), p. 10; United Nations, op. cit. note 1, p. 260.

41. Brown and Carter, op. cit. note 40, p. 7.

42. Ibid.

43. Sunita Narain, "The Flush Toilet is Ecologically Mindless," *Down to Earth*, 28 February 2002, pp. 28–32; dead zones in R. J. Diaz, J. Nestlerode, and M. L. Diaz, "A Global Perspective on the Effects of

Eutrophication and Hypoxia on Aquatic Biota," in G. L. Rupp and M. D. White (eds.), *Proceedings of the 7th Annual Symposium on Fish Physiology, Toxicology and Water Quality*, Estonia, 12–15 May 2003 (Athens, GA: U.S. Environmental Protection Agency (EPA), Ecosystems Research Division: 2003).

44. Narain, op. cit. note 43.

45. Ibid.

46. Ibid.

47. EPA, "Water Efficiency Technology Factsheet—Composting Toilets," information sheet (Washington, DC: September 1999); Jack Kieffer, Appalachia—Science in the Public Interest, *Humanure: Preparation of Compost from the Toilet for Use in the Garden,* ASPI Technical Series TP 41 (Mount Vernon, KY: ASPI Publications, 1998).

48. EPA, op. cit. note 47.

49. Tony Sitathan, "Bridge Over Troubled Waters," *Asia Times*, 23 August 2002; "Singapore Opens Fourth Recycling Plant to Turn Sewage into Water," *Associated Press*, 12 July 2005.

50. Peter H. Gleick, *The World's Water 2004-2005: The Biennial Report on Freshwater Resources* (Washington, DC: Island Press, 2004), p. 149.

51. Ibid., pp. 106, 113–15.

52. United Nations, *World Population Prospects, The 2004 Revision: Highlights,* op. cit. note 9, p. 1; United Nations, op. cit. note 1, pp. 1, 4.

53. Hari Srinivas, "Defining Squatter Settlements," Global Development Research Center Web site, www.gdrc.org/uem/define-squatter.html, viewed 9 August 2005.

54. Ibid.

55. O'Meara, op. cit. note 6, p. 49.

56. Rasna Warah, *The Challenge of Slums: Global Report on Human Settlements 2003* (New York: U.N. Human Settlements Programme, 2003).

57. Srinivas, op. cit. note 53.

58. E. O. Wilson, *Biophilia* (Cambridge, MA: Harvard University Press, 1984); S. R. Kellert and E. O. Wilson, eds., *The Biophilia Hypothesis* (Washington, DC: Island Press, 1993).

59. Theodore Roszak, Mary Gomes, and Allen Kanner, eds., *Restoring the Earth, Healing the Mind* (San Francisco: Sierra Club Books, 1995).

60. Public transport ridership growth rate calculated from American Public Transportation Association, *APTA Transit Ridership Report*, at www.apta.com/research/stats/ridershp/riderep/documents/history.pdf, viewed 10 August 2005; Justin Blum, "Oil Prices Spike As Storm Nears," *Washington Post*, 20 September 2005.

61. Ding Guangwei and Li Shishun, "Analysis of Impetuses to Change of

Agricultural Land Resources in China," *Bulletin of the Chinese Academy of Sciences*, vol. 13, no. 1 (1999).

62. Molly O'Meara Sheehan, *City Limits: Putting the Breaks on Sprawl*, Worldwatch Paper 156 (Washington, DC: Worldwatch Institute, June 2001), p. 11; Schrank and Lomax, op. cit. note 2.
63. Jim Motavalli, "The High Cost of Free Parking," *E: The Environmental Magazine*, March–April 2005.
64. O'Meara, op. cit. note 6, p. 49; Donald C. Shoup, "Congress Okays Cash Out," *Access*, fall 1998, pp. 2–8.
65. "Paris To Cut City Centre Traffic," *BBC News*, 15 March 2005; J. H. Crawford, "Existing Carfree Places," at www.carfree.com; see also J. H. Crawford, *Carfree Cities* (Utrecht, Netherlands: International Books, July 2000).
66. Lyndsey Layton, "Mass Transit Popularity Surges in U.S.," *Washington Post*, 30 April 2000; Bruce Younkin, Manager of Fleet Operations at Penn State University, State College, PA, discussion with Janet Larsen, Earth Policy Institute, 4 December 2000.

Chapter 12. Building a New Economy

1. Expansion in world economy from International Monetary Fund (IMF), *World Economic Outlook Database*, at www.imf.org/external/pubs/ft/weo, updated April 2005; Angus Maddison, *The World Economy: A Millennial Perspective* (Paris: Organisation for Economic Co-operation and Development (OECD), 2001).
2. Øystein Dahle from discussion with author at State of the World Conference, Aspen, CO, 22 July 2001.
3. Redefining Progress, *The Economists' Statement on Climate Change* (Oakland, CA: 1997); ECOTEC Research and Consulting, *Study on the Economic and Environmental Implications of the Use of Environmental Taxes and Charges in the European Union and its Member States* (Brussels: 2001), pp. 24–25; David Malin Roodman, "Environmental Tax Shifts Multiplying," in Lester R. Brown et al., *Vital Signs 2000* (New York: W.W. Norton & Company, 2000), pp. 138–39.
4. Roodman, op. cit. note 3; German Federal Environment Ministry, *Environmental Effects of the Ecological Tax Reform* (Bonn: 2002); Donald W. Aitken, "Germany Launches Its Transition: How One of the Most Advanced Industrial Nations Is Moving to 100 Percent Energy from Renewable Sources," *Solar Today*, March/April 2005, pp. 26–29.
5. Ministry of Finance, Sweden, "The Budget for 2005: A Commitment to More Jobs and Increased Welfare," press release (Stockholm: 20 September 2004); Ministry of Finance, Sweden, "Taxation and the Environment," press release (Stockholm: 25 May 2005); household size from Target Group Index, "Household Size," *Global TGI Barometer* (Miami: 2005); population from United Nations, *World Population*

Prospects: The 2004 Revision (New York: 2005).

6. Andrew Hoerner and Benoît Bosquet, *Environmental Tax Reform: The European Experience* (Washington, DC: Center for a Sustainable Economy, 2001); European Environment Agency (EEA), *Environmental Taxes: Recent Developments in Tools for Integration*, Environmental Issues Series No. 18 (Copenhagen: 2000); U.S. chlorofluorocarbon tax from Elizabeth Cook, *Ozone Protection in the United States: Elements of Success* (Washington, DC: World Resources Institute, 1996); city of Victoria from David Malin Roodman, "Environmental Taxes Spread," in Lester R. Brown et al., *Vital Signs 1996* (New York: W.W. Norton & Company, 1996), pp. 114–15.

7. Tom Miles, "London Drivers to Pay UK's First Congestion Tax," *Reuters*, 28 February 2002; Randy Kennedy, "The Day the Traffic Disappeared," *New York Times Magazine*, 20 April 2003, pp. 42–45; Sarah Blaskovich, "London Hikes Congestion Charge to Force More Cars off the Streets," *Associated Press*, 3 July 2005.

8. World Energy Council, *Energy Efficiency Policies and Indicators* (London: 2001), Annex 1; Howard W. French, "A City's Traffic Plans Are Snarled by China's Car Culture," *New York Times*, 12 July 2005.

9. U.S. Department of Agriculture, Economic Research Service, "Cigarette Price Increase Follows Tobacco Pact," *Agricultural Outlook*, January–February 1999.

10. Centers for Disease Control and Prevention, "Annual Smoking-Attributable Mortality, Years of Potential Life Lost, and Economic Costs—United States, 1995–1999," *Morbidity and Mortality Weekly Report*, 12 April 2002; Campaign for Tobacco-Free Kids et al., *Show Us the Money: A Report on the States' Allocation of the Tobacco Settlement Dollars* (Washington, DC: 2003); New York from Jodi Wilgoren, "Facing New Costs, Some Smokers Say 'Enough,'" *New York Times*, 17 July 2002; cigarette death toll from World Health Organization, *World Health Report 2002* (Geneva: 2002), p. 10.

11. International Center for Technology Assessment, *The Real Price of Gasoline*, Report No. 3 (Washington, DC: 1998), p. 34; U.S. Department of Energy (DOE), Energy Information Administration (EIA), *This Week in Petroleum* (Washington, DC: various issues).

12. Mick Corliss, "Carbon Tax Stuck in Detour to Kyoto," *Japan Times*, 17 January 2002; "China Studying Energy Conservation Taxes," *Asia Times*, 22 April 2005.

13. Peter P. Wrany and Kai Schlegelmilch, "The Ecological Tax Reform in Germany," prepared for the UN/OECD Workshop on Enhancing the Environment by Reforming Energy Prices, Pruhonice, Czech Republic, 14–16 June 2000.

14. OECD, European Commission, and EEA, *Environmentally Related Taxes Database*, at www.oecd.org/env/tax-database, updated 13 May 2003.

15. "BTM Predicts Continued Growth for Wind Industry," in Soren

Krohn, *Wind Energy Policy in Denmark: Status 2002* (Copenhagen: Danish Wind Energy Association, 2002), p. 8.

16. N. Gregory Mankiw, "Gas Tax Now!" *Fortune*, 24 May 1999, pp. 60–64.

17. "Addicted to Oil," *The Economist*, 15 December 2001; environmental tax support from David Malin Roodman, *The Natural Wealth of Nations* (New York: W.W. Norton & Company, 1998), p. 243.

18. Roodman, op. cit. note 17, pp. 15–27.

19. Australia in John Tierney, "A Tale of Two Fisheries," *New York Times Magazine*, 27 August 2000; South Australian Southern Zone Rock Lobster Fishery Management Committee, *Southern Zone Rock Lobster Annual Report 2003–2004* (Adelaide, South Australia: May 2005), pp. 2–5.

20. Richard Schmalensee et al., "An Interim Evaluation of Sulfur Dioxide Emissions Trading," in Robert N. Stavins, ed., *Economics of the Environment* (New York: W.W. Norton & Company, 2000), pp. 455–71.

21. Edwin Clark, letter to author, 25 July 2001.

22. André de Moor and Peter Calamai, *Subsidizing Unsustainable Development* (San José, Costa Rica: Earth Council, 1997); Barbara Crossette, "Subsidies Hurt Environment, Critics Say Before Talks," *New York Times*, 23 June 1997.

23. World Bank, *World Development Report 2003* (New York: Oxford University Press, 2003), pp. 30, 142.

24. Belgium, France, and Japan from Seth Dunn, "King Coal's Weakening Grip on Power," *World Watch*, September/October 1999, pp. 10–19; coal subsidy reduction in Germany from Robin Pomeroy, "EU Ministers Clear German Coal Subsidies," *Reuters*, 10 June 2002; subsidy cut figures in China from Roodman, op. cit. note 17, p. 109; sulfur coals tax from DOE, EIA, *China: Environmental Issues* (Washington, DC: 2001).

25. John Whitelegg and Spencer Fitz-Gibbon, *Aviation's Economic Downside*, 3rd ed. (London: Green Party of England & Wales, 2003); dollar conversion based on December 2003 exchange rate in IMF, "Representative Exchange Rates for Selected Currencies in December 2003," *Exchange Rate Archives by Month*, at www.imf.org/external/np/fin/rates/param_rms_mth.cfm, viewed 1 October 2005.

26. Erich Pica, ed., *Running On Empty: How Environmentally Harmful Energy Subsidies Siphon Billions from Taxpayers*, A Green Scissors Report (Washington, DC: Friends of the Earth, 2002), pp. 2–3.

27. Internet's start from Barry M. Leiner et al., "A Brief History of the Internet," at www.isoc.org/internet/history/brief.shtml, viewed 4 August 2000; wind power in California from Peter H. Asmus, *Wind Energy, Green Marketing, and Global Climate Change* (Sacramento, CA: California Regulatory Research Project, 1999), and from California Energy Commission, "Wind Energy in California," at

www.energy.ca.gov/wind/overview.html, viewed 15 January 2003.

28. Marine Stewardship Council, "World's First Sustainable Seafood Products Launched," press release (London: 3 March 2000); Marine Stewardship Council, "Marine Stewardship Council Awards Sustainability Label to Alaska Salmon," press release (London: 5 September 2000).
29. Marine Stewardship Council, "Sustainability Label to Alaska Salmon," op. cit. note 28; Marine Stewardship Council, "Certified Fisheries," at www.msc.org, viewed 15 August 2005.
30. World Wide Fund for Nature (WWF), *The Forest Industry in the 21st Century* (Surrey, U.K.: 2001).
31. Ibid.
32. Ibid.
33. WWF, *Certification: A Future for the World's Forests* (Surrey, U.K.: WWF Forests for Life Campaign, 2000), p. 4; Forest Stewardship Council, *FSC Certified Forests* (Bonn, Germany: 2005), p. 53.
34. WWF, op. cit. note 30; Natural Resources Defense Council (NRDC), "Good Wood: How Forest Certification Helps the Environment," at www.nrdc.org/land/forests/qcert.asp, viewed 15 August 2005; Rainforest Alliance, "Profiles in Sustainable Forestry: IKEA—Furniture for Better Forestry," at www.rainforest-alliance.org/programs/profiles/documents/IKEAProfile.pdf, viewed 24 August 2005; Rainforests.net, "The Forest Industry in the 21st Century: Top 5 Wood Buyers," factsheet, at www.rainforests.net/top5woodbuyers.htm, viewed 27 September 2005.
35. "Russia Set to Begin Certification of Forests," and "Russia Works Out System for Mandatory Wood Certification," *Interfax*, 5 June 2001; Russia now has 4 million hectares of FSC certified forest and another 10–15 million hectares are actively seeking accreditation, according to Forest Stewardship Council, "FSC takes off in Russia," *FSC News*, 30 June 2005.
36. National Renewable Energy Laboratory, *Summary of Green Pricing Programs* (Golden, CO: updated 12 July 2001).
37. Global Green USA, "Santa Monica Unanimously Approves RFP Process to Switch All City Facilities to Green Power," press release (Los Angeles: 14 October 1998); Oakland from Peter Asmus, *Reaping the Wind* (Washington, DC: Island Press, 2000); New American Dream, "Institutional Purchasing Program: What's Happening Around the Country," at www.newdream.org/procure/categories.php, updated January 2004.
38. Environmental Protection Agency (EPA), Green Power Partnership, "Top 25 Partners," factsheet, at www.epa.gov/greenpower/partners/top25.htm, viewed 28 June 2005; EPA, Green Power Partnership, "Our Partners," factsheet, at www.epa.gov/greenpower/partners/gpp_partners.htm, viewed 19 September 2005.
39. Junko Edahiro, e-mail to author, 8 October 2005.

40. Consumers Union, "In Time for Earth Day, Consumers Union Launches www.eco-labels.org," press release (Yonkers, NY: 10 April 2001); Federal Environmental Agency (Germany), "Information Sheet for Submission of New Proposals for the 'Blue Angel' Environmental Label" (Berlin: Federal Environmental Agency, 2001); Canada Environmental Choice from www.environmentalchoice.com; U.S. Energy Star program information from www.energystar.gov.

41. Ernst Ulrich von Weizsäcker, Amory B. Lovins, and L. Hunter Lovins, *Factor Four: Doubling Wealth, Halving Resource Use* (London: Earthscan, 1997); Friedrich Schmidt-Bleek et al., *Factor 10: Making Sustainability Accountable, Putting Resource Productivity into Praxis* (Carnoules, France: Factor 10 Club, 1998), p. 5.

42. William McDonough and Michael Braungart, *Cradle to Cradle: Remaking the Way We Make Things* (New York: North Point Press, 2002); Rebecca Smith, "Beyond Recycling: Manufacturers Embrace 'C2C' Design," *Wall Street Journal*, 3 March 2005.

43. U.S. Geological Survey (USGS), "Iron and Steel Scrap," in *Mineral Commodity Summaries* (Reston, VA: U.S. Department of the Interior, 2005), pp. 88–89; Steel Recycling Institute, "Recycling Scrapped Automobiles: Recycling Steel And Iron Used In Automobiles," brochure (Pittsburgh, PA: no date).

44. Recycling rates from USGS, op. cit. note 43.

45. USGS, "Recycling—Metals," in *Minerals Yearbook 2003: Volume I—Metals and Minerals* (Reston, VA: U.S. Department of the Interior, 2004), pp. 61.5–61.6; Italy and Spain from Hal Kane, "Steel Recycling Rising Slowly," in Lester R. Brown et al., *Vital Signs 1992* (New York: W.W. Norton & Company, 1992), p. 98.

46. "Recycling Taken to a New Level: Buildings," *Associated Press*, 1 November 2004.

47. Tim Burt, "VW is Set for $500m Recycling Provision," *Financial Times*, 12 February 2001; Mark Magnier, "Disassembly Lines Hum in Japan's New Industry," *Los Angeles Times*, 13 May 2001.

48. Finland in Brenda Platt and Neil Seldman, *Wasting and Recycling in the United States 2000* (Athens, GA: GrassRoots Recycling Network, 2000); Prince Edward Island Government, "PEI Bans the Can," Prince Edward Island official Web site, at www.gov.pe.ca/index.php3?number=43924, viewed 15 August 2005.

49. Brenda Platt and Doug Rowe, *Reduce, Reuse, Refill!* (Washington, DC: Institute for Local Self-Reliance, 2002); energy in David Saphire, *Case Reopened: Reassessing Refillable Bottles* (New York: INFORM, Inc., 1994).

50. Dupont will cut all material waste and emission of toxic substances to the environment, according to its "Safety, Health, and Environmental Commitment," as reported 15 April 1998 by University of California at Berkeley, "People Product Strategy" program, at best.me.berkeley.edu/~pps/pps/dupont_dfe.html; "How High the Moon—The Chal-

lenge of 'Sufficient' Goals," *The New Bottom Line*, 30 June 2004.

51. NEC Corporation, *Annual Environmental Report 2000: Ecology and Technology* (Tokyo: 2000), pp. 24–27.
52. John E. Young, "The Sudden New Strength of Recycling," *World Watch*, July/August 1995, p. 24.
53. "China is No. 1 in Asian Cell Phone Market," *International Herald Tribune*, 17 August 2000.
54. Catherine Ferrier, *Bottled Water: Understanding a Social Phenomenon* (Surrey, U.K.: WWF, 2001).
55. Ibid.
56. Ibid.
57. Leanne Farrell et al., *Dirty Metals: Mining, Communities and the Environment* (Washington, DC: Earthworks and Oxfam America, 2004), pp. 4–5; gold, iron, and steel production data from USGS, "Gold," "Iron Ore," and "Iron and Steel," in *Mineral Commodity Summaries* (Reston, VA: U.S. Department of the Interior, 2005), pp. 72–73, 84–87; ratios of ore mined to metal produced from Lester Brown, *Eco-Economy* (Washington, DC: Earth Policy Institute, 2001), p. 130.
58. Share of gold to jewelry from Earthworks, "Valentine's Gold Jewelry Sales Generate 34,000,000 Tons of Mine Waste," press release (Washington, DC: 11 February 2005); Lemke from "Don't Mine Gold for Jewels," *Reuters*, 10 December 2000.
59. Clive Hamilton and Hal Turton, *Subsidies to the Aluminum Industry and Climate Change, Background Paper No. 21,* Submission to Senate Environment References Committee Inquiry into Australia's Response to Global Warming (Canberra, Australia: The Australia Institute, 1999), pp. 3–4; Hal Turton, *The Aluminium Smelting Industry: Structure, Market Power, Subsidies and Greenhouse Gas Emissions*, Discussion Paper Number 44 (Canberra, Australia: The Australia Institute, 2002), p. vii; dollar conversion based on January 2002 exchange rate in IMF, op. cit. note 25; John E. Young, "Aluminum's Real Tab," *World Watch*, March/April 1992, pp. 26–33.
60. Weizsäcker quoted in John Young, "The New Materialism: A Matter of Policy," *World Watch*, September/October 1994, p. 34.
61. Coal and natural gas consumption from BP, *BP Statistical Review of World Energy* (London: Group Media & Publishing, 2005), pp. 26, 33; European Wind Energy Association, *Wind Energy—The Facts: An Analysis of Wind Energy in the EU-25*, Executive Summary (Brussels: 2004), pp. 2, 7.
62. Aquaculture growth calculated from U.N. Food and Agriculture Organization (FAO), *FISHSTAT Plus*, electronic database, at www.fao.org/fi/statist/FISOFT/FISHPLUS.asp, updated March 2005; fish protein conversion from Rosamond L. Naylor et al., "Effect of Aquaculture on World Fish Supplies," *Nature*, vol. 405 (29 June 2000), p. 1,022.

63. Michael Renner, "Vehicle Production Sets New Record," and Gary Gardner, "Bicycle Production Recovers," both in Worldwatch Institute, *Vital Signs 2001* (New York: W.W. Norton & Company, 2001), pp. 68–71; Ward's Communications, *Ward's World Motor Vehicle Data 2004* (Southfield, MI: 2004), p. 216; John Crenshaw, Bicycle Retailer and Industry News, email to Danielle Murray, Earth Policy Institute, 19 August 2005.

64. Gary Gardner, "Bicycle Production Rolls Forward," in Worldwatch Institute, *Vital Signs 2002* (New York: W.W. Norton & Company, 2002), pp. 76–77.

65. United Nations, "New UN Report Outlines Indicators for Sustainable Energy Use," press release (New York: 15 April 2005).

66. Total land area is 13.1 billion hectares, arable land is 1.4 billion hectares, according to FAO, *FAOSTAT Statistics Database*, at apps.fao.org, updated 4 April 2005.

67. Oil expenditures calculated from petroleum demand and price per barrel from DOE, EIA, *Short-Term Energy Outlook—August 2005* (Washington, DC: 2005).

68. Aquaculture growth calculated from FAO, op. cit. note 62.

69. Plantation area from FAO, *Forest Resources Assessment (FRA) 2000* (Rome: 2001), p. 402.

Chapter 13. Plan B: Building a New Future

1. United Nations, *World Population Prospects: The 2004 Revision* (New York: 2005).

2. 1998 data from U.N. Food and Agriculture Organization (FAO), *The State of Food Insecurity in the World 2000* (Rome: 2000); 2002 data from FAO, *The State of Food Insecurity in the World 2004* (Rome: 2004).

3. Fund for Peace and the Carnegie Endowment for International Peace, "The Failed States Index," *Foreign Policy*, July/August 2005, pp. 56–65.

4. Munich Re, *Topics Annual Review: Natural Catastrophes 2001* (Munich, Germany: 2002), pp. 16–17; "Katrina May Cost as Much as Four Years of War: Government Certain to Pay More than $200 Billion Following Hurricane," *Associated Press*, 10 September 2005; P. J. Webster et al., "Changes in Tropical Cyclone Number, Duration, and Intensity in a Warming Environment," *Science*, vol. 309 (16 September 2005), pp. 1844–46.

5. Centers for Disease Control and Prevention (CDC), "Heat-Related Deaths—Chicago, Illinois, 1996–2001, and United States, 1979–1999," *Morbidity and Mortality Weekly Report*, 4 July 2003; estimate of deaths across Europe compiled in Janet Larsen, "Record Heat Wave in Europe Takes 35,000 Lives," *Eco-Economy Update* (Washington, DC: Earth Policy Institute, 9 October 2003), updated with

Istituto Nazionale di Statistica, *Bilancio Demografico Nazionale: Anno 2003* (Rome: 15 July 2004).

6. Silvia Aloisi, "Senegal Mulls 'Green Wall' to Stop Desert Advance," *Reuters*, 1 August 2005.

7. Richard Black, "Arctic Ice 'Disappearing Quickly,'" *BBC News*, 28 September 2005; National Snow and Ice Data Center/University of Washington, "Sea Ice Decline Intensifies," press release (Boulder, CO: 28 September 2005); Steve Connor, "Global Warming 'Past the Point of No Return,'" *Independent* (London), 16 September 2005; R. Warrick et al., "Changes in Sea-Level," in J. T. Houghton et al., eds., *Climate Change, 1995: The Science of Climate Change* (Cambridge, U.K.: Cambridge University Press, 1995), pp. 359–405, cited in Dorthe Dahl-Jensen, "The Greenland Ice Sheet Reacts," *Science*, 21 July 2000, pp. 404–05.

8. Jared Diamond, *Collapse: How Societies Choose to Fail or Succeed* (New York: Penguin Group, 2005).

9. Record oil prices in U.S. Department of Energy (DOE), Energy Information Administration, "This Week in Petroleum" press release, (Washington, DC: 28 September 2005).

10. Joseph Tainter, *The Collapse of Complex Societies* (Cambridge, U.K.: Cambridge University Press, 1988).

11. World Business Academy, "Interface's Ray Anderson: Mid-Course Correction," *Global Reconstruction*, vol. 19, issue 5 (2 June 2005); Ray Anderson, "A Call for Systemic Change," speech at the National Conference on Science, Policy, & the Environment: Education for a Secure and Sustainable Future, Washington, DC, 31 January 2003.

12. For information on mobilization, see Francis Walton, *Miracle of World War II: How American Industry Made Victory Possible* (Macmillan: New York, 1956).

13. Franklin Roosevelt, "State of the Union Address," 6 January 1942, at www.ibiblio.org/pha/7-2-188/188-35.html.

14. Harold G. Vatter, *The US Economy in World War II* (New York: Columbia University Press, 1985), p. 13.

15. Doris Kearns Goodwin, *No Ordinary Time—Franklin and Eleanor Roosevelt: The Home Front in World War II* (New York: Simon & Schuster, 1994), p. 316; "Point Rationing Comes of Age," *Business Week*, 19 February 1944.

16. "War Production—The Job 'That Couldn't Be Done,'" *Business Week*, 5 May 1945; Donald M. Nelsen, *Arsenal of Democracy: The Story of American War Production* (New York: Harcourt, Brace and Co., 1946), p. 243.

17. Goodwin, op. cit. note 15.

18. Sir Edward Grey quoted in Walton, op. cit. note 12.

19. Jeffrey Sachs, "One Tenth of 1 Percent to Make the World Safer," *Washington Post,* 21 November 2001.

20. See Table 7–1 and associated discussion in Chapter 7 for more information.

21. Ibid.

22. See Tables 7–1 and 8–1 and associated discussion for more information on basic social goals (Chapter 7) and earth restoration goals (Chapter 8).

23. Table 13–2 compiled by Earth Policy Institute from Center for Arms Control and Non-Proliferation, "Highlights of Senate Armed Services Committee Action on the Fiscal Year 2006 Defense Authorization Bill (S. 1042)," factsheet, at www.armscontrolcenter.org/archives/001919.php, 22 July 2005; Christopher Hellman, "U.S. Military Budget is the World's Largest, and Still Growing," Center for Arms Control and Non-Proliferation, at www.armscontrolcenter.org/archives/001221.php, 7 February 2005, based on data from the International Institute for Strategic Studies and the U.S. Department of Defense; Elisabeth Sköns et al., "Military Expenditure," in Stockholm International Peace Research Institute, *SIPRI Yearbook 2005: Armaments, Disarmament and International Security* (Oxford: Oxford University Press, 2005); U.S. Department of Defense, Office of the Under Secretary of Defense (Comptroller), *National Defense Budget Estimates for FY 2006* (Washington, DC: 2005); Eugene Carroll from Christopher Hellman, "Last of the Big Time Spenders: U.S. Military Budget Still the World's Largest and Growing," Center for Defense Information, at www.cdi.org/issues/wme/spendersFY03.html, 4 February 2002.

24. For more information on tax restructuring, see Chapter 12.

25. For more information on energy efficiency, see Chapter 10.

26. Gordon Brown, "Marshall Plan for the Next 50 Years," *Washington Post*, 17 December 2001.

27. Gerard Bon, "France's Chirac Backs Tax to Fight World Poverty," *Reuters*, 4 September 2002.

28. "A Long Decade of Negotiations: The Difficult Birth of the Kyoto Protocol," *European Affairs*, summer 2002.

29. J. Andrew Hoerner and Benoît Bosquet, *Environmental Tax Reform: The European Experience* (Washington, DC: Center for a Sustainable Economy, 2001), pp. 17–18.

30. Ministry of Finance, Sweden, "The Budget for 2005: A Commitment to More Jobs and Increased Welfare," press release (Stockholm: 20 September 2004); Ministry of Finance, Sweden, "Taxation and the Environment," press release (Stockholm: 25 May 2005).

31. Fred Pearce, "Cities Lead the Way to a Greener World," *New Scientist*, 4 June 2005; Office of the Mayor, Greg Nickels, Seattle, "U.S. Mayors' Climate Protection Agreement," at www.seattle.gov/mayor/climate, updated 3 October 2005.

32. John Richardson, "States Poised to Set Limits on Emissions," *Portland Press Herald*, 21 September 2005; Kathy Belyeu, "States of Motion:

Not Content to Wait for Federal Action, More U.S. States Act to Develop Renewable Energy," *Solar Today*, May/June 2005.

33. Jared Diamond, "The Ends of the World as We Know Them," *New York Times*, 1 January 2005; Diamond, op. cit. note 8.

34. Diamond, op. cit. note 8.

35. Geoffrey Dabelko, "Nobel of the Ball: Kenyan Eco-Activist Wangari Maathai Wins Nobel Peace Prize," *Grist Magazine*, 8 October 2004.

36. For more information about the United Nations Foundation, see www.unfoundation.org.

37. For more information about the Bill & Melinda Gates Foundation, see www.gatesfoundation.org.

38. Diamond, op. cit. note 8; Ronald Wright, *A Short History of Progress* (New York: Carroll and Graf Publishers, 2005); Jeffrey Sachs, "Can Extreme Poverty Be Eliminated?" *Scientific American,* September 2005, pp. 56–65; Amory Lovins, "More Profit with Less Carbon," *Scientific American,* September 2005, pp. 74–82.

Index

Acknowledgments

If it takes a village to raise a child, then it takes the entire world to do a book. It begins with the work of thousands of scientists and research teams in many fields whose analyses we draw on for information, insights, and understanding. The process ends with the teams who translate the book into other languages. We are indebted to the thousands of researchers, all the translation teams, and countless others.

When dealing with such a broad range of issues and their interrelationships, an author necessarily relies heavily on the work of other analysts. There are three individuals in particular whose work has shaped my thinking: Jared Diamond and his work on the survival or decline of earlier civilizations, Jeffrey Sachs and his tireless efforts to analyze and eradicate poverty, and Amory Lovins and his ingenious work on the potential for raising energy efficiency. In drawing on their work, I feel as though I am standing on the shoulders of giants.

There are many other individuals whose work inspired and informed this book, including Herman Daly on the relationship between the economy and the environment, Gene Sperling on education in developing countries, Enrique Peñalosa and Richard Register on cities, Sandra Postel and Peter Gleick on water, Bill McDonough and Michael Braungart on materials recycling, and Ernst von Weizsäcker, with his pioneering work on environmental tax reform and reducing materials use.

Our research team, led by Janet Larsen, went through literally thousands of research reports, articles, and books, gathering, organizing, and analyzing the information that fed into this book. Janet also helped conceptualize *Plan B 2.0*. She was the first, for instance, to see the need for the earth restoration budget. In research and writing, Janet is my alter ego, my best critic, and a sounding board for new ideas.

Elizabeth Mygatt not only made a strong research contribution, but she was invaluable in bringing the manuscript down the homestretch. Several others helped with research at various times as the book evolved, including Viviana Jiménez, Danielle Murray (who was especially helpful with the research and analysis on biofuels), Emily Arnold, and Erin Greenfield.

Reah Janise Kauffman, our Vice President, manages the Institute, coordinates our worldwide network of publishers, and serves as my special assistant. Perhaps most important, she has succeeded in boosting our earned income from publication sales, royalties, and speaking fees to where it now covers over half of our budget. And, as if this were not enough, she reviewed the manuscript twice, aiding in its evolution at every step of the way.

Millicent Johnson, our Director of Publications Sales, not only manages our publications department, she also serves as our office quartermaster and librarian. Millicent, who cheerfully handles the thousands of book orders, takes pride in her one-day turnaround policy.

Reviewers who helped shape the final product include my colleagues, each of whom reviewed the manuscript at least twice, and more than a dozen talented individuals from outside the Institute. Peter Goldmark, for many years publisher of the International Herald Tribune, used his rich experience to help us identify the strengths and weakness of the manuscript. Peter is simultaneously one of the book's strongest supporters and one of its most able critics. His comments on the structure of last chapter were particularly helpful.

Edwin (Toby) Clark, an engineer and economist by training, brought his decades of environmental experience as an environmental analyst at the Council on Environmental Quality and as an administrator at the U.S. Environmental Protection Agency to bear on the manuscript, providing both broad structural suggestions and detailed page-by-page commentary.

Doug and Debra Baker contributed their wide-ranging scientific knowledge, ranging from physics to meteorology, to chapter-by-chapter critiques that were both constructive and encouraging. Maureen Kuwano Hinkle drew on her 26 years of experience working on agricultural issues with the Environmental Defense Fund and the Audubon Society to review the book twice, providing valuable comments and encouragement along the way. Gail Gorham provided general feedback and was particularly helpful in structuring Chapter 1. Reed Funk, Rutgers University biologist, read the entire manuscript and made particularly useful comments on the chapters dealing with agriculture and natural systems.

Three Board members—Scott McVay, joined by his wife Hella; William Mansfield; and Raisa Scriabine—read the entire manuscript and provided numerous useful comments.

Those who read selected chapters include University of Nebraska agronomist Kenneth Cassman, who provided substantive suggestions for improving the chapters dealing with agriculture. Brian Brown, my son, suggested several steps to focus the book more sharply. William Brown at the U.S. Department of Energy was particularly helpful with Chapter 2. Hadan Kauffman offered a very positive commentary on Chapter 1.

As always, we are in debt to our editor, Linda Starke, who brings nearly 30 years of international experience in editing environmental books and reports to the table. She has brought her sure hand to the editing of not only this book, but all of my books during this period.

The book was produced in record time thanks to the efforts of Elizabeth Doherty, who gave up evenings and weekends to prepare the page proofs. The index has again been ably prepared under deadline pressure by Ritch Pope.

We are supported by a network of dedicated publishers for our books and Eco-Economy Updates in some 22 languages—Arabic, Catalan, Chinese, Czech, English, Danish, French, Indonesian, Italian, Japanese, Korean, Marathi, Persian, Polish, Portuguese (in Brazil), Romanian, Russian, Spanish, Swedish, Thai, Turkish, and Ukrainian. There are three publishers in English (U.S.A./Canada, U.K./Commonwealth, and India/South Asia) and two in Spanish (Spain and Latin America).

These translations are often the work of environmentally committed individuals. In Iran, Hamid Taravati and his wife,

Farzaneh Bahar, are both medical doctors who head an environmental NGO. In recognition of their fine translations, the Peka Institute, an association of Iranian publishers, selected the Farsi edition of *Eco-Economy* as the best nonfiction book in 2003.

In China, Lin Zixin has arranged the publication of my books in Chinese for more than 20 years. For the most recent books, he has personally led the team of translators. I am also grateful to him for arranging a trip to Inner Mongolia and Gansu provinces that helped me better understand the pressures on the land in China's northwest.

In Japan, Soki Oda, who started Worldwatch Japan some 20 years ago, leads our publication efforts and arranges book promotional tours in Japan. He is already planning the outreach effort for the Japanese edition of *Plan B 2.0*. Junko Edahiro, founder and head of Japan for Sustainability, also supports us in Japan. For many years my interpreter, she now also organizes highly successful fundraisers for the Earth Policy Institute and is, in addition, our leading individual donor.

Gianfranco Bologna, with whom I've had a delightful relationship for nearly 25 years, arranges for the publishing of our books in Italian. As head of World Wildlife Fund Italy, he is uniquely positioned to assist in this effort.

In Brazil, Eduardo Athayde is responsible for translating our books into Portuguese for electronic downloading. He also does an excellent job of organizing fast-paced promotional tours in that country's major cities.

In Romania, former President Ion Iliescu, who started publishing our books some 18 years ago when he headed Editura Tehnica, personally assumes responsibility for getting our books out quickly. He is aided by Roman Chirila.

In Turkey, TEMA, the leading environmental NGO—which works especially on reforesting the countryside—has for many years published my books.

In South Korea, Yul Choi, founder of the Korean Federation for Environmental Movement (KFEM), has published my books and overseen their launching through KFEM's publishing house, Doyosae.

We have been fortunate in India to have the highly respected Orient Longman publishing our books for distribution throughout South Asia.

A small organization like EPI relies on the goodwill and dedication of many people. We are grateful to the growing number of individuals and organizations who translate and post our Eco-Economy Updates on their Web sites, including Leif Ohlsson (Swedish), Li Kangmin (Chinese), Gianfranco Bologna (Italian), Soki Oda (Japanese), Professor Sankar Sen (Bengali), Ole Hansen (Danish), Eduardo Athayde (Portuguese), Joseph Robertson (Spanish), François Kornmann (French), and Misha Jones (Russian).

I would also like to thank personally the members of our Plan B Team—the 650 or so individuals who purchased five or more copies of the first edition of *Plan B: Rescuing a Planet Under Stress and a Civilization in Trouble* and distributed it to friends, colleagues, and opinion leaders.

We are also indebted to our funders, without whose support this book would not exist. Among these are the United Nations Population Fund and several foundations, including the Appleton, Farview, McBride Family, Shenandoah, Summit, and Wallace Genetic foundations. A special thanks to the Lannan Foundation, whose generous and timely three-year grant has allowed us to acquire additional staff as well as the luxury of time for completing this book.

Earth Policy is supported by individual donors, including Ray Anderson, Doug and Debra Baker, Susan Brown, Don Collins and Sarah Epstein, Junko Edahiro, Judy Gradwohl, Paul Growald, Maureen Hinkle, Judy Hyde, Scott and Hella McVay, Peter Seidel, and Geraldine Wang of the William Penn Foundation, as well as the Cultural Vision Fund. This unwavering support from foundations and individuals enables us to offer a plan of hope for future generations.

And finally, saving the best for last, my thanks to the team at W.W. Norton & Company: Amy Cherry, our editor; Heather Goodman, Amy's assistant; Andrew Marasia, who put the book on a fast-track production schedule; Ingsu Liu, Art Director for the book jacket; Bill Rusin, Marketing Director; and Drake McFeely, President, with special thanks for his support. It is a delight to work with such a talented team and to have been published by W.W. Norton for more than 30 years.

Lester R. Brown

About the Author

Lester R. Brown is President of Earth Policy Institute, a nonprofit, interdisciplinary research organization based in Washington, D.C., which he founded in May 2001. The purpose of the Earth Policy Institute is to provide a vision of an environmentally sustainable economy and a roadmap of how to get from here to there.

Brown has been described as "one of the world's most influential thinkers" by the *Washington Post*. The *Telegraph* of Calcutta called him "the guru of the environmental movement." In 1986, the Library of Congress requested his papers for their archives.

Some 30 years ago, Brown helped pioneer the concept of environmentally sustainable development, a concept he uses in his design of an eco-economy. He was the Founder and President of the Worldwatch Institute during its first 26 years. During a career that started with tomato farming, Brown has authored or coauthored many books and been awarded over 20 honorary degrees. His books have appeared in more than 40 languages.

Brown is a MacArthur Fellow and the recipient of countless prizes and awards, including the 1987 United Nations Environment Prize, the 1989 World Wide Fund for Nature Gold Medal, and the 1994 Blue Planet Prize for his "exceptional contributions to solving global environmental problems." In 1995, Marquis *Who's Who*, on the occasion of its fiftieth edition, selected Lester Brown as one of 50 Great Americans. Most recently he was awarded the Presidential Medal of Italy and the Borgström Prize by the Royal Swedish Academy of Agriculture and Forestry, and he was appointed an honorary professor of the Chinese Academy of Sciences. He lives in Washington, D.C.

If you have found this book useful and would like to share it with others, consider joining our

Plan B Team.

To do so, order five or more copies at our bulk discount rate at www.earthpolicy.org/Books/PB2/index.htm.

This book is not the final word. We will continue to unfold new issues and update the analysis in our

Eco-Economy Updates.

Follow this progress by subscribing to our free, low-volume electronic listserv. Please sign up at www.earthpolicy.org to get these four-page Updates by e-mail as they are released.

Past Eco-Economy Updates and all of the Earth Policy Institute's research, including this book, are posted on our Web site www.earthpolicy.org for free downloading.

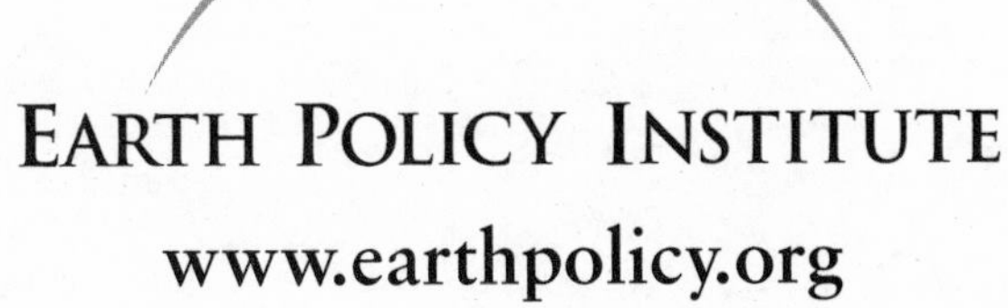